Texts in Theoretical Computer Science
An EATCS Series

Springer
Berlin
Heidelberg
New York
Hong Kong
London
Milan
Paris
Tokyo

Juraj Hromkovič

Theoretical Computer Science

Introduction to Automata, Computability, Complexity, Algorithmics, Randomization, Communication, and Cryptography

With 73 Figures and 6 Tables

Springer

Author

Prof. Dr. Juraj Hromkovič
RWTH Aachen
Computer Science I, Algorithms and Complexity
Ahornstr. 55
52074 Aachen, Germany
jh@i1.informatik.rwth-aachen.de

Series Editors

Prof. Dr. Wilfried Brauer
Institut für Informatik, Technische Universität München
Arcisstraße 21, 80333 München, Germany
Brauer@informatik.tu-muenchen.de

Prof. Dr. Grzegorz Rozenberg
Leiden Institute of Advanced Computer Science
University of Leiden
Niels Bohrweg 1, 2333 CA Leiden, The Netherlands
rozenber@liacs.nl

Prof. Dr. Arto Salomaa
Turku Centre for Computer Science
Lemminkäisenkatu 14 A, 20520 Turku, Finland
asalomaa@utu.fi

Library of Congress Cataloging-in-Publication Data applied for

Die Deutsche Bibliothek – CIP-Einheitsaufnahme

Bibliographic information published by Die Deutsche Bibliothek
Die Deutsche Bibliothek lists this publication in the Deutsche Nationalbibliografie;
detailed bibliographic data is available in the Internet at <http://dnb.ddb.de>.

ACM Subject Classification (1998):
F.1.1-3, F.2.2, F.4.3, I.4.2, E.3, E.4, C.2.1

ISBN 978-3-642-05729-8

Springer-Verlag Berlin Heidelberg New York,
a member of BertelsmannSpringer Science+Business Media GmbH

Printed in Germany

Cover Design: KünkelLopka, Heidelberg
Illustrations: Ingrid Zámečniková

Printed on acid-free paper 45/3142PS - 5 4 3 2 1 0

To my parents

Keep me away from the wisdom
which does not cry,
the philosophy which does not laugh,
and the greatness
which does not bow before children ...

KHALIL GIBRAN
Handful of Beach Sand

Preface

This textbook is an introduction to theoretical computer science with a focus on the development of its algorithmic concepts. It is based on a substantially extended translation of the German textbook "Algorithmische Konzepte der Informatik" written for the first introductory course to theoretical fundamentals of computer science at the University of Aachen. The topics have been chosen to strike a balance between classical fundamentals related to automata, computability, and NP-completeness and modern topics such as approximation algorithms, randomization, cryptography, and interconnection networks.

In contrast to the technical and applied areas of computer science, theoretical computer science is strongly related to fundamental questions about the existence of algorithmic solutions, physical limits of computing, methodology of algorithm design, etc. Since these topics are strongly connected with mathematics and not always directly related to applications, students often consider theoretical computer science too difficult and not motivating enough. The main goal of this book is to change this negative opinion on the theory. Theoretical computer science is a fascinating scientific discipline. Through its spectacular, deep results and high interdisciplinarity, it has made great contributions to our view of the world. On the other hand, there is no doubt about its relevance to practice. It provides methodology as well as particular concepts and techniques that can be applied throughout the entire process of design, implementation, and analysis of software systems. Moreover, without the know-how of theoretical fundamentals many everyday applications of enormous economic importance (such as E-commerce) would be impossible.

The right choice of topics and related motivations is not the only effort we make to stimulate the reader's interest for theoretical computer science. Despite the fact that there is no easy way to develop a deep understanding and mastery of methods that have significant and impressive applications, we try to provide an easy route into the fundamentals of computer science, and show that matters strongly related to mathematical rigor can be readily accessible even for beginners. Simplicity and transparency are the main educational features of this book. All ideas, concepts, analysis, and proofs are first explained in an informal way in order to build the right intuition. They are then carefully specified in rigorous detail. Following this strategy we choose to illustrate the main concepts and ideas using the most transparent, simple examples rather than to present the best, but too technical results. Presenting the main concepts of the theory in the order of their historical discovery, we aim to teach the development of the computer scientist's kind of thinking from the early days to the present.

I would like to express my deepest thanks to Dirk Bongartz, Hans-Joachim Böckenhauer, Alexander Ferrein, and Yvonne Moh for carefully reading the whole manuscript and their numerous comments and suggestions. Very special thanks go to Yvonne Moh for her help during the whole preparation of the English manuscript. I am also indebted to Volker Claus, Galina Jirásková, Bagdat El Abdouni Khayari, Georg Schnitger, Karol Tauber, Erich Valkema, and Manuel Wahle for various comments and support during the work on the book.

I would like to cordially thank Ingrid Zámečniková for her illustrations, the only perfect part of this book. The excellent cooperation with Alfred Hofmann and Ingeborg Mayer from Springer-Verlag is gratefully acknowledged. Last but not least I would like to express my deepest thanks to Tanja for her collection of citations.

Aachen,
September 2003

Juraj Hromkovič

Contents

When you want to build a ship,
then do not drum the men together
in order to procure wood,
to give instructions or to distribute the work
but teach them longing for the wide endless sea.

A. de Saint-Exupéry

1

Introduction

1.1 What Is Computer Science?

Computer science and informatics are the most common terms used to name the science, to whose fundamentals this textbook is devoted. Everybody studying or practicing this scientific discipline should now and then think about how one would define computer science, and contemplate its contributions to science, education and daily life. It is important to realize that learning more and more about a scientific discipline and going deeper and deeper into the understanding of its nature always results in the development of our opinion on the role of this particular science in the context of all sciences. Hence, it is especially important for students to consistently review their perception of computer science. Here, we do not hesitate to provoke a conflict between your current opinion of computer science and the viewpoints presented in this introduction and so to initialize a discussion that could lead to the development of your understanding of informatics.

Let us first attempt to answer the question

> *"What is computer science?"*

It is difficult to provide an exact and complete definition of a scientific discipline. A commonly accepted definition is:

> *Computer science is the science of algorithmic processing, representation, storage and transmission of information.*

This definition presents information and algorithm as the main objects investigated in computer science. However, it neglects to properly reveal the nature and methodology of computer science. Another question regarding the substance of computer science is

> *"To which scientific discipline does computer science belong? Is it a meta science such as mathematics and philosophy, a natural science or an engineering discipline?"*

An answer to this question serves not only to clarify the objects of the investigation, it also must be determined by the methodology and contributions of computer science. The answer is that computer science cannot be uniquely assigned to any of these disciplines. Computer science includes aspects of mathematics, and natural sciences as well as of engineering. We will briefly explain why.

Similar to philosophy and mathematics, computer science investigates general categories such as

> *determinism, nondeterminism, randomness, information, truth, untruth, complexity, language, proof, knowledge, communication, approximation, algorithm, simulation, etc.*

and contributes to the understanding of these categories. Computer science has shed new light on and brought new meaning to many of these categories.

A natural science, in contrast to philosophy and mathematics, studies concrete natural objects and processes, determines the border between possible and impossible and investigates quantitative rules of natural processes. It models, analyzes, and confirms the credibility of hypothesized models through experiments. These aspects are similarly prevalent in computer science. The objects are information and algorithms (programs, computers) and the investigated processes are the physically existing computations. Let us try to document this by looking at the development of computer science. Historically, the first important research question in computer science was the following one with philosophical roots.

> *"Are there well-defined problems that cannot be automatically (by a computer, regardless of the computational powers of contemporary computers or futuristic ones) solved?"*

Efforts to answer this question led to the founding of computer science as an independent science. The answer to this question is positive. We are now aware of many practical problems that we would like to solve algorithmically, but which are not algorithmically solvable. This conclusion is based on a sound mathematical proof of algorithmic nonsolvability (i.e., on a proof of the nonexistence of algorithms solving the given problem), and not on the fact that no algorithmic solution has been discovered so far.

After developing methods for classifying problems according to their algorithmic solvability, one asks the following scientific question:

> *"How difficult are concrete algorithmic problems?"*

This difficulty is not measured in the difficulty of developing an algorithmic solution, or in the size of the designed program. Rather, this difficulty is measured in the amount of work necessary and sufficient to algorithmically compute the solution for a given problem instance. One learns of the existence of hard problems, for which computing solutions needs energy exceeding that of the entire universe. There are algorithmically solvable problems such that the execution of any program solving them would require more time than has passed since the Big Bang. Hence the mere existence of a program for a particular problem is not an indication that this problem is solvable within practical limits.

Efforts to classify problems into practically solvable (tractable) and practically insolvable led to the most fascinating scientific discoveries of theoretical computer science.

As an example, let us consider randomized algorithms. Most programs (algorithms), as we know them, are deterministic. By deterministic, we mean that the program and the input completely determine all steps of the work on the problem. At every moment, the next action of the program is unambiguously determined and depends only on the current data. Randomized programs may have several options for the next action. Which option is taken is randomly chosen. The work of a randomized algorithm may be viewed as if the algorithm tosses a coin from time to time to determine its next action, i.e., to choose the next strategy in its search for the correct answer. Hence, a randomized program can have many different computations for an input. In contrast to deterministic programs that reliably deliver the right solution for any input, randomized programs may give erroneous results. The aim is to suppress the probability of such false computations, which under some circumstances means to decrease the proportion of false computations.

At first sight, randomized programs may seem unreliable, as opposed to their deterministic counterparts. Why then the necessity for randomized programs? There are important problems whose solution by the best known deterministic algorithm require more computer work than one can realistically execute. Such problems appear to be practically insolvable. But a miracle can happen: this miracle can be a randomized algorithm that solves the problem within minutes, with a miniscule error probability of one in a trillion. Can one ban such a program as unreliable? A deterministic program that requires a day's computer work is more unreliable than a randomized program running in a few minutes, because the probability that a hardware error occurs during this 24 hours of computation is much higher than the error probability of the fast randomized program. A concrete example of utmost practical significance is primality testing. In the ubiquitous use of cryptographic public-key protocols, huge prime numbers (approximately 500 digits long) must be generated. The first deterministic algorithms for primality testing were based on testing the divisibility of the input n. Alone, the number of primes smaller than $\sqrt{n}$ for such huge values of n exceeds the number of protons in the universe. Hence, such deterministic algorithms are practically useless. Recently, a new deterministic algorithm for primality testing running in time $O(m^{12})$ for n of binary length m was developed. But it needs to execute more than 10^{32} computer instructions in order to test a 500-digit number and so the amount of time since the Big Bang is not sufficient to execute such a computation on the fastest computers. However, there are several randomized algorithms that test primality of such large numbers within minutes or even seconds on a regular PC.

Another spectacular example is a communication protocol for comparison of the contents of two databases, stored in two distant computers. It is mathematically provable that every deterministic communication protocol that tests the equivalence of these contents, needs to exchange as many bits as those within the databases. For a database with 10^{16} bits, this would prove to be tedious. A randomized communication protocol can test this equivalence using a message of merely 2000 bits. The error probability of this test is less than one in a trillion.

How is this possible? It is difficult to explain this without some basic knowledge of computer science. The search for the explanation behind the strengths of randomized algorithms is a fascinating research project, going into the deepest fundamentals of mathematics, philosophy, and natural sciences. Nature is our best teacher, and randomness plays a larger role in nature than one would believe. Computer scientists can cite many systems where the required characteristics and behaviors of such systems are achievable only through the concept of randomization. In such examples, every deterministic reliable system is made up of billions of subsystems, and these subsystems must interact correctly. Such a complex system, highly dependent on numerous subcomponents, is not practical. In the case that an error occurs, it would be almost impossible to detect it. Needless to say, the costs of developing such a system is also astronomical. On the other hand, one can develop small randomized systems with the required behavior. Because of their small size, such systems are inexpensive and the work of their components is easily verifiable. And the crucial point is that the probability of a wrong behavior is so miniscule that it is negligible.

Despite its above-illustrated scientific aspects, computer science is a typical problem-oriented and practical engineering discipline for many scientists. Computer science not only includes the technical aspects of engineering such as:

> *organization of development processes (phases, milestones, documentation), formulation of strategic goals and limits, modeling, description, specification, quality assurance, tests, integration into existing systems, reuse, and tool support,*

it also encompasses the management aspects such as:

> *team organization and team leadership, costs estimation, planning, productivity, quality management, estimation of time plans and deadlines, product release, contractual obligations, and marketing.*

A computer scientist should also be a true pragmatic practitioner. When constructing complex software or hardware systems, one must often make decisions based on one's experience, because one does not have any opportunity to model and analyze the highly complex reality.

Considering our definition of computer science, one may get the impression that the study of computer science is too difficult. One needs mathematical knowledge as well as the understanding of the way of thinking in natural sciences, and on top of that, to be able to work like an engineer. This may really be a strong requirement, but it is also the greatest advantage of this education. The main drawback of current science is in its overspecialization, which leads to an independent development of small subdisciplines. Each branch has developed its own language, often incomprehensible even for researchers in a related field. It has gone so far that the standard way of arguing in one branch is perceived as superficial and inadmissible in another branch. This slows down the development of interdisciplinary research. Computer science is interdisciplinary at heart. It is focused on the search for solutions for problems in all areas of sciences and in everyday life, wherever the use of computers is imaginable. While doing so, it employs a wide spectrum of methods, ranging from precise formal mathematical methods to experience-based "know-how" of engineering. The opportunity to concurrently learn the different languages of different areas and the different ways of thinking, all in one discipline, is the most precious gift conferred on a computer science student.

1.2 A Fascinating Theory

This book is devoted to an elementary introduction to the fundamentals of theoretical computer science. Theoretical computer science is a fascinating scientific discipline. Through its spectacular results and high interdisciplinarity, it has made great contributions to our view of the world. However, theoretical computer science is not the favorite subject of students, as statistics would confirm. Many students even view theoretical computer science as a hurdle that one has to overcome in order to graduate. There are several reasons for this widespread opinion. One reason is that amongst all areas of computer science, theoretical computer science is the mathematically most demanding part and hence the lectures on theoretical fundamentals belong to the hardest courses in computer science. Not to forget, many computer science students start their study with a wrong impression of computer science, and many lecturers of theoretical computer science do not present their courses in a sufficiently attractive way. Excessive pressure for precise representation of the minute technical details of mathematical proofs plus a lack of motivation, a lack of relevance, a lack of informal development of ideas within the proper framework and a lack of direct implementation and usage, can ruin the image of any fascinating field of science.

In our previous depiction of computer science as a science with many faces, we have already indirectly brought attention to the importance of their theoretical fundamentals. Because there are several important reasons for the indispensability of theoretical fundamentals in the study of computer science, we would like to list them in what follows.

1. *Philosophical depth*
   Theoretical computer science explores knowledge and develops new concepts and notions that influence science in its very core. Theoretical computer science gives partial or complete answers to philosophical questions such as:
   - Are there problems that are not automatically (algorithmically) solvable? If so, where does the boundary lie between automatic solvability and automatic insolvability?
   - Are nondeterministic and randomized processes capable of what deterministic processes are incapable? Is nondeterminism and randomization better (more efficient) than determinism?
   - How does one define the difficulty (hardness) of problems?
   - Where are the limits of "practical" algorithmic solvability?
   - What is a mathematical proof? Is it more difficult to find mathematical proofs algorithmically than to verify the correctness of a given proof algorithmically?
   - How to define a random object?

   It is important to note that many of these questions cannot be properly formulated without the formal concepts of algorithm and computation. Thus, theoretical computer science has enriched the language of science through these new terms, contributing to its development. Many known basic categories of science, such as determinism, chance, and nondeterminism have gained new meanings, and through this, our general view of the world has been influenced.

2. *Applicability and spectacular results*
   Theoretical computer science is relevant to practice. On one hand, it provides methodological insights that influence our first strategic decision over the processing of algorithmic problems. On the other hand, it provides particular concepts and methods that can be applied during the whole process of design and implementation. Moreover, without the knowledge and concepts of theoretical computer science many applications would be impossible. Besides the concept of randomized algorithms (as already mentioned), there are many other "miracles" that were born in theoretical computer science.
   There are difficult optimization problems for which a simple relaxation of their constraints and requirements decreases the hardness of the problem so much that this decrease corresponds to a gigantic leap from an unrealistic computational demand to that of a few minutes. Often, this relaxation is so small that it is practically negligible.
   Do you believe that it is possible to convince somebody of the knowledge of a secret (password), without having to reveal a single bit of this secret? Do you believe that two persons can determine who of them is older without revealing their ages to the other party? Do you believe that one can almost with certainty check the correctness of mathematical proofs

of several thousand pages, without reading it, only by looking at a few randomly chosen bits of them? All these things are possible. This not only shows that, thanks to theory, things are made possible though previously they were believed to be impossible, it also shows that research in theoretical computer science is exciting and full of surprises, and so one can be inspired and enthused by theoretical computer science.

3. *Lifespan of knowledge*
Through the rapid development of technology, the world of applied computer science continuously evolves. Half of the existing information about software and hardware products is obsolete after 5 years. Hence, an education that is disproportionately devoted to system information and current technologies, does not provide appropriate job prospects. Whereas the concepts and methodology in theoretical computer science have a longer average lifespan of several decades. Such knowledge will serve its owner well for a long period of time.

4. *Interdisciplinary orientation*
Theoretical computer science is interdisciplinary in its own right and can take part in many exciting frontiers of research and development – genome projects, medical diagnostics, optimization in all areas of economy and technical sciences, automatic speech recognition, and space exploration, just to name a few.
As much as computer science contributes to all other fields, it also benefits from the contributions from other fields. The study of computations on the level of elementary particles, whose behavior follows the rules of quantum mechanics, focuses on the efficient execution of computations in the microworld whose execution in the macroworld has failed. The theoretical model of a quantum computer already exists, but its implementation is a huge challenge to physicists. Currently, nobody can overview all consequences of a successful construction of quantum computers. Independent of the success of this project, the rules of the microworld are so surprising and counterintuitive for those who have gained their experience in the macroworld, that one expects many more "miracles" from the use of quantum theory. It is already obvious today that reliable and secure communication exists in the microworld, since every attempt to learn the message submitted will be detected and warded off by the sender. Another exciting area is computing with DNA molecules. DNA molecules are information carriers, and hence, it is not surprising that one can exploit them for information storage and transmission. Today, we are aware that DNA molecules are capable of imitating the work of electronic computers. This is not only theoretically obvious, several simulations of computations by chemical operations over DNA molecules have been performed in laboratories. One cannot exclude the emergence of DNA molecular computers,

where a few DNA molecules can take over the work of an electronic computer.

5. *Way of thinking*
   Mathematicians attribute the special role mathematics play in education through development, enrichment and shaping the way of thinking, i.e., through contributing to the general development of one's personality. If this contribution by mathematics is so highly regarded, then one must also acknowledge the importance of computer science for the general education and the enrichment of the way of thinking.
   Theoretical computer science encourages creating and analyzing mathematical models of real systems and searching for concepts and methods to solve concrete problems. Remember that precisely understanding which features of a real system are exactly captured by one's model and which characteristics are only approximated or even neglected is the main assumption for a success in science and engineering. Because of this, theoretical computer science calls attention to teaching the evolution of mathematical concepts and models in a strong relation to real problems. Thus, studying computer science, one has the chance of learning how to combine theoretical knowledge with practical experience and so to develop a way of thinking that is powerful enough to attack complex real-world problems.

1.3 To the Student

This book has been written primarily for you. The aim of this book is not only to introduce some basic concepts of computer science, it also attempts to inspire. We leave it to you to decide how successful we are in our endeavor.

In the first sections of this chapter we have attempted to convince you that computer science is a fascinating science, full of fun, excitement, and joy. As a proof that there are people who truly enjoy the work in theoretical computer science, I would like to quote one of my friends from various lectures and research seminars.

> "And if, after our pressure, this stupid error probability still hesitates to go below 0.5, then we will consider other levers. It will deeply repent for having the courage to steal into my model."
>
> "Today, we have forced an ϵ to its knees. Let me tell you, this was a high pressure steam. First, we set an approximation on it. It nervously twitched, but unfortunately only a bit. And then the heavy blows of a semidefinite matrix made it impassive. It was wonderful!"
>
> "You say it! No stingy, around knocking λ will profane our μ without paying for it. We will dress it down pretty well. It will never understand how this happened to it."

Do you consider such emotions for a rigorous topic too excessive? Personally, I do not. It is great fun to participate in such lectures where one realizes immediately the difficulties one has to surmount. Passion is the main driving force behind education and research. Once you have developed a passion for a topic, you have already won half the race. If you have not found a passion till now, it is high time to start looking. Should this search come to a dead end, then it would be appropriate to extend your search to other scientific disciplines or activities beyond computer science.

This textbook is devoted to some fundamental areas of theoretical computer science. Why is the study of theoretical computer science considered difficult? There is no easy way to the development of a deep understanding and mastery of methods that have significant and impressive applications. This should not come as a surprise. Should one hope to accomplish the 100 meter dash under 10 s, or jump over 8 m, one has to invest years of hard training. To achieve something exceptional, one has to put in exceptional effort. The acquisition of knowledge is no exception. Perhaps you may even face more resistance during this pursuit, because in contrast to sports your motivation may falter as you lose sight of your aims.[1] It also demands endurance, especially readiness to repetitively explore every topic in order to gain a deeper understanding for the interconnection between them for the context of the whole theory.

This book seeks to facilitate the entry into some fundamental parts of theoretical computer science. For this purpose we use the following three concepts.

1. *Simplicity and transparency*
   We will explain simple notions in simple terms. We avoid the use of unnecessary mathematical abstractions. Hence, we attempt to be as concrete as possible. Through this, we build the introduction on elementary mathematical knowledge. Presenting complicated arguments and proofs, we will first explain the ideas in a simple and transparent way before providing the formal proofs.
   Sections marked with a "*" are more involved and technical. Undergraduates are advised to skip these sections. The technical discussion of the Turing machine in the computability section is also optional, since the understanding of computability can be established on any programming language.
   Clarity takes priority over the presentation of the best known results. When a transparent argument of a weaker result can bring across the idea succinctly, then we will opt for it instead of presenting a strong but technically demanding and confusing argument of the best known result.

[1] Maybe, you even do not see your aims clearly at the beginning of the process of learning.

Throughout this book, we will work systematically, taking small steps to journey from the simple through to the complicated. We will avoid any interruptions in thoughts.

2. *Less is sometimes more or a context-sensitive presentation*
   Many study guides and textbooks falsely assume that the first and foremost aim is the delivery of a quantum of information to the reader. Hence they often go down the wrong track: maximum knowledge within minimum time, presented in minimal space. This haste usually results in the presentation of a great amount of individual results, and thus neglecting the context of the entire course.
   The philosophy behind this book is different. We would like to build and influence the student's ways of thinking. We are interested in the historical development of computer science concepts and ways of thinking, and the presentation of definitions, results, proofs, and methods is only a means to the end. Hence, we are not overly concerned about the amount of information, preferring to sacrifice 10 to 20% of the teaching material. In return, we dedicate more time to the motivation, aims, connection between practice and theoretical concepts, and especially to the internal context of the presented theory. We place special emphasis on the creation of new terms. The notions and definitions do not appear out of the blue, as seemingly so in some lectures using the formal language of mathematics. The formally defined terms are always an approximation or an abstraction of intuitive ideas. The formalization of these ideas enables us to make accurate statements and conclusions about certain objects and events. They also allow for formal and direct argumentation. We strive to explain our choice of the formalization of terms and models used and to point out the limitations of their usage. To learn to work on the level of terms creation (basic definitions) is very important, because most of the essential progress happens exactly on this level.
3. *Support of iterative teaching*
   The strategy of this book is also tailored to cultivate repetitive reconsideration of presented concepts. Every chapter opens with a section "Aims", in which the motivations, teaching objectives, and relations to the topics of previous chapters are presented. The core of the chapter is dedicated to the formalization of informal ideas by theoretical concepts and to the study within the framework of these concepts. At every essential development, we will pinpoint its relevance to our aims. We end each chapter with a short summary and an outlook. Here, the major highlights of the chapter are informally summarized and the relevance to other parts of theory is once again reviewed. We also briefly mention some further developments of the presented theoretical concepts, survey more advanced results and discuss the gap between the achieved knowledge and the research objectives. As usual, the learning process is supported by exercises. The exercises are

not moved to some special subsections but they are distributed in the text with our recommendation to deal with them immediately after one has reached them while reading this textbook. They serve to learn to successfully apply the presented concepts and methods as well as to deepen the reader's understanding of the material.

Our aim is not only to introduce to you the exciting world of computer science, but also to offer a "simply applicable, context-sensitive ticket" into the fundamentals of computer science. The simplicity of the presentation does not imply that this book is not sufficiently rigorous, but that the matters presented in this introductory material are presented so transparently that you can absorb the knowledge of the book in an unusually short time. The assumptions for a successful use of this ticket are minimal – some experience with programming (equivalent to the course work of one semester) and some basic mathematical knowledge. Standard course work such as *Computer Architectures, Algorithms and Data Structures* are not necessary, although they can be helpful for understanding of conceptual relationships.

1.4 Structure of the Book

This book is divided into nine chapters, counting this introduction. Chapter 2 serves as a springboard. Here, the formal language of computer science and the ways of representing objects and problems are introduced. Any computation of a computer can always be viewed as the transformation of a text into another text, since we represent the input data and output data as texts. Chapter 2 provides the fundamentals for working with texts and uses them to develop a formal specification of algorithmic problems. Additionally, Chapter 2 contemplates the questions how one can measure the information content of a text, and when a text can be considered to be random.

Chapter 3 introduces the finite automaton as the simplest computing model. The aim is not to provide an introduction to automata theory. Instead, it is to prepare the reader for the complex definition of a formal model of algorithms (programs). We use finite automata for a simple introduction of the key terms of computer science, such as states and configurations of computing model, computation, computation step, determinism, nondeterminism, descriptional complexity, and simulation. This helps to simplify the understanding of these terms in the general framework of the Turing-machine model discussed later.

Chapter 4 is devoted to the *Turing machine*, which is the formal model of the intuitive notion *algorithm*. Since the technical behavior of Turing machines mirrors that of programming in computer machine code, we will try to restrict this section to the bare minimum required for understanding and dealing with the above-mentioned basic terms.

Chapter 5 is an introduction to computability theory. Here we pose the question

"Which problems are algorithmically solvable and which are not?"

We present some methods that one can successfully apply for answering this question for concrete problems of interest. Here, we work on two levels. The first level performs the argumentation only by using an intuitive understanding of the term program. The second level involves the formal proofs based on the model of Turing machines.

Chapter 6 is an introduction to the complexity theory. The main questions posed here are:

"How to measure the hardness of algorithmic problems?"

and

"Do there exist arbitrarily hard problems?"

The hardness of a computing problem is measured in its computational complexity which is the amount of computer work sufficient and necessary for solving the problem by an algorithm. First we present fundamental results about complexity measures and complexity classes. Some of the proofs of these results are too technical for an introductory material and so we omit them. We learn here that there exist problems of such a high computational complexity that the energy of the whole universe does not suffice to solve them by an algorithm. The kernel of this chapter is the presentation of the NP-completeness concept that can be viewed as a method for classifying problems into "practically solvable" (tractable) and "practically unsolvable" problems. This concept is based on the study of the relation between nondeterministic and deterministic computations, one of the core research topics in theoretical computer science. Any known deterministic simulation of a nondeterministic algorithm requires an exponential increase of time complexity and one does not believe that there exists an efficient simulation of nondeterminism by determinism. Here, we provide an important argument for this belief, which touches the philosophical fundamentals of mathematics. We show in some framework that the complexity of solving a problem by a deterministic algorithm corresponds to the complexity of creating a proof of a mathematical theorem, while the complexity of solving a problem in a nondeterministic way corresponds to the complexity of verifying the correctness of a given mathematical proof. Hence, the question of whether nondeterminism is more powerful than determinism is equivalent to the question of whether it is easier to verify given proofs than to find them.

Chapter 7 presents some jewels of algorithmics.[2] This chapter is a continuation of Chapter 6 and asks

"What can one do with difficult problems, for which the best algorithms take years to solve?"

[2] Here, we are not referring to parts of the lectures algorithm and data structures or classical lectures on algorithms, but only to algorithm design for hard problems.

We present here concepts such as the pseudopolynomial algorithm, local search, approximation algorithms, and heuristics such as simulated annealing. We will explain the methodology of these concepts, which are based on the fact that a slight relaxation on the constraints can lead to an enormous leap from unrealistic computational complexity of the original problem to a matter of a few minutes of computer work. For instance, if one relaxes the requirement of computing an optimal solution of an optimization problem to the requirement to compute a feasible solution whose quality does not differ too much from that of an optimal solution, then one can substantially reduce the computational complexity of the optimization problem considered.

In the case of randomized algorithms, as previously mentioned in Section 1.1, we relax the demand for guaranteeing the correct solution to compute one that has a large probability of being correct. The concept of randomization belongs undoubtedly to the basic concepts of algorithmics for hard problems and from this point of view it should be a part of Chapter 7. However, since it is of essential importance for many theoretical and practical core areas of computer science, we dedicate an entire chapter to it. In Chapter 8, we shall not limit ourselves to merely presenting some impressive examples of efficient randomized algorithms (e.g. randomized primality testing). We, furthermore, attempt to highlight some of the reasons leading to the success of randomized algorithms. Within this framework, we introduce methods such as abundance of witness and fingerprinting to illustrate the basic design paradigms of randomized algorithms.

The final chapter concentrates on communication problems. Thanks to the recent technological breakthroughs over the past years, we are now able to transfer large amounts of data and information. Hence, the algorithmics to solve communication problems has become a dynamic and flourishing part of computer science. Chapter 9 first presents an overview of secure communication. We start with a few examples of cryptosystems and introduce the concept of public-key cryptosystems. We use these concepts to show the forgery-safe usage of signatures. We then present the concept of zero-knowledge proof systems, which plays an important role in cryptography. We conclude with the design of an interconnection network to illustrate the problems in the area of network communication.

The revelations of mysticism are
the deepest one can experience.
They are the most fundamental feeling,
which lies in the cradle of all true art and science.
Who does not know it
is not able to wonder anymore;
will never have a deep surprise.
One is more dead than alive...
as an extinguished candle...

A. Einstein

2

Alphabets, Words, Languages, and Algorithmic Problems

2.1 Objectives

Looking more carefully at machine computations we observe that computers work on texts that are nothing more than sequences of symbols over a given alphabet. The programs are texts over the alphabet of the computer keyboard, all data are saved in a computer as sequences of zeros and ones, inputs and outputs are texts or can at least be represented as texts over a suitable alphabet. From this point of view every program transforms input texts into output texts.

The first aim of this chapter is to introduce a formalism that is suitable for working with texts viewed as information representations. The basic terms introduced here are **alphabet**, **word**, and **language**. These terms are necessary as a starting point for formally specifying the fundamental notions of computer science such as algorithm (program), computer, computation, etc. In Section 2.2 we learn to work with these basic terms, use them to represent data, and practice some basic operations over texts.

The second aim of this chapter is to learn how to use the terms introduced, and how they can be applied to get formal representations of algorithmic problems. Here we focus especially on two problem classes – **decision problems** and **optimization problems**.

The third and last aim of this chapter is to deal with the questions related to text compression. We introduce the notion of Kolmogorov complexity. This notion can be used not only to define the shortest representation of texts (data), but also to measure the amount of information in a text and to find a reasonable definition of the notion of **random** text. This is a contribution of computer science to science on the philosophical (very fundamental) level because it reasonably explains when an object or an event can be declared random. Another very important point is that Kolmogorov complexity is a powerful instrument for investigating computations and we will be using it several times in the later chapters.

2.2 Alphabets, Words, and Languages

In information processing one represents data and objects by sequences of symbols. Exactly in the same way as in the development of natural languages we start by fixing the set of symbols used to represent data. In what follows $\mathbb{N} = \{0, 1, 2, \ldots\}$ denotes the set of natural numbers.

Definition 2.1. *Any nonempty finite set is called an* **alphabet**. *Every element of an alphabet Σ is called a* **symbol** *of Σ.*

The meaning of an alphabet is the same as for natural languages. The alphabet is applied to create a script (written representation of a language). For our purposes it is only important to know that one can choose arbitrary, but finitely many, symbols in order to create representations of the studied objects.

In what follows we list some of the more frequently used alphabets.

- $\Sigma_{\text{bool}} = \{0, 1\}$ is the Boolean alphabet used in computers.
- $\Sigma_{\text{lat}} = \{a, b, c, \ldots, z\}$ is the Latin alphabet.
- $\Sigma_{\text{keyboard}} = \Sigma_{\text{lat}} \cup \{A, B, \ldots, Z, \sqcup, >, <, (,), \ \ldots, !\}$ is the alphabet of all symbols of the computer keyboard, where $\sqcup$ denotes the blank.
- $\Sigma_m = \{0, 1, 2, \ldots, m-1\}$ for any $m \geq 1$ is an alphabet for the m-ary representation of numbers.
- $\Sigma_{\text{logic}} = \{0, 1, x, (,), \wedge, \vee, \neg\}$ is an alphabet that can be used to represent Boolean formulae.

In what follows we define words as sequences of symbols. In contrast to natural languages, where a word is a verbal unit of a language, a word in computer-science terminology corresponds to an arbitrary text.

Definition 2.2. *Let Σ be an alphabet. A* **word** *over Σ is any finite sequence of symbols of Σ. The* **empty word** $\boldsymbol{\lambda}$ *is the only word consisting of zero symbols*[1].

The **length of a word** *w over Σ, denoted by $|w|$, is the number of symbols in w (i.e., the length of w as a sequence).*

The set of all words over Σ is denoted by $\boldsymbol{\Sigma^*}$. $\boldsymbol{\Sigma^+} = \Sigma^* - \{\lambda\}$ denotes the set of words without the empty word.

The sequence $0, 1, 0, 0, 1, 1$ is a word over the alphabets Σ_{bool} and Σ_{keyboard}. The length of this word is $|0, 1, 0, 0, 1, 1| = 6$. The empty word λ is a word over every alphabet and $|\lambda| = 0$.

[1] In some literature, ϵ is used in place of λ.

Agreement. In what follows we omit the commas in the representations of words, i.e, we write

$$x_1x_2 \ldots x_n \text{ instead of } x_1, x_2, \ldots, x_n.$$

Thus we use 010011 to represent $0, 1, 0, 0, 1, 1$.

The symbol blank ␣ over Σ_{keyboard} is different from λ because ␣ is a symbol of Σ_{keyboard}, and hence $|␣| = 1$. Using ␣ the content of a book or a program can be considered as a word over Σ_{keyboard}.

$$\begin{aligned}(\Sigma_{\text{bool}})^* &= \{\lambda, 0, 1, 00, 01, 10, 11, 000, 001, 010, 100, 011, \ldots\} \\ &= \{\lambda\} \cup \{x_1x_2 \ldots x_i \mid i \in \mathbb{N},\ x_j \in \Sigma_{\text{bool}} \text{ for } j = 1, \ldots, i\}.\end{aligned}$$

From the above example we see that a possibility of enumerating all words over a given alphabet is to write all words of lengths $i = 0, 1, 2, \ldots$, one after the other.

Exercise 2.3. Let Σ be an alphabet. Estimate, for every $i \in \mathbb{N}$, how many words over Σ of length i exist.

Exercise 2.4. Let $\Sigma = \{0, 1, \#\}$, and let k and n be positive integers, such that $k \leq n$.

(i) Estimate the number of words of length n that contain exactly k occurrences of the symbol 0.
(ii) Estimate the number of words of length n that contain at most k occurrences of the symbol 0.

Words can be used to represent different objects such as numbers, formulae, graphs, and programs. A word

$$x = x_1x_2 \ldots x_n \in (\Sigma_{\text{bool}})^*, x_i \in \Sigma_{\text{bool}} \text{ for } i = 1, \ldots, n$$

can be viewed as the binary representation of the nonnegative integer

$$Number(x) = \sum_{i=1}^{n} 2^{n-i} \cdot x_i.$$

For any nonnegative integer m, we denote by

$$Bin(m) \in \Sigma_{\text{bool}}^*$$

the shortest binary[2] representation of m. Hence,

$$Number(Bin(m)) = m.$$

---

[2] The requirement, that $Bin(m)$ is the shortest representation of $Number(Bin(m)) = m$, means nothing more than that the first symbol of $Bin(m)$ must be 1.

Exercise 2.5. The binary representation of every positive integer m begins with 1. What is the length of $Bin(m)$ for a positive integer m?

Exercise 2.6. Let $x \in (\Sigma_m)^*$ for a positive integer m. Let x be the m-ary representation of a nonnegative integer $Number_m(x)$. How does one compute $Number_m(x)$ from x?

A sequence of integers $a_1, a_2, \ldots, a_m$, with $m \in \mathbb{N}$ and $a_i \in \mathbb{N}$ for $i = 1, \ldots, m$, can be represented as

$$Bin(a_1)\#Bin(a_2)\# \cdots \#Bin(a_m) \in \{0, 1, \#\}^*.$$

In what follows we usually use this representation for sets and sequences of integers.

Let $G = (V, E)$ be a directed graph with the set V of vertices and the set $E \subseteq \{(u, v) \mid u, v \in V,\ u \neq v\}$ of edges. Let $|V| = n$ be the cardinality of V. Recall that one can represent G by its adjacency matrix M_G. The Boolean matrix $M_G = [a_{ij}]$ has the size $n \times n$ and

$$a_{ij} = 1 \iff (v_i, v_j) \in E.$$

Hence, $a_{ij} = 1$ means that G has an edge (v_i, v_j) from v_i to v_j and $a_{ij} = 0$ means that there is no edge leading from v_i to v_j in G. A Boolean matrix M can be represented as a word over the alphabet $\{0, 1, \#\}$ as follows. One writes the rows of M one after another and the symbol # is used to mark the end of every row. Consider the graph in Figure 2.1.

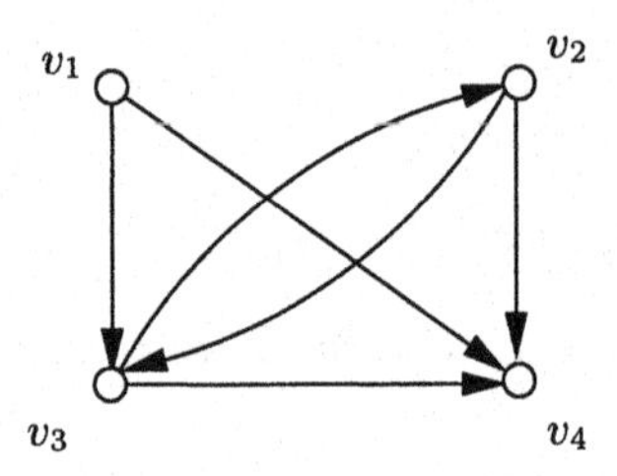

Fig. 2.1.

The corresponding adjacency matrix is

$$\begin{pmatrix} 0\,0\,1\,1 \\ 0\,0\,1\,1 \\ 0\,1\,0\,1 \\ 0\,0\,0\,0 \end{pmatrix}.$$

The proposed representation of this matrix is the word (over $\{0, 1, \#\}$)

0011#0011#0101#0000#.

It is obvious that this representation is unambiguous, meaning that given a graph representation,[3] one can unambiguously reconstruct the corresponding graph.

Exercise 2.7. The proposed representation of a graph as a word over $\{0, 1, \#\}$ has the length $n(n+1)$ for any graph of n vertices. Can you design a shorter (and different) unambiguous representation of graphs?

Exercise 2.8. Design a representation of graphs over the alphabet Σ_{bool}.

The inputs of many algorithmic problems are often weighted, undirected graphs $G = (V, E, h)$ where h is a function from E to $\mathbb{N} - \{0\}$. This means that a weight (cost) $h(e)$ is assigned to every edge $e \in E$. We know that such graphs can also be represented by adjacency matrices. Analogously $a_{ij} = 0$ implies that the edge[4] $\{v_i, v_j\}$ is not present in G. If $\{v_i, v_j\} \in E$, then the matrix element

$$a_{ij} = h(\{v_i, v_j\}),$$

is the weight of the edge $\{v_i, v_j\}$.

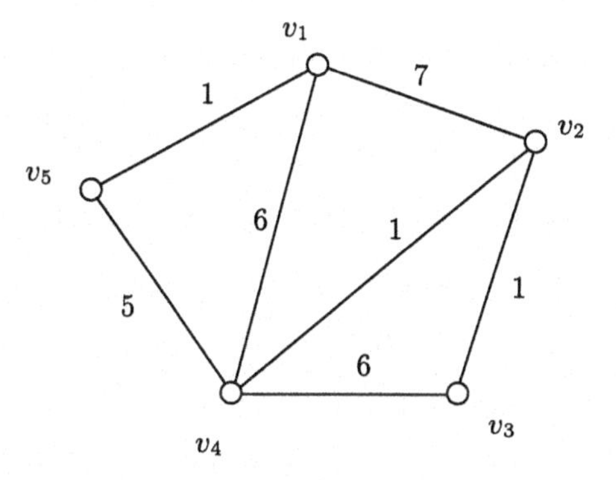

Fig. 2.2.

In this case one can take the binary representations of the weights $a_{ij} = h(\{v_i, v_j\})$ and separate them by the symbol #. To mark the end of a row of the adjacency matrix one can choose to use the word ##. The resulting adjacency matrix of the graph in Figure 2.2 is

$$\begin{pmatrix} 0 & 7 & 0 & 6 & 1 \\ 7 & 0 & 1 & 1 & 0 \\ 0 & 1 & 0 & 6 & 0 \\ 6 & 1 & 6 & 0 & 5 \\ 1 & 0 & 0 & 5 & 0 \end{pmatrix}$$

[3] In what follows the term representation is always connected with unambiguity.

[4] The undirected edge between u and v is denoted by $\{u, v\}$. For the directed edge from u to v we use the usual notation (u, v).

and its representation as a word over $\{0, 1, \#\}$ is

0#111#0#110#1##111#0#1#1#0##0#1#0#110#0##110#1#110#0#101##1#0#0#101#0.

For the adjacency matrix $M_G = [a_{ij}]$ of any undirected graph G we have obviously $a_{ij} = a_{ji}$ for all i, j. This implies a redundancy in the above-proposed representation of G because all edges are recorded twice. Therefore it is sufficient to consider only the elements above the main diagonal of M_G. The resulting representation of G over $\{0, 1, \#\}$ is

$$111\#0\#110\#1\#\#1\#1\#0\#\#110\#0\#\#101.$$

The last example that we consider here is the representation of Boolean formulae over the Boolean operators negation ($\neg$), disjunction ($\vee$) and conjunction ($\wedge$). In what follows we denote Boolean variables in formulae by $x_1, x_2, x_3, \ldots$. The number of possible variables is infinite, therefore we cannot use the symbols $x_1, x_2, x_3, \ldots$ as symbols of the alphabet. Instead we use the alphabet

$$\Sigma_{\text{logic}} = \{0, 1, x, (,), \wedge, \vee, \neg\}$$

to code the Boolean variable x_i as the word

$$x Bin(i)$$

for all $i \in \mathbb{N} - \{0\}$. All other symbols of the formula are projected one to one to its representation. Therefore, the formula

$$(x_1 \vee x_7) \wedge \neg(x_{12}) \wedge (x_4 \vee x_8 \vee \neg(x_2))$$

has the following representation

$$(x1 \vee x111) \wedge \neg(x1100) \wedge (x100 \vee x1000 \vee \neg(x10)).$$

A useful operation over words is the simple concatenation of two words.

Definition 2.9. *Let Σ be an alphabet. A* **concatenation** *with respect to Σ is the mapping $K : \Sigma^* \times \Sigma^* \to \Sigma^*$ given by*

$$K(x, y) = x \cdot y = xy$$

for all $x, y \in \Sigma^*$.

Let $x = 0aa1bb$ and $y = 111b$ for $\Sigma = \{0, 1, a, b\}$. Then $K(x, y) = x \cdot y = 0aa1bb111b$.

Remark 2.10. The concatenation K for Σ is an associative operation over Σ^*, because

$$K(u, K(v, w)) = u \cdot (v \cdot w) = uvw = (u \cdot v) \cdot w = K(K(u, v), w)$$

for all $u, v, w \in \Sigma^*$. Furthermore, for every $x \in \Sigma^*$:

$$x \cdot \lambda = \lambda \cdot x = x.$$

Therefore (Σ^*, K, λ) is a monoid with the neutral element λ.

Clearly, the concatenation is commutative only for alphabets of cardinality 1.

Remark 2.11. For all $x, y \in \Sigma^*$,

$$|xy| = |x \cdot y| = |x| + |y|.$$

In what follows we prefer the notation xy over the notations $K(x, y)$ and $x \cdot y$.

Definition 2.12. *Let Σ be an alphabet. For all $x \in \Sigma^*$ and all positive integers i, we define the i-th iteration x^i of x as*

$$x^i = xx^{i-1},$$

where $x^0 = \lambda$.

Thus, for instance, $K(aabba, aaaaa) = aabbaaaaa = a^2b^2a^6 = a^2b^2(aa)^3$. We see that the above-introduced notation allows us to find a shorter representation of some words.

In what follows we define subwords of a word x as connected parts of x.

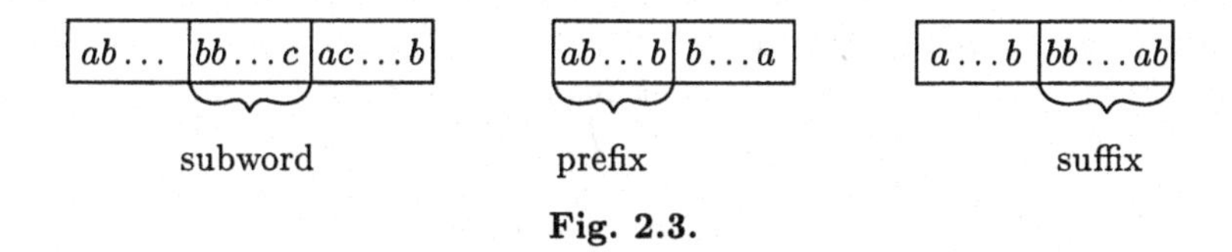

Fig. 2.3.

Definition 2.13. *Let $v, w \in \Sigma^*$ for an alphabet Σ.*

- *v is a* **subword** *of $w \Leftrightarrow \exists x, y \in \Sigma^* : w = xvy$.*
- *v is a* **suffix** *of $w \Leftrightarrow \exists x \in \Sigma^* : w = xv$.*
- *v is a* **prefix** *of $w \Leftrightarrow \exists y \in \Sigma^* : w = vy$.*
- *$v \neq \lambda$ is a* **proper** *subword [suffix, prefix] of w iff $v \neq w$ and v is a subword [suffix, prefix] of w (Figure 2.3).*

We have $(abc)^3 = abcabcabc$, and the word abc is a proper prefix of $(abc)^3$. The word bc is a proper suffix of $(abc)^3$.

Exercise 2.14. Let Σ be an alphabet and let $x \in \Sigma^*$ with $|x| = n$ for an $n \in \mathbb{N} - \{0\}$. Which is the maximal possible number of different subwords of x? Count all different subwords of the word $abbcbbab$.

Definition 2.15. *Let $x \in \Sigma^*$ and let $a \in \Sigma$ for an alphabet Σ. We define $|x|_a$ as the number of occurrences of a in x.*

For each set A, $|\boldsymbol{A}|$ denotes the cardinality of A and $\boldsymbol{\mathcal{P}(A)} = \{S \mid S \subseteq A\}$ is the powerset of A.

Hence $|(abbab)|_a = 2$ and $|(11bb0)|_0 = 1$. For all $x \in \Sigma^*$ we have

$$|x| = \sum_{a \in \Sigma} |x|_a.$$

There are several parts of this textbook where one needs a fixed order of all words over a given alphabet. The most convenient way is to consider the canonical order defined below.

Definition 2.16. *Let $\Sigma = \{s_1, s_2, \ldots, s_m\}$, $m \geq 1$, be an alphabet and let $s_1 < s_2 < \cdots < s_m$ be a linear order of elements of Σ. We define the* **canonical order** *over Σ^* for all $u, v \in \Sigma^*$ by:*

$$\begin{aligned} u < v \Leftrightarrow \quad & (|u| < |v|) \\ & \vee\ (|u| = |v| \wedge u = x \cdot s_i \cdot u' \wedge v = x \cdot s_j \cdot v' \\ & \textit{for some } x, u', v' \in \Sigma^* \textit{ and } i < j). \end{aligned}$$

Now, we define the term language as any set of words over a fixed alphabet.

Definition 2.17. *A* **language** *over an alphabet Σ is any subset of Σ^*. The complement* $\boldsymbol{L^C}$ *of the language L with respect to Σ is the language $\Sigma^* - L$.*
$\boldsymbol{L_\emptyset} = \emptyset$ *is the* **empty language**.
$\boldsymbol{L_\lambda} = \{\lambda\}$ *is the language that contains only the empty word.*
Let L_1 and L_2 be languages over Σ. Then

$$\boldsymbol{L_1 \cdot L_2} = \boldsymbol{L_1 L_2} = \{vw \mid v \in L_1 \textit{ and } w \in L_2\}$$

is the concatenation of L_1 and L_2. Let L be a language over Σ. We define

$$\begin{aligned} \boldsymbol{L^0} &:= L_\lambda, \\ \boldsymbol{L^{i+1}} &= L^i \cdot L \textit{ for all } i \in \mathbb{N}, \\ \boldsymbol{L^*} &= \bigcup_{i \in \mathbb{N}} L^i, \\ \boldsymbol{L^+} &= \bigcup_{i \in \mathbb{N} - \{0\}} L^i = L \cdot L^*. \end{aligned}$$

L^* is called the **Kleene star** *of L.*

The following sets are examples of languages over $\Sigma = \{a, b\}$:

$$\begin{aligned} L_1 &= \emptyset, \\ L_2 &= \{\lambda\}, \\ L_3 &= \{\lambda, ab, abab\}, \\ L_4 &= \Sigma^* = \{\lambda, a, b, aa, \ldots\}, \end{aligned}$$

$$\begin{aligned}
L_5 &= \Sigma^+ = \{a, b, aa, \ldots\},\\
L_6 &= \{a\}^* = \{\lambda, a, aa, aaa, \ldots\} = \{a^i \mid i \in \mathbb{N}\},\\
L_7 &= \{a^p \mid p \text{ is prime }\},\\
L_8 &= \{a^i b^{2\cdot i} a^i \mid i \in \mathbb{N}\},\\
L_9 &= \Sigma,\\
L_{10} &= \Sigma^3 = \{aaa, aab, aba, abb, baa, bab, bba, bbb\}.
\end{aligned}$$

The set of all grammatically correct texts in English is a language over Σ_{keyboard}, and the set of all syntactically correct programs in JAVA is also a language over Σ_{keyboard}.

Observe that

$$\Sigma^i = \{x \in \Sigma^* \mid |x| = i\},$$

and that

$$L_\emptyset L = L_\emptyset = \emptyset$$

and

$$L_\lambda \cdot L = L.$$

Exercise 2.18. Let $L_1 = \{\lambda, ab, b^3a^4\}$ and $L_2 = \{ab, b, ab^2, b^4\}$. Which words belong to the language L_1L_2?

Our next aim is to practice working with languages. Since languages are sets, the standard set operations union ($\cup$) and intersection ($\cap$) are used. We add the concatenation and the Kleene star to these set operations. The first question we pose is whether the distributive laws with respect to $\cup$ and concatenation (with respect to $\cap$ and concatenation) hold. For $\cup$ and concatenation the following lemma provides a positive answer. To prove the equality of two sets A and B we use the standard methods of set theory, where we usually show $A \subseteq B$ and $B \subseteq A$ separately, which then implies $A = B$. To show $A \subseteq B$ it is sufficient to prove for every element $x \in A$, that x belongs to B ($x \in B$).

Lemma 2.19. *Let L_1, L_2, and L_3 be languages over an alphabet Σ. Then*

$$L_1L_2 \cup L_1L_3 = L_1(L_2 \cup L_3).$$

Proof. First we show $L_1L_2 \cup L_1L_3 \subseteq L_1(L_2 \cup L_3)$. The comments in brackets explain the step executed.

$L_1L_2 \subseteq L_1(L_2 \cup L_3)$ holds because

$$\begin{aligned}
L_1L_2 &= \{xy \mid x \in L_1 \wedge y \in L_2\} \quad \{\text{definition of concatenation}\}\\
&\subseteq \{xy \mid x \in L_1 \wedge y \in L_2 \cup L_3\} \quad \{\text{since } L_2 \subseteq L_2 \cup L_3\}\\
&= L_1 \cdot (L_2 \cup L_3) \quad \{\text{definition of concatenation}\}.
\end{aligned}$$

In the same way one can show $L_1L_3 \subseteq L_1(L_2 \cup L_3)$. Therefore $L_1L_2 \cup L_1L_3 \subseteq L_1(L_2 \cup L_3)$.

Now we prove the inclusion $L_1(L_2 \cup L_3) \subseteq L_1L_2 \cup L_1L_3$.

Let $x \in L_1(L_2 \cup L_3)$. Then

$$\begin{aligned}
& x \in \{yz \mid y \in L_1 \wedge z \in L_2 \cup L_3\} \\
& \qquad \{\text{definition of concatenation}\} \\
\Rightarrow\ & \exists y \in L_1 \wedge \exists z \in L_2 \cup L_3, \text{ such that } x = yz \\
\Rightarrow\ & \exists y \in L_1 \wedge (\exists z \in L_2 \vee \exists z \in L_3), \text{ such that } x = yz \\
& \qquad \{\text{definition of } \cup\} \\
\Leftrightarrow\ & (\exists y \in L_1 \wedge \exists z \in L_2 : x = yz) \vee (\exists y \in L_1 \wedge \exists z \in L_3 : x = yz) \\
& \qquad \{\text{distributive law for } \wedge, \vee\} \\
\Leftrightarrow\ & (x \in \underbrace{\{yz \mid y \in L_1 \wedge z \in L_2\}}_{L_1L_2}) \vee (x \in \underbrace{\{yz \mid y \in L_1 \wedge z \in L_3\}}_{L_1L_3}) \\
& \qquad \{\text{definition of concatenation}\} \\
\Leftrightarrow\ & x \in L_1L_2 \cup L_1L_3 \quad \{\text{definition of } \cup\}.
\end{aligned}$$

□

Now, we would like to deal with the question of whether the distributive law holds for concatenation and intersection. It may be a surprise at the first glance that the answer to this question is negative. Only one inclusion of the distributive law equality holds and we show it in the next lemma.

Lemma 2.20. *Let L_1, L_2, L_3 be languages over an alphabet Σ. Then*

$$L_1(L_2 \cap L_3) \subseteq L_1L_2 \cap L_1L_3.$$

Proof. Let $x \in L_1(L_2 \cap L_3)$. This is equivalent to

$$\begin{aligned}
& x \in \{yz \mid y \in L_1 \wedge z \in L_2 \cap L_3\} \\
& \qquad \{\text{definition of concatenation}\} \\
\Leftrightarrow\ & \exists y, z \in \Sigma^*, y \in L_1 \wedge (z \in L_2 \wedge z \in L_3), \text{ such that } x = yz \\
& \qquad \{\text{definition of } \cap\} \\
\Leftrightarrow\ & \exists y, z \in \Sigma^*, (y \in L_1 \wedge z \in L_2) \wedge (y \in L_1 \wedge z \in L_3) : x = yz \\
\Rightarrow\ & \exists y, z \in \Sigma^*, (yz \in L_1L_2) \wedge (yz \in L_1L_3) : x = yz \\
& \qquad \{\text{definition of concatenation}\} \\
\Leftrightarrow\ & x \in L_1L_2 \cap L_1L_3 \quad \{\text{definition of } \cap\}.
\end{aligned}$$

□

To show that $L_1(L_2 \cap L_3) \supseteq L_1L_2 \cap L_1L_3$ is not always true, it is sufficient to find three concrete languages U_1, U_2, and U_3 such that

$$U_1(U_2 \cap U_3) \subset U_1U_2 \cap U_1U_3.$$

When searching for convenient languages U_1, U_2 and U_3, one can base one's strategy on the fact that there is only one implication in the proof of Lemma 2.20 that cannot be converted (exchanged by equivalence). If a word x belongs to L_1L_2 as well as to L_1L_3, then one may not conclude in general that $x = yz$ with $x \in L_1$ and $z \in L_2 \cap L_3$. This happens, for instance, when x is written as

$$x = y_1z_1 = y_2z_2 \text{ with } y_1 \neq y_2 (\text{i.e.}, z_1 \neq z_2)$$

where $y_1, y_2 \in L_1$, $z_1 \in L_2$, and $z_2 \in L_3$. Here we have $x \in L_1L_2 \cap L_1L_3$, but there is no evidence that x has to be in $L_1(L_2 \cap L_3)$.

Lemma 2.21. *There exist* U_1, U_2, $U_3 \in (\Sigma_{\text{bool}})^*$, *such that*

$$U_1(U_2 \cap U_3) \subsetneq U_1U_2 \cap U_1U_3.$$

Proof. First, we choose $U_2 = \{0\}$ and $U_3 = \{10\}$. Hence, we obtain $U_2 \cap U_3 = \emptyset$, and thus

$$U_1(U_2 \cap U_3) = \emptyset$$

for any language U_1. Now, it is sufficient to find a U_1 such that $U_1U_2 \cap U_1U_3$ would not be empty. We set $U_1 = \{\lambda, 1\}$. Then

$$U_1U_2 = \{0, 10\}, U_1U_3 = \{10, 110\}$$

and hence

$$U_1U_2 \cap U_1U_3 = \{10\} \neq \emptyset.$$

□

Exercise 2.22. Let L_1, L_2 and L_3 be languages over the alphabet $\{0\}$. Is the equality

$$L_1(L_2 \cap L_3) = L_1L_2 \cap L_1L_3$$

valid?

Exercise 2.23. Let $L_1 \subseteq \Sigma_1^*$ and $L_2, L_3 \subset \Sigma_2^*$ for alphabets Σ_1 and Σ_2 with $\Sigma_1 \cap \Sigma_2 = \emptyset$. Is the equality

$$L_1(L_2 \cap L_3) = L_1L_2 \cap L_1L_3$$

valid?

Exercise 2.24. Do languages L_1, L_2, and L_3 exist, such that $L_1(L_2 \cap L_3)$ is finite and $L_1L_2 \cap L_1L_3$ is infinite?

In what follows we will work with the Kleene star.

Example 2.25. The following equality holds,

$$\{a\}^*\{b\}^* = \{a^ib^j \mid i, j \in \mathbb{N}\}.$$

Proof. First we show $\{a\}^*\{b\}^* \subseteq \{a^ib^j \mid i,j \in \mathbb{N}\}$.

Let $x \in \{a\}^*\{b\}^*$. Then

$$\begin{aligned}
& x = yz, \text{ where } y \in \{a\}^* \wedge z \in \{b\}^* \\
& \quad \{\text{definition of concatenation}\} \\
\Rightarrow\ & x = yz, \text{ where } (\exists k \in \mathbb{N} : y \in \{a\}^k) \wedge (\exists m \in \mathbb{N} : z \in \{b\}^m) \\
& \quad \{\text{definition of the Kleene star}\} \\
\Leftrightarrow\ & x = yz, \text{ where } (\exists k \in \mathbb{N} : y = a^k) \wedge (\exists m \in \mathbb{N} : z = b^m) \\
\Leftrightarrow\ & \exists k, m \in \mathbb{N}, \text{ such that } x = a^k b^m \\
\Rightarrow\ & x \in \{a^ib^j \mid i,j \in \mathbb{N}\}.
\end{aligned}$$

Next, we show $\{a^ib^j \mid i,j \in \mathbb{N}\} \subseteq \{a\}^*\{b\}^*$.

Let $x \in \{a^ib^j \mid i,j \in \mathbb{N}\}$. Then

$$\begin{aligned}
& x = a^r b^l \text{ for some integers } r, l \in \mathbb{N} \\
\Rightarrow\ & x \in \{a\}^*\{b\}^*, \text{ since } a^r \in \{a\}^*, b^l \in \{b\}^*.
\end{aligned}$$

□

Exercise 2.26. Prove or disprove the truth of the equality

$$(\{a\}^*\{b\}^*)^* = \{a,b\}^*.$$

Definition 2.27. *Let Σ_1 and Σ_2 be two arbitrary alphabets. A* **homomorphism** *from Σ_1 to Σ_2 is any function $h : \Sigma_1^* \to \Sigma_2^*$ that satisfies the following conditions:*

(i) $h(\lambda) = \lambda$ and
(ii) $h(uv) = h(u) \cdot h(v)$ for all $u, v \in \Sigma_1^*$.

One can easily observe that to specify a homomorphism it is sufficient to fix $h(a)$ for all symbols $a \in \Sigma_1$.

Exercise 2.28. Let h be a homomorphism from Σ_1 to Σ_2. Prove by induction that for all words $x = x_1x_2 \ldots x_m$, $x_i \in \Sigma_1$ for $i = 1, \ldots, m$,

$$h(x) = h(x_1)h(x_2)\ldots h(x_m).$$

Consider a mapping h given by

$$h(\#) = 10, h(0) = 00 \text{ and } h(1) = 11.$$

Clearly h specifies a homomorphism from $\{0,1,\#\}$ to Σ_{bool}. For instance,

$$\begin{aligned}
h(011\#101\#) &= h(0)h(1)h(1)h(\#)h(1)h(0)h(1)h(\#) \\
&= 0011111011001110.
\end{aligned}$$

One can use h to transfer a representation of some objects over $\{0,1,\#\}$ to a new representation of these objects over Σ_{bool}.

Exercise 2.29. Define a homomorphism from $\{0, 1, \#\}$ to Σ_{bool} that maps infinitely many words over $\{0, 1, \#\}$ to one word from $(\Sigma_{\text{bool}})^*$.

Exercise 2.30. Define an injective homomorphism from Σ_{logic} to Σ_{bool} that provides an unambiguous representation of Boolean formulae over Σ_{bool}.

Exercise 2.31. Let Σ_1 and Σ_2 be alphabets. Let h be a homomorphism from Σ_1 to Σ_2. For any language $L \subseteq \Sigma_1^*$ we define

$$\mathbf{h(L)} = \{h(w) \mid w \in L\}.$$

Let $L_1, L_2 \subseteq \Sigma_1^*$. Prove or disprove the following equality:

$$h(L_1)h(L_2) = h(L_1L_2).$$

2.3 Algorithmic Problems

Before giving the formal definition of the notion "algorithm" by the Turing-machine model in Chapter 4 we view algorithms as programs. We assume that the reader knows what a program is. For our purposes, the specific programming language is irrelevant. When using the synonym "program" for "algorithm" we require that the program computes a correct output for each feasible input. This means that an algorithm is considered to be a program that halts for any input (i.e., does not have any infinite computation) and solves the given problem. Given this assumption a program (an algorithm) A performs a mapping

$$A : \Sigma_1^* \to \Sigma_2^*$$

for some alphabets Σ_1 and Σ_2. This means that

(i) the inputs are represented as words over an alphabet Σ_1,
(ii) the outputs are represented as words over an alphabet Σ_2, and
(iii) A unambiguously assigns an output to every input.

For any algorithm A and any input x we denote by $\boldsymbol{A(x)}$ the output of the algorithm A for the input x. We say that two algorithms (programs) A and B are **equivalent** if they work over the same alphabet Σ and $A(x) = B(x)$ for all $x \in \Sigma^*$.

Definition 2.32. *The* **decision problem** $(\boldsymbol{\Sigma}, \boldsymbol{L})$ *for a given alphabet* Σ *and a given language* $L \subseteq \Sigma^*$ *is to decide for any* $x \in \Sigma^*$*, whether*

$$x \in L \text{ or } x \notin L.$$

An algorithm A **solves** *the decision problem* (L, Σ)*, if, for all* $x \in \Sigma^*$*:*

$$A(x) = \begin{cases} 1, & \text{if } x \in L, \\ 0, & \text{if } x \notin L. \end{cases}$$

We also say that A **recognizes** L.

If, for a language L, there exists an algorithm that recognizes L, we say that the language L is **recursive**[5]. We often use a language $L \subseteq \Sigma^*$ to specify a concrete property of words from Σ^* (or objects that are represented by the words). Words in L satisfy this property and words from $L^{\complement} = \Sigma^* - L$ do not have this property.

Usually we describe a decision problem (Σ, L) in the following way:

(Σ, L)
Input: $x \in \Sigma^*$.
Output: $A(x) \in \Sigma_{\text{bool}} = \{0, 1\}$, where

$$A(x) = \begin{cases} 1, \text{ if } & x \in L \text{ (Yes, } x \text{ has the property),} \\ 0, \text{ if } & x \notin L \text{ (No, } x \text{ does not have the property).} \end{cases}$$

For instance, $(\{a, b\}, \{a^n b^n \mid n \in \mathbb{N}\})$ is a decision problem that can be also specified as follows:

$(\{a, b\}, \{a^n b^n \mid n \in \mathbb{N}\})$
Input: $x \in \{a, b\}^*$.
Output: Yes, if $x = a^n b^n$ for an $n \in \mathbb{N}$.
No, otherwise.

Example 2.33. A well-known decision problem of large practical importance is **primality testing**

$$(\Sigma_{\text{bool}}, \{x \in (\Sigma_{\text{bool}})^* \mid \textit{Number}(x) \text{ is a prime}\}).$$

The usual representation is

$(\Sigma_{\text{bool}}, \{x \in (\Sigma_{\text{bool}})^* \mid \textit{Number}(x) \text{ is a prime}\})$
Input: $x \in (\Sigma_{\text{bool}})^*$.
Output: Yes, if $\textit{Number}(x)$ is a prime,
No, otherwise.

Example 2.34. Let $L = \{x \in (\Sigma_{\text{keyboard}})^* \mid x \text{ is a syntactically correct program in C++}\}$. We consider the following problem that is a subproblem of any compiler for C++.

Input: $x \in (\Sigma_{\text{keyboard}})^*$.
Output: Yes, if $x \in L$,
No, otherwise.

[5] Recursion is one of the fundamental terms of computer science. Therefore we use later a formal model of computation (algorithm) in order to give a formally precise definition of this term.

Example 2.35. The **Hamiltonian Cycle Problem (HC)** is (Σ, HC), where $\Sigma = \{0, 1, \#\}$ and

$$\mathrm{HC} = \{x \in \Sigma^* \mid x \text{ represents an undirected graph that contains a Hamiltonian cycle}^6\}.$$

Example 2.36. The **Satisfiability Problem (SAT)** is $(\Sigma_{\mathrm{logic}}, \mathrm{SAT})$ with

$$\mathrm{SAT} = \{x \in (\Sigma_{\mathrm{logic}})^* \mid x \text{ represents a satisfiable Boolean formula}\}.$$

An important subclass of decision problems is the class of equivalence problems. For instance, the equivalence problem for programs is to decide whether two given programs A and B of the same programming language (i.e., the input is $(A, B) \in (\Sigma_{\mathrm{keyboard}})^*$) are equivalent. Another example of an equivalence problem is to decide whether two given Boolean formulae represent the same Boolean function.

Definition 2.37. *Let Σ and Γ be two alphabets. We say that an algorithm A* **computes a function $f : \Sigma^* \to \Gamma^*$**, *if for all $x \in \Sigma^*$*

$$A(x) = f(x).$$

Decision problems are a special case of computing functions, because solving a decision problem is equivalent to computing the characteristic function of a language.[7]

At first glance one may think that computing functions is the most general representation of algorithmic problems. The following definition shows that this is not true.

Definition 2.38. *Let Σ and Γ be alphabets and let $R \subseteq \Sigma^* \times \Gamma^*$ be a relation in Σ^* and Γ^*. An algorithm* **A computes R** *(or A* **solves the relation problem R**), *if, for every $x \in \Sigma^*$:*

$$(x, A(x)) \in R.$$

From Definition 2.38 we see that to solve a relation problem R for a given input x, it suffices to find one y from a possibly infinite set of ys with the property $(x, y) \in R$. The following examples show that the relation problems are not only an abstract generalization of computing functions,[8] but that many practical problems are problems about computing a relation.

---

6 A Hamiltonian cycle of a graph G is a cycle (a closed path) that contains every vertex of G exactly once.

7 The characteristic function f_L of a language $L \subseteq \Sigma^*$ is a function from Σ^* to $\{0, 1\}$ with $f_L(x) = 1$ iff $x \in L$.

8 Remember that functions are special relations with the property that for every x there exists exactly one y with $(x, y) \in R$.

Let $R_{\text{fac}} \subseteq (\Sigma_{\text{bool}})^* \times (\Sigma_{\text{bool}})^*$, where $(x, y) \in R_{\text{fac}}$ if and only if $Number(y)$ is a factor[9] of $Number(x)$ or $y = 1$ when $Number(x)$ is a prime. A transparent representation of this relation problem follows.

$\underline{R_{\text{fac}}}$
Input: $x \in (\Sigma_{\text{bool}})^*$.
Output: $y \in (\Sigma_{\text{bool}})^*$, where $y = 1$, if $Number(x)$ is a prime, and $Number(y)$ is a factor of $Number(x)$ if x is composite.

Another hard problem is the search for a proof to a theorem. Let $R_{\text{Proof}} \subseteq (\Sigma_{\text{keyboard}})^* \times (\Sigma_{\text{keyboard}})^*$, where $(x, y) \in R_{\text{Proof}}$ if

- either x is a code of a true assertion in a mathematical theory considered and y is the representation of a proof of this assertion, or
- $y = \sqcup$ and x does not represent any true assertion of the considered theory.

In what follows we introduce optimization problems that can be viewed as a special case of relation problems. Optimization problems are of central interest in practice as well as in theory. In order to give a transparent representation of optimization problems, we use the following description instead of the representation of a relation problem. Informally, an input instance of an optimization problem determines a set $\mathcal{M}(x)$ of feasible solutions for x. Thus, we can say we have a relation R with

$$(x, y) \in R \text{ iff } y \text{ is a feasible solution for } x.$$

But this R is not the relation problem one has to solve. The input x additionally determines the cost of every y in $\mathcal{M}(x)$. The output for x must be one of the feasible solutions with the most favorable (maximal or minimal) cost.

Definition 2.39. *An* **optimization problem** *is a 6-tuple* $\mathcal{U} = (\Sigma_I, \Sigma_O, L, \mathcal{M}, cost, goal)$, *where*

(i) Σ_I *is an alphabet, called the* **input alphabet**.
(ii) Σ_O *is an alphabet, called the* **output alphabet**.
(iii) $L \subseteq \Sigma_I^*$ *is the language of* **feasible inputs** *(as inputs one allows only words that have a reasonable interpretation).*
An $x \in L$ *is called a* **problem instance of** $\mathcal{U}$.
(iv) $\mathcal{M}$ *is a function from* L *to* $\mathcal{P}(\Sigma_O^*)$, *and for each* $x \in L$, *the set* $\mathcal{M}(x)$ *is the set of* **feasible solutions for** x.
(v) $cost$ *is a function* $cost : \bigcup_{x \in L}(\mathcal{M}(x) \times \{x\}) \to \mathbb{R}^+$, *called the* **cost function**.
(vi) $goal \in \{minimum, maximum\}$ *is the objective.*

---

[9] A positive integer a is a factor of an integer b if a divides b and $a \notin \{1, b\}$.

A feasible solution $\alpha \in \mathcal{M}(x)$ is called **optimal** *for the problem instance x of U, if*

$$cost(\alpha, x) = \mathbf{Opt}_{\mathcal{U}}(x) = goal\{cost(\beta, x) \mid \beta \in \mathcal{M}(x)\}.$$

We say that an algorithm A **solves** *$\mathcal{U}$, if for any $x \in L$*

(i) $A(x) \in \mathcal{M}(x)$ $\big(A(x)$ is a feasible solution for the problem instance x of $\mathcal{U}\big)$, and
(ii) $cost(A(x), x) = goal\{cost(\beta, x) \mid \beta \in \mathcal{M}(x)\}$.

If goal = minimum, $\mathcal{U}$ is called a **minimization problem,**
if goal = maximum, $\mathcal{U}$ is called a **maximization problem.**

To understand why we define optimization problems in this way, let us have a look at the above formal definition of an optimization problem as a 6-tuple $\mathcal{U} = (\Sigma_I, \Sigma_O, L, \mathcal{M}, cost, goal)$. The input alphabet Σ_I has the same meaning as the input alphabet of decision problems, i.e., Σ_I is used for the representation of input instances of $\mathcal{U}$. Analogously, the output alphabet Σ_O is used for representing the outputs (feasible solutions). The language $L \subseteq \Sigma_I^*$ is the set of correct representations of problem instances. We assume that no word from $L^{\mathsf{C}} = \Sigma_I^* - L$ will occur as an input. This means that we focus on determining the complexity of the optimization and not on solving the decision problem (Σ_I, L).

A problem instance x usually specifies a set of constraints and $\mathcal{M}(x)$ is the set of objects (feasible solutions for x) that satisfy these constraints. In the typical case the problem instance x also determines the $cost(\alpha, x)$ for every solution $\alpha \in \mathcal{M}(x)$. The task is to find an optimal solution in the set $\mathcal{M}(x)$ of feasible solutions for x. The typical difficulty of solving $\mathcal{U}$ lies in the fact that the set $\mathcal{M}(x)$ has such a large cardinality that it is practically[10] impossible to generate all feasible solutions from $\mathcal{M}(x)$ in order to pick the best one.

To make the specification of optimization problems transparent, we often omit the specification of Σ_I and Σ_O and the specification of coding the data over Σ_I and Σ_O. We simply assume that the typical data such as integers, graphs, and formulae are represented in the way described above. This also simplifies the situation in that we can now address these objects directly instead of working with their formal representation. Therefore one can transparently describe an optimization problem by specifying

- the set L of problem instances,
- the constraints given by any problem instance $x \in L$ and the corresponding set $\mathcal{M}(x)$ for any $x \in L$,
- the cost function, and
- the goal.

[10] In an efficient way

Example 2.40. **Traveling Salesman Problem (TSP)**

TSP

Input: A weighted complete graph (G, c), where $G = (V, E)$, $V = \{v_1, \ldots, v_n\}$ for an $n \in \mathbb{N} - \{0\}$, and $c : E \to \mathbb{N} - \{0\}$.
{More precisely, an input is a word $x \in \{0, 1, \#\}^*$ that codes (represents) a weighted complete graph (G, c)}.

Constraints: For any problem instance (G, c), $\mathcal{M}(G, c)$ is the set of all Hamiltonian cycles of G. Each Hamiltonian cycle can be represented as a sequence $v_{i_1}, v_{i_2}, \ldots, v_{i_n}, v_{i_1}$ of vertices,[11] where $(i_1, i_2, \ldots, i_n)$ is a permutation of $(1, 2, \ldots, n)$.
{A strictly formal representation of $\mathcal{M}(G, c)$ is the set of all words $y_1\#y_2\#\ldots\#y_n \in \{0, 1, \#\}^* = \Sigma_O^*$ where $y_i \in \{0, 1\}^+$ for $i = 1, 2, \ldots, n$, with

$$\{Number(y_1), Number(y_2), \ldots, Number(y_n)\} = \{1, 2, \ldots, n\},$$

and $Number(y_1) = 1$.}

Costs: For every Hamiltonian cycle $H = v_{i_1}, v_{i_2}, \ldots, v_{i_n}, v_{i_1} \in \mathcal{M}(G, c)$,

$$cost((v_{i_1}, \ldots, v_{i_n}, v_{i_1}), (G, c)) = \sum_{j=1}^{n} c\left(\{v_{i_j}, v_{i_{(j \bmod n)+1}}\}\right),$$

i.e., the cost of every Hamiltonian cycle is the sum of the weights of all its edges.

Goal: minimum.

For the problem instance of TSP in Figure 2.4 we have

$$cost((v_1, v_2, v_3, v_4, v_5, v_1), (G, c)) = 8 + 1 + 7 + 2 + 1 = 19$$
$$cost((v_1, v_5, v_3, v_2, v_4, v_1), (G, c)) = 1 + 1 + 1 + 1 + 1 = 5.$$

The Hamiltonian cycle $v_1, v_5, v_3, v_2, v_4, v_1$ is the only optimal solution for this problem instance of TSP.

TSP is a hard optimization problem. But in many applications, we are only interested in problem instances that have some special nice properties that can make searching for an optimal solution easier. We say that an optimization problem $\mathcal{U}_1 = (\Sigma_I, \Sigma_O, L', \mathcal{M}, cost, goal)$ is a **subproblem** of an optimization problem $\mathcal{U}_2 = (\Sigma_I, \Sigma_O, L, \mathcal{M}, cost, goal)$ if $L' \subseteq L$, i.e., when $\mathcal{U}_1$ can be obtained from $\mathcal{U}_2$ by restricting the set of feasible (allowed) inputs. We define the **metric TSP (Δ-TSP)** as a subproblem of TSP. This means that the constraints, costs, and goal of Δ-TSP remain the same as in TSP and that only the set of input instances is reduced to a special subclass of inputs.

[11] Note, that this formalism assigns several different correct representations to any Hamiltonian cycle.

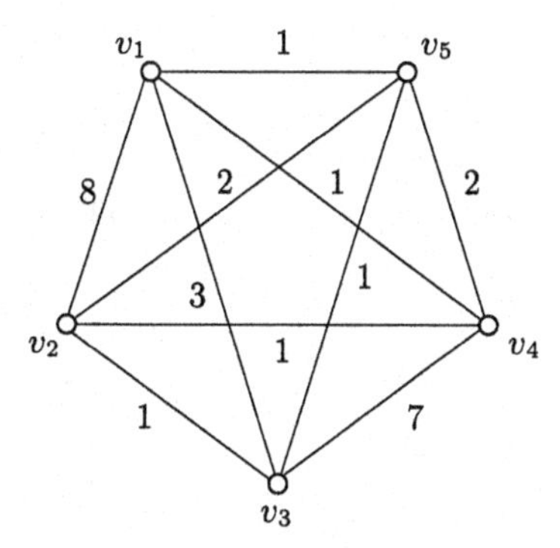

Fig. 2.4.

We require that every input instance (G, c) of Δ-TSP satisfies the so-called **triangle inequality**, i.e.,

$$c(\{u, v\}) \leq c(\{u, w\}) + c(\{w, v\})$$

for all vertices u, v, w of G. The triangle inequality can be viewed as a natural property of problem instances because it states that the direct connection between the vertices u and v may not be more expensive than any other path (detour) between u and v. Thus, if the vertices of G represent places (towns), the edges represent connections (roads) between these places, and the costs correspond to the distances, the triangle inequality may be a natural property of this model of reality. Observe, that the problem instance in Figure 2.4 does not satisfy the triangle inequality.

Exercise 2.41. Prove that $|\mathcal{M}((G, c))| = (n - 1)!/2$ for any graph G with n vertices where $n > 2$.

A **vertex cover** of a graph $G = (V, E)$ is any set U of vertices of G (i.e., $U \subseteq E$) such that every edge from E is incident[12] to at least one vertex from U. For instance, the set $\{v_2, v_4, v_5\}$ is a vertex cover of the graph in Figure 2.5, because each edge of this graph is incident with at least one of there three vertices. The set $\{v_1, v_2, v_3\}$ is not a vertex cover of the graph in Figure 2.5, because the edge $\{v_4, v_5\}$ is not covered by any of the vertices v_1, v_2, and v_3.

Exercise 2.42. The **minimum vertex cover problem, MIN-VCP**, is a minimization problem, where one searches for a vertex cover of minimal cardinality for a given graph G.

(i) Estimate the set of all vertex covers of the graph in Figure 2.5.
(ii) Give a formal specification of MIN-VCP as a 6-tuple. Use the alphabet $\{0, 1, \#\}$ to represent the input instances and the feasible solutions.

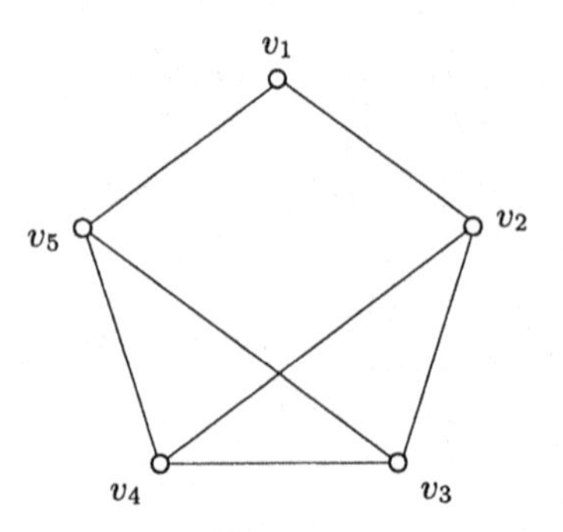

Fig. 2.5.

Example 2.43. **The maximum clique problem (MAX-CL)**
A clique of a graph $G = (V, E)$ is any set $U \subseteq V$, such that

$$\{\{u, v\} \mid u, v \in U, u \neq v\} \subseteq E,$$

i.e., any set U whose vertices build the complete subgraph of G of size $|U|$. The maximum clique problem is to find a clique of the maximal size in a given G. The formal representation of MAX-CL as an optimization problem follows.

MAX-CL
Input: A graph $G = (V, E)$.
Constraints: $\mathcal{M}(G) = \{S \subseteq V \mid \{\{u, v\} \mid u, v \in S, u \neq v\} \subseteq E\}$, i.e., $\mathcal{M}(G)$ contains all cliques of G.
Costs: For every $S \in \mathcal{M}(G)$, $cost(S, G) = |S|$.
Goal: maximum.

Exercise 2.44. A graph $T = (V, E')$ is called a **spanning tree** of a graph $G = (V, E)$ if T is a tree (a cycle-free connected graph) and $E' \subseteq E$. The weight of a spanning tree $T = (V, E')$ of G is $\sum_{e \in E'} c(e)$, i.e., the sum of the weights of all edges in E'. The minimum spanning tree problem is to find a spanning tree of the minimal weight in a given graph G. Give a formal specification of the minimum spanning tree problem as an optimization problem.

Example 2.45. **The maximum satisfiability problem (MAX-SAT)**
Let $X = \{x_1, x_2, \ldots\}$ be the set of Boolean variables. The set of all literals over X is $Lit_X = \{x, \overline{x} \mid x \in X\}$, where $\overline{x}$ is the negation of x for every variable x. The values 0 and 1 are called Boolean values (constants). A **clause** is any finite disjunction over literals (for instance, $x_1 \vee \overline{x}_3 \vee x_4 \vee \overline{x}_7$). A (Boolean) formula F is in the **conjunctive normal form (CNF)**, if F is a finite conjunction of clauses.

[12] An edge is incident to a vertex if this vertex is one of the two endpoints of this edge, i.e., the edge $\{u, v\}$ is incident to the vertices u and v.

An example of a formula over X in CNF is:

$$\Phi = (x_1 \vee x_2) \wedge (\overline{x}_1 \vee \overline{x}_2 \vee \overline{x}_3) \wedge \overline{x}_2 \wedge (x_2 \vee x_3) \wedge x_3 \wedge (\overline{x}_1 \vee \overline{x}_3).$$

The maximum satisfiability problem, MAX-SAT, is to find an input assignment to the variables of a given formula in CNF such that the number of satisfied clauses is maximized.

MAX-SAT

Input: A formula $\Phi = F_1 \wedge F_2 \wedge \cdots \wedge F_m$ over X in CNF, where F_i is a clause for $i = 1, \ldots, m$, $m \in \mathbb{N} - \{0\}$.

Constraints: For every formula Φ over a set $\{x_{i_1}, x_{i_2}, \ldots, x_{i_n}\}$ of n Boolean variables, the set of feasible solutions is

$$\mathcal{M}(\Phi) = \{0,1\}^n.$$

{Every $\alpha = \alpha_1 \ldots \alpha_n \in \mathcal{M}(\Phi)$, $\alpha_j \in \{0,1\}$ for $j = 1, \ldots, n$, represents an assignment where the value α_j is assigned to the variable x_{i_j}.}

Costs: For every formula Φ and any $\alpha \in \mathcal{M}(\Phi)$, $cost(\alpha, \Phi)$ is the number of clauses of Φ satisfied by α.

Goal: maximum.

For the formula Φ described above, Table 2.1 presents all eight assignments to the variables x_1, x_2, x_3 and we can easily observe that the assignments 001, 011 and 101 satisfy five clauses each and hence they are optimal solutions for Φ.

Table 2.1.

x_1 x_2 x_3	$x_1 \vee x_2$	$\overline{x}_1 \vee \overline{x}_2 \vee \overline{x}_3$	$\overline{x}_2$	$x_2 \vee x_3$	x_3	$\overline{x}_1 \vee \overline{x}_3$	# of satisfied clauses
0 0 0	0	1	1	0	0	1	3
0 0 1	0	1	1	1	1	1	5
0 1 0	1	1	0	1	0	1	4
0 1 1	1	1	0	1	1	1	5
1 0 0	1	1	1	0	0	1	4
1 0 1	1	1	1	1	1	0	5
1 1 0	1	1	0	1	0	1	4
1 1 1	1	0	0	1	1	0	3

Example 2.46. **Integer linear programming (ILP)**

Given a system of linear equations and a linear function over variables of this linear equation system, the task is to find a solution to the system of equations such that the value of the linear function is minimized. ILP can be phrased as an optimization problem as follows.

ILP

Input: A $m \times n$ matrix

$$A = [a_{ij}]_{i=1,\ldots m, j=1,\ldots,n}$$

and two vectors

$$b = (b_1, \ldots, b_m)^{\mathsf{T}} \text{ and } c = (c_1, \ldots, c_n)$$

for $n, m \in \mathbb{N} - \{0\}$, where a_{ij}, b_i, c_j are integers for $i = 1, \ldots, m$ and $j = 1, \ldots, n$.

Constraints: $\mathcal{M}(A, b, c) = \{X = (x_1, \ldots, x_n)^{\mathsf{T}} \in \mathbb{N}^n \mid AX = b\}$.

$\{\mathcal{M}(A, b, c)$ is the set of all solutions (vectors) that satisfy the system $AX = b$ of linear equations determined by A and b.

Costs: For every $X = (x_1, \ldots, x_n) \in \mathcal{M}(A, b, c)$,

$$cost(X, (A, b, c)) = c \cdot X = \sum_{i=1}^{n} c_i x_i.$$

Goal: minimum.

Aside from the decision problems and optimization problems introduced above we also consider algorithmic problems of another nature. These problems require no input. Their only task is to generate a word or an infinite sequence of symbols.

Definition 2.47. *Let Σ be an alphabet and let $x \in \Sigma^*$. We say that an algorithm* **generates** *the word x, if A outputs x for the input λ.*

The following program A generates the word 100111.

```
A :    begin
         write(100111);
       end
```

For every positive integer n, the following program A_n generates the word $(01)^n$.

```
A_n :  begin
         for i = 1 to n do
           write(01);
       end
```

A program that generates a word x can be viewed as an alternative representation of x. Thereby one can save some words in memory as programs that generate these words.

Definition 2.48. *Let Σ be an alphabet, and let $L \subseteq \Sigma^*$. A is an algorithm* **enumerating L**, *if, for every positive integer n, A outputs $x_1, x_2, \ldots, x_n$, where $x_1, x_2, \ldots, x_n$ are the first n words of L with respect to the canonical order.*

Example 2.49. Let $\Sigma = \{0\}$ and let $L = \{0^p \mid p \text{ is a prime}\}$.

Input: n
Output: $0^2, 0^3, 0^5, 0^7, \ldots, 0^{p_n}$, where p_n is the n-th smallest prime.

Exercise 2.50. Prove that a language is recursive if and only if there exists an algorithm that enumerates L.

2.4 Kolmogorov Complexity

In this section we consider words as information carriers and we focus on finding a reasonable way to measure the information contents of words. We do this only for words over the basic alphabet Σ_{bool}. An informal idea can be to say that a word w has a small information content if there is a short representation of this word (i.e., if it is compressible), and that a word w has a large information content if there does not exist any short representation of w (i.e., there is no representation shorter than $|w|$). The intuition behind this is that a word with small information content is regular and hence easy to describe, and a word with high information content is irregular.[13] Therefore the only way of representing it is to write it completely bit by bit. Based on this idea, the word

$$011011011011011011011011$$

with a short representation $(011)^8$ has a smaller information content than the word

$$0101101000101101001110110010.$$

The process of producing a representation of w that is shorter than $|w|$ or the previous representation of w is called the **compression**[14] of w.

The next idea could be to fix a convenient compression method and use the length of the resulting compressed representation of w as a measure of its information content. Clearly, one has to assert that the compressed representation of w is again a word over Σ_{bool} because the use of a larger alphabet to obtain a shorter representation of a given word yields no true compression.

Exercise 2.51. Find an injective mapping H from $(\Sigma_{\text{bool}})^*$ to $\{0,1,2,3,4\}^* = \Sigma_5^*$, such that

$$|x| \geq 2 \cdot |H(x)|$$

[13] A chaotic distribution of 0s and 1s

[14] Formally, a compression can be viewed as an injective mapping from $(\Sigma_{\text{bool}})^*$ to $(\Sigma_{\text{bool}})^*$.

for every $x \in (\Sigma_{\text{bool}})^*$, $|x| \geq 4$. Which compression factor can be achieved, if instead of Σ_5 one uses the alphabet Σ_m for an integer $m > 5$?

If one uses the representations for the compression of words, then one can start with the alphabet $\{0, 1, (,)\}$ and represent

$$w^a \text{ by } (w)Bin(a)$$

for any $w \in (\Sigma_{\text{bool}})^*$. Thus, for instance, (011)1000 stands for $(011)^8$ or

$$(0)1010(010)1(01)1101 \text{ stands for } (0)^{10}(010)^1(01)^{13}.$$

To get finally a representation over Σ_{bool}, one can use the homomorphism from $\{0, 1, (,)\}$ to Σ_{bool} defined by

$$h(0) = 00, \quad h(1) = 11, \quad h(\,(\,) = 10, \quad \text{and } h(\,)\,) = 01.$$

In this way, the compressed representation of $(011)^8$ becomes

$$10001111011000000.$$

Note, that this compression method is correct because each compressed representation of w unambiguously determines the original word w.

The problem is that one can propose infinitely many different compression mechanisms. Which is the right one? For instance, one can further improve the compression method introduced above by compressing the representation of powers. Thus, $(011)^{2^{20}}$ can be used as a shorter representation of $(011)^{1048576}$. Using this strategy, the compression can be arbitrarily improved by generating the representations such as

$$(01)1^{2^{2^n}}, (01)^{2^{2^{2^n}}}, \ldots$$

for regular words over Σ_{bool}. This means that for any of our compression methods M, there exists another compression method that is better than M for infinitely many words over Σ_{bool}. Therefore, if one wants to have an objective and robust measure of information content in words, one cannot take any of the above strategies.

Matters can get worse. Let us consider the following compression method.

For every $x \in (\Sigma_{\text{bool}})^*$, the nonnegative integer $Number(x)$ can be represented as its factorization

$$p_1^{i_1} \cdot p_2^{i_2} \cdot \cdots \cdot p_k^{i_k}$$

for primes $p_1 < p_2 < \cdots < p_k$, $i_1, i_2, \ldots, i_k \in \mathbb{N} - \{0\}$ for $j = 1, 2, \ldots, k$. A possible representation $p_1^{i_1} \cdot p_2^{i_2} \cdot \cdots \cdot p_k^{i_k}$ over $\{0, 1, (,), \}$ can be

$$Bin(p_1)(Bin(i_1))Bin(p_2)(Bin(i_2)) \ldots Bin(p_k)(Bin(i_k)).$$

Applying the above-introduced homomorphism h one obtains a binary representation of x. The bad news is that this compression method is incomparable with the method based on subword repetitions in the sense that compression by subword repetitions can perform better than compression by factorization and vice versa.

Exercise 2.52. Find two words $x, y \in (\Sigma_{\text{bool}})^*$ such that

1. the compression by subword repetitions provides a substantially shorter representation of x than the compression method by factorization, and
2. the compression by factorization results in a substantially shorter representation of y than the compression by repetitions.

A definition of a complexity measure must be robust, in the sense that the measured complexity has to have a broad validity, to be applicable in a general framework. Considering the dependence of the size of word representations on a particular compression method, it is obvious that fixing one compression method for the definition of the size of the information content of a word leads to a situation where one does not have any possibility to formulate generally valid assertions about the information content of words. The following definition by Kolmogorov provides a way out of this apparent deadlock. It is important to observe that the introduction of the algorithm (program) is the crucial point that enables one to find a way of information-content measurement.

Definition 2.53. *For any word $x \in (\Sigma_{\text{bool}})^*$, the* **Kolmogorov complexity $K(x)$ of the word x** *is the binary length*[15] *of the shortest Pascal program that generates x.*

We know that any compiler for Pascal generates the machine code of every syntactically correct program written in Pascal and that the machine code of a program is nothing other than a word over $(\Sigma_{\text{bool}})^*$. Hence, for every word $x \in (\Sigma_{\text{bool}})^*$, we consider all (infinitely many) machine codes of programs that generate x and the length of the shortest one is the Kolmogorov complexity of x.

Is $K(x)$ a good candidate for the size of the information content of x? When one wants to include all compression methods, then surely, yes. For every compression method M that computes a compression $M(x)$ to any word x, one can write a program that contains $M(x)$ as a parameter (a constant of the program) and generates x from $M(x)$. But before analyzing the definition of $K(x)$ in depth, we present some basic properties of Kolmogorov complexity to deepen our understanding of this concept.

The first property of $K(x)$ guarantees that $K(x)$ cannot be substantially larger that $|x|$. Obviously, this property is highly valued.

[15] More precisely, the length of the shortest binary code of a program generating x

Lemma 2.54. *There exists a constant d, such that for every $x \in (\Sigma_{\text{bool}})^*$*

$$K(x) \leq |x| + d.$$

Proof. For any $x \in (\Sigma_{\text{bool}})^*$ we consider the following program[16] A_x:

```
begin
  write(x);
end
```

The parts begin, write, end and commas of the program A_x are the same for every $x \in (\Sigma_{\text{bool}})^*$ and the length of their representation in the machine code is bounded by a small constant d independent of x. The word x is represented in A_x as x and therefore its contribution to the length of the (binary) machine code of A_x is exactly $|x|$. □

Obviously, regular words with many subword repetitions have small Kolmogorov complexity. Let $y_n = 0^n \in \{0,1\}^*$ for any $n \in \mathbb{N} - \{0\}$. The following program Y_n generates y_n.

```
begin
  for i = 1 to n do
    write(0);
end
```

All parts of programs Y_n are the same, except the number n. The length of $Bin(n)$ is exactly $\lceil \log_2(n+1) \rceil$ and thus the contribution of n to the binary representation of Y_n is at most $\lceil \log_2 n \rceil + 1$. Hence, there exists a constant c such that

$$K(y_n) \leq \lceil \log_2 n \rceil + c = \lceil \log_2 |y_n| \rceil + c$$

for all $n \in \mathbb{N} - \{0\}$.

Now, consider $z_n = 0^{n^2} \in \{0,1\}^*$ for any positive integer n. The following program Z_n generates the word z_n.

```
begin
  m := n;
  m := m × m;
  for i = 1 to m do
    write(0);
end
```

All programs Z_n are similar, except for the integer n. If the length of the binary code of all parts of Z_n except n is d, then

[16] For simplicity we use a Pascal-like programming language, where, for instance, the declaration of variables is omitted.

$$K(z_n) \leq \lceil \log_2(n+1) \rceil + d \leq \lceil \log_2(\sqrt{|z_n|} \rceil + d + 1.$$

Note, that the crucial point in our calculation is that the program Z_n does not need to store the value n^2 as a constant. Instead, Z_n stores the smaller value n, and the value n^2 is computed during the execution of the program. Since the size of the memory during the computation of Z_n is not included in the length of the description of Z_n, we have saved approximately $\lceil \log_2 n \rceil$ representation bits in this way.

Exercise 2.55. Prove the following assertion. There exists a constant c such that, for every positive integer n,

$$K\left((01)^{2^n}\right) \leq \lceil \log_2(n+1) \rceil + c = \left\lceil \log_2 \log_2 \left(\left|(01)^{2^n}\right|/2\right)\right\rceil + c.$$

Exercise 2.56. Find an infinite sequence of words $y_1, y_2, y_3, \ldots$ over Σ_{bool} that satisfies the following conditions:

(i) $|y_i| < |y_{i+1}|$ for all $i \in \mathbb{N} - \{0\}$, and
(ii) there exists a constant c such that, for all $i \in \mathbb{N} - \{0\}$,

$$K(y_i) \leq \lceil \log_2 \log_2 \log_2 |y_i| \rceil + c.$$

Exercise 2.57. Prove that for each positive integer m there exists a word w_m such that

$$|w_m| - K(w_m) > m.$$

One can also measure the information content of positive integers by simply taking the Kolmogorov-complexity of their binary representations.

Definition 2.58. *The* **Kolmogorov complexity of a positive integer** n *is*

$$\boldsymbol{K(n)} = K(Bin(n)).$$

Exercise 2.59. Let $n = pq$ be a positive integer. Prove

$$K(n) \leq K(p) + K(q) + c$$

for a constant c, that does not depend on n, p, and q.

The next fundamental result shows that there exist words that are not compressible with respect to Kolmogorov complexity.

Lemma 2.60. *For every positive integer n there exists a word $w_n \in (\Sigma_{\text{bool}})^n$, such that*

$$K(w_n) \geq |w_n| = n,$$

i.e., for any positive integer n there is a noncompressible word of length n.

Proof. The proof is based on the following simple combinatorial idea. We have exactly 2^n words $x_1, \ldots, x_{2^n}$ in $(\Sigma_{\text{bool}})^n$. Let, for $i = 1, 2, \ldots, 2^n$, C-Prog$(x_i) \in \{0,1\}^*$ be the machine code of the program Prog(x_i) that generates x_i and $K(x_i) = |\text{C-Prog}(x_i)|$, i.e., Prog$(x_i)$ is one of the shortest[17] programs that generate x_i.

Clearly, for $i \neq j$, C-Prog(x_i) and C-Prog(x_j) must be different because x_i and x_j are different. This means that we have 2^n different machine codes

$$\text{C-Prog}(x_1), \text{C-Prog}(x_2), \ldots, \text{C-Prog}(x_{2^n})$$

of shortest programs for $x_1, x_2, \ldots, x_{2^n}$. It is sufficient to show that at least one of these program machine codes has length at least n.

Our combinatorial argument simply says that one cannot have 2^n different machine codes (words), all shorter than n. Each machine code is a word over $(\Sigma_{\text{bool}})^*$. The number of words of length i over Σ_{bool} is exactly 2^i. Hence the number of all nonempty words over Σ_{bool} of length at most $n-1$ is

$$\sum_{i=1}^{n-1} 2^i = 2^n - 2 < 2^n.$$

Therefore, there is at least one word among C-Prog$(x_1), \ldots,$ C-Prog(x_{2^n}) with a length of at least n. Let C-Prog(x_j) be such a word with

$$|\text{C-Prog}(x_j)| \geq n.$$

Since

$$|\text{C-Prog}(x_j)| = K(x_j),$$

x_j is not compressible. □

Exercise 2.61. Prove that for all $i, n \in \mathbb{N} - \{0\}$, $i < n$, there exist $2^n - 2^{n-i}$ different words x in $(\Sigma_{\text{bool}})^n$ such that

$$K(x) \geq n - i.$$

Exercise 2.62. Prove that there are infinitely many positive integers m such that

$$K(m) \geq \lceil \log_2 m \rceil - 1.$$

Let us now return to the question of whether Kolmogorov complexity is a sufficiently robust measure for the information content of words. Instead of choosing a special compression method we have taken a formal model of programs that implicitly involves all possible compression methods. But one can regard fixing of the programming language Pascal as a restriction. Is it not possible that another language, for instance Java or C++, would provide

[17] Note, that there may exist several shortest programs that generate x_i and we simply fix one of them.

shorter representation than the representation by Pascal programs for some words? Does the commitment to Pascal decrease the robustness of measuring the information content of words?

The answers to these questions are negative. In the following theorem we show that fixing a programming language exerts only a limited influence on the Kolmogorov complexity of words and hence the Pascal-based definition of Kolmogorov complexity is a reasonable formalization for the intuitive notion information content.

For every word x over Σ_{bool} and any programming language A, let $K_A(x)$ be the Kolmogorov complexity of x with respect to the programming language A, i.e., the length of the shortest machine code of a program in A that generates x.

Theorem 2.63. *Let A and B be programming languages. There exists a constant $c_{A,B}$ depending only on A and B, such that*

$$|K_A(x) - K_B(x)| \leq c_{A,B}$$

for all $x \in (\Sigma_{\text{bool}})^*$.

Proof. We know that we can construct an interpreter $U_{A \to B}$ for any arbitrary programming languages A and B. $U_{A \to B}$ is a program in B that translates any program P_A (written in the programming language A) into an equivalent program P_B of the programming language B. $U_{A \to B}$ then executes program P_B on the input x of P_A. Considering the word generation we can simplify this by assuming that $U_{A \to B}$ receives a program P_A in the programming language A as an input parameter and that $U_{A \to B}$ executes the same computational (generation) as P_A does.

Hence, $U_{A \to B}$ with input P_A is a program in the programming language B that generates the same words as P_A. Let $c_{A \to B}$ be the binary length of the program $U_{A \to B}$ (without its input). Let P_x be a program in A that generates x. Then $U_{A \to B}$ with input P_x generates x, too. Since $U_{A \to B}$ is a program in B we have

$$K_B(x) \leq K_A(x) + c_{A \to B}. \tag{2.1}$$

If one takes an interpreter $U_{B \to A}$ from the programming language B to A with the binary length $c_{B \to A}$, then one obtains

$$K_A(x) \leq K_B(x) + c_{B \to A} \tag{2.2}$$

for every $x \in (\Sigma_{\text{bool}})^*$. Taking $c_{A,B}$ as the maximum of $c_{A \to B}$ and $c_{B \to A}$, the inequalities (2.1) and (2.2) imply

$$|K_A(x) - K_B(x)| \leq c_{A,B}.$$

□

Now that we have accepted Kolmogorov complexity as a reasonable measure of the information content of words, let us show its usefulness as an instrument for investigating important relationships in mathematics and computer science. The power of Kolmogorov complexity lies especially in the derivation of transparent combinatorial arguments. In what follows we present three applications of Kolmogorov complexity that can be understood without any special previous knowledge. Further fundamental applications are provided in the chapters on automata and computability.

The first application is on the fundamental level of the notion formation. "Chance" is one of the basic notions of science and we will deal with this term in Chapter 8 about randomized algorithms. Here, we would like to pose the following question:

> *"Which object or its word representation can be considered to be random?"*

The classical probability theory cannot help us to answer this question because it only assigns the occurrence probability to every possible event. If one has a random choice from the set of all words in $(\Sigma_{\text{bool}})^n$, then each word of this set has equal probability of being chosen. Could one relate the probability of choosing a word to its degree of randomness? If yes, then 0^n would be exactly as random as any irregular word and this does not correspond to the intuitive meaning of the adjective "random". According to the BBC English dictionary

> *"something is random if it is done without a definitive plan or pattern",*

i.e., a random object has a chaotic structure, it is completely irregular. Now our considerations about compression and information content come in handy. A word is random when it does not have any representation shorter than its length, i.e., it does not allow for any compression. In other words,

> *a word is random if any description of its creation is at least as large as its full representation bit by bit.*

Therefore, the following definition is considered to be the best-known formalization of the word "random".

Definition 2.64. *A word $x \in (\Sigma_{\text{bool}})^*$ is said to be* **random,**[18] *if*

$$K(x) \geq |x|.$$

A positive integer n is said to be **random,** *if*

$$K(n) = K(Bin(n)) \geq \lceil \log_2(n+1) \rceil - 1.$$

[18] Note that we subtract 1 from $\lceil \log_2(n+1) \rceil$ because we know that each binary representation of an integer begins with the digit 1.

The next application shows that the existence of a program (algorithm) solving a decision problem $(\Sigma_{\text{bool}}, L)$ provides some information about the Kolmogorov complexity of words in L. For instance, when no two words in L have the same length, then the Kolmogorov complexity of each word in L is at most $\log_2 |x|$.

Theorem 2.65. *Let L be a language over Σ_{bool}. Let z_n be the n-th word in L with respect to the canonical order for any positive integer n. Furthermore, assume that there exists a program (algorithm) A_L that decides $(\Sigma_{\text{bool}}, L)$. Then there exists a constant c independent of n, such that*

$$K(z_n) \leq \lceil \log_2(n+1) \rceil + c,$$

for all positive integers n.

Proof. For any positive integer n we design a program C_n that generates z_n. Each C_n contains A_L as a subroutine. The program C_n can be described as follows.

```
begin
  i = 0;
  x := λ;
  while i < n do begin
    ⟨Compute A_L(x) with the program A_L⟩;
    if A_L(x) = 1 then begin
      i := i + 1;
      z := x;
    end;
    x := the successor of x in the canonical order;
  end
  write(z);
end
```

We observe that C_n generates the words from $(\Sigma_{\text{bool}})^*$ in the canonical order. For each x, C_n checks whether $x \in L$ with the help of subroutine A_L. While doing this C_n counts the number of words accepted by A_L. Clearly, the output of C_n is the n-th word z_n of L.

We again observe that all programs C_n are identical except the parameter n. Let c be the length of the binary machine code of C_n except for the part coding n. Then, the binary length of C_n is exactly

$$c + \lceil \log_2(n+1) \rceil$$

for all positive integers n. Since C_n generates z_n, $K(z_n)$ is at most the binary length of C_n. □

A typical misunderstanding of the Kolmogorov complexity is that the binary length of C_n is dependent on the size of the memory needed to save the values of the variables c, i, and z during the computation of C_n. We refute this by pointing out the fact that saving the values of the variables of C_n may need more bits (a longer binary representation) and that the binary representation of C_n does not matter, since these values are not a part of the representation of C_n. In the description of C_n we only need to code the names of the variables. The only value that is a part of C_n is the value n. For a program C_n, the value n is a constant whose binary representation is a part of the machine code of C_n. In the next chapter we will develop the proof idea of Theorem 2.65 to get a method for proving that some languages cannot be recognized by finite automata.

Exercise 2.66. Let p be a polynomial of one variable. Let $L \subseteq (\Sigma_{\text{bool}})^*$ be an infinite recursive language with the property $|L \cap (\Sigma_{\text{bool}})^n| \leq p(n)$ for all positive integers n. Let, for any positive integer m, z_m be the m-th word in L with respect to the canonical order. What is the upper bound on $K(z_m)$ in terms of $|z_m|$ for any m?

The last application of Kolmogorov complexity presented in this section is related to a fundamental result of the number theory of enormous importance for the design of randomized algorithms. For any positive integer n, let *Prim* (n) denote the number of primes smaller than or equal to n. The following fundamental theorem says that primes are densely distributed among the integers. One can observe the density of primes among the positive integers in the following sequence of smallest natural numbers:

$$1, \underline{2}, \underline{3}, 4, \underline{5}, 6, \underline{7}, 8, 9, 10, \underline{11}, 12, \underline{13}, 14, 15, 16, \underline{17}, 18, \underline{19}, 20, 21, 22, \underline{23}, \ldots .$$

Theorem 2.67. Prime Number Theorem

$$\lim_{n \to \infty} \frac{\mathit{Prim}\,(n)}{n / \ln n} = 1.$$

The prime number theorem is one of the most remarkable discoveries in the whole of mathematics. It tells us that the number of primes[19] grows approximately as fast as the function $n/\ln n$. For "small" values of n the exact value *Prim* (n) can be computed. Table 2.2 shows some of these values.

The following inequality shows how close *Prim* (n) and $n/\ln n$ are for $n \geq 67$.

$$\ln n - \frac{3}{2} < \frac{n}{\mathit{Prim}\,(n)} < \ln n - \frac{1}{2}.$$

[19] Note, that the prime number theorem provides only this average density value and we do not know anything more about the distribution of the individual primes among the integers, except that this distribution is extremely irregular.

Table 2.2.

n	$Prim\,(n)$	$\frac{Prim(n)}{(n/\ln n)}$
10^3	168	~ 1.161
10^6	78 498	~ 1.084
10^9	50 847 478	~ 1.053

Exercise 2.68. Let p_i denote the i-th smallest prime for $i = 1, 2, \ldots$. Use Theorem 2.67 to prove that

$$\lim_{n\to\infty} \frac{p_n}{n \ln n} = 1.$$

Gauss, based on an empirical study of prime number tables, was the first to conjecture that $Prim\,(n)\,/n$ is approximately $1/\ln n$ especially for large n. But a rigorous proof of Gauss' conjecture was far beyond the power of mathematical methods during his time. The first strategic ideas for attacking this open problem were drawn by Riemann in the mid-19th century and the final, very complicated proof was given independently by Hadamard and Vallée Pousson in 1896. An interesting point is that they needed the "unnatural" concept of complex numbers in order to prove this theorem about natural numbers and it took yet another half century before Norbert Wiener was able to modify the proof so as to avoid the use of complex numbers. Another important point is that the question as to how fast $Prim\,(n)\,/(n/\ln n)$ converges to 1 is strongly related to the famous extended Riemann's hypothesis, one of the most spectacular open problems in mathematics.

Now, we arrive at the point where we want to show the usefulness of the concept of Kolmogorov complexity by giving a simple combinatorial proof of a weaker version of the prime number theorem. This weaker version of the prime number theorem is still strong enough for all our applications in the design of randomized algorithms, and the presented simple proof helps us to understand why there are so many primes. The crucial idea is to show that if the density of primes is essentially smaller than $n/\ln n$, then one could represent any integer in such a short way that it would contradict the injectivity of this representation.[20]

First, we show a fundamental result that says there are infinitely many primes with a special property.

Lemma 2.69. *Let $S = n_1, n_2, n_3, \ldots$ be an increasing infinite sequence of positive integers with $K(n_i) \geq \lceil \log_2 n_i \rceil / 2$. Let, for $i = 1, 2, \ldots$, q_i be the largest prime factor of n_i. Then, the set*

$$Q = \{q_i \mid i \in \mathbb{N} - \{0\}\}$$

is infinite.

[20] We have already seen that it is impossible to provide a compressed representation for each word (number).

Proof. We prove Lemma 2.69 by contradiction. Assume that $Q = \{q_i \mid i \in \mathbb{N} - \{0\}\}$ is a finite set. Let p_m be the largest prime in Q. Then, every positive integer $n_i, i \in \mathbb{N} - \{0\}$, of the sequence S can be unambiguously expressed as

$$n_i = p_1^{r_{i,1}} \cdot p_2^{r_{i,2}} \cdot \dots \cdot p_m^{r_{i,m}}$$

for some nonnegative integers $r_{i,1}, r_{i,2}, \dots, r_{i,m} \in \mathbb{N}$. Since $p_1, p_2, \dots, p_m$ are fixed for the representation of every integer in S, n_i is determined by the exponents $r_{i,1}, r_{i,2}, \dots, r_{i,m}$. Hence, one can write a program A_i that generates n_i for given $r_{i,1}, \dots, r_{i,m}$. Let c be the binary length of the program A_i, excluding the representation of the parameters $r_{i,1}, \dots, r_{i,m}$ (i.e., the binary length of the representation of that part of A_i that is the same for all programs A_j). Since the binary length of A_i gives an upper bound on $K(n_i)$, we obtain for all $i \in \mathbb{N} - \{0\}$,

$$K(n_i) \leq c + \lceil \log_2(r_{i,1}+1) \rceil + \lceil \log_2(r_{i,2}+1) \rceil + \dots + \lceil \log_2(r_{i,m}+1) \rceil.$$

Since $r_{i,j} \leq \log_2 n_i$ for all $j \in \{1, 2, \dots, m\}$ we have, for all $i \in \mathbb{N} - \{0\}$,

$$K(n_i) \leq c + m \cdot \lceil \log_2(\log_2 n_i + 1) \rceil.$$

Since m and c are constants with respect to i, the inequality

$$\lceil \log_2 n_i \rceil / 2 \leq c + m \cdot \lceil \log_2(\log_2 n_i + 1) \rceil$$

can be true for only finitely many i. However, this contradicts our assumption that

$$K(n_i) \geq \lceil \log_2 n_i \rceil / 2$$

for all positive integers n. □

Exercise 2.70. In Lemma 2.69 we assume that there exists an increasing infinite sequence of positive integers $n_1, n_2, n_3, \dots$ with the property $K(n_i) \geq \lceil \log_2 n_i \rceil / 2$. How far can you weaken this assumption without touching the validity of the conclusion of Lemma 2.69?

Exercise 2.71. Let t be a positive integer. How big is the subset of integers in the interval $[2^t, 2^t + 1, \dots, 2^{t+1} - 1]$, that does not satisfy

$$K(n) \geq \lceil \log_2 n \rceil / 2?$$

Lemma 2.69 not only shows that there exist infinitely many primes, but also that the set of the largest prime factors of any infinite sequence of positive integers with nontrivial Kolmogorov complexity is infinite. We will use this assertion for proving the following lower bound for *Prim* (n).

Theorem 2.72.* *For infinitely many* $k \in \mathbb{N}$,

$$Prim\,(k) \geq \frac{k}{64 \log_2 k \cdot (\log_2 \log_2 k)^2}.$$

Proof. Recall that p_j denotes the j-th smallest prime for $j = 1, 2, \ldots$. Let $n \geq 2$ be an arbitrary integer, and let p_m be the largest prime factor of n. Obviously, n can be generated from $(p_m, n/p_m)$ by simply multiplying p_m with n/p_m. This can even be done with the smaller information $(m, n/p_m)$ because there is a program that, for a given m, generates p_m by simply enumerating the first m primes.

Let us now think about how to (unambiguously) represent the pair $(m, n/p_m)$ as a word over Σ_{bool}. One cannot simply concatenate $Bin(m)$ and $Bin(n/p_m)$ because the boundary between $Bin(m)$ and $Bin(n/p_m)$ cannot be identified, resulting in several interpretations of $Bin(m)Bin(n/p_m)$. The solution is to consider coding $Bin(m)$ in the following way. Let

$$Bin(m) = a_1 a_2 a_3 \ldots a_{\lceil \log_2(m+1) \rceil}$$

for $a_i \in \Sigma_{\text{bool}}$ for $i = 1, 2, \ldots, \lceil \log_2(m+1) \rceil$. Then we define

$$\overline{Bin}(m) = a_1 0 a_2 0 a_3 0 \ldots a_{\lceil \log_2(m+1) \rceil - 1} 0 a_{\lceil \log_2(m+1) \rceil} 1.$$

The word $\overline{Bin}(m)Bin(n/p_m)$ is an unambiguous representation of $(m, n/p_m)$, and hence of n, because the end of the code $\overline{Bin}(m)$ of m is unambiguously indicated by the first 1 on an even position in $\overline{Bin}(m)Bin(n/p_m)$. The length of this representation is

$$2 \cdot \lceil \log_2(m+1) \rceil + \lceil \log_2(\frac{n}{p_m} + 1) \rceil.$$

But this length is still too large for us and we try to decrease it by coding $(m, n/p_m)$ by

$$\overline{Bin}(\lceil \log_2(m+1) \rceil) Bin(m) Bin(n/p_m).$$

This representation is unambiguous because the end of $\overline{Bin}(\lceil \log_2(m+1) \rceil)$ is marked by the first 1 on an even position and $\overline{Bin}(\lceil \log_2(m+1) \rceil)$ says that the following $\lceil \log_2(m+1) \rceil$ bits belong to the representation of $Bin(m)$. The length of this coding of $(m, n/p_m)$ is

$$2\lceil \log_2 \lceil \log_2(m+1) \rceil \rceil + \lceil \log_2(m+1) \rceil + \lceil \log_2(\frac{n}{p_m} + 1) \rceil.$$

Using the above-introduced strategy one can perform infinitely many improvements, but for our theorem it is sufficient to make only one more. Definitively, we represent $(m, n/p_m)$ as

$$\begin{aligned}
&Word(m, n/p_m) = \\
&\overline{Bin}(\lceil \log_2 \lceil \log_2(m+1) \rceil \rceil) Bin(\lceil \log_2(m+1) \rceil) Bin(m) Bin(n/p_m).
\end{aligned}$$

Hence,

$$\begin{aligned}
|\,Word(m, n/p_m)| = {} & 2 \cdot \lceil \log_2 \lceil \log_2 \lceil \log_2(m+1) \rceil \rceil \rceil \\
& + \lceil \log_2 \lceil \log_2(m+1) \rceil \rceil \\
& + \lceil \log_2(m+1) \rceil + \lceil \log_2(\frac{n}{p_m} + 1) \rceil.
\end{aligned} \tag{2.3}$$

Now, the word $Word(m, n/p_m)$ is considered as a compression of $Bin(n)$. Since we consider a fixed, uniform compression strategy for every n, using the same arguments as in Lemma 2.60 and Exercise 2.61, we obtain, for all $i \in \mathbb{N} - \{0\}$, that more than half of the integers n in

$$S_i = \{2^i, 2^i + 1, \ldots, 2^{i+1} - 1\}$$

have

$$|Word(m, n/p_m)| \geq \lceil \log_2(n+1) \rceil - 1.$$

Analogously, more than half of the integers n in S_i have

$$K(n) \geq \lceil \log_2(n+1) \rceil - 2.$$

This implies that, for every positive integer i, there exists an integer $n_i \in S_i$ (i.e., $2^i \leq n_i \leq 2^{i+1} - 1$) such that

$$|Word(m, n_i/p_m)| \geq \lceil \log_2(n_i + 1) \rceil - 1$$

and

$$K(n) \geq \lceil \log_2(n_i + 1) \rceil - 2.$$

In other words there are infinitely many positive integers n satisfying the inequalities

$$|Word(m, n/p_m)| \geq \lceil \log_2(n+1) \rceil - 1 \tag{2.4}$$

and

$$K(n) \geq \lceil \log_2(n+1) \rceil - 2. \tag{2.5}$$

Combining inequalities (2.4) and (2.3), one obtains

$$\begin{aligned}\lceil \log_2(n+1) \rceil - 1 \leq\ & 2\lceil \log_2 \lceil \log_2 \lceil \log_2(m+1) \rceil \rceil \rceil \\ & + \lceil \log_2 \lceil \log_2(m+1) \rceil \rceil + \lceil \log_2(m+1) \rceil + \lceil \log_2 (\frac{n}{p_m} + 1) \rceil.\end{aligned}$$

Because $\log_2(n/p_m) = \log_2 n - \log_2 p_m$ we get

$$\begin{aligned}\log_2 p_m &\leq 2 \cdot \lceil \log_2 \log_2 \log_2(m+1) \rceil + \lceil \log_2 \log_2(m+1) \rceil + \lceil \log_2(m+1) \rceil + 2 \\ &\leq 2 \cdot \log_2 \log_2 \log_2 m + \log_2 \log_2 m + \log_2 m + 6.\end{aligned}$$

Therefore

$$p_m \leq 64 \cdot m \cdot \log_2 m \cdot (\log_2 \log_2 m)^2. \tag{2.6}$$

Because of inequality (2.5) the infinite sequence $n_1, n_2, n_3, \ldots$ satisfies the assumptions of Lemma 2.69. Therefore inequality (2.6) holds for infinitely many m (infinitely many primes p_m). This means, that the m-th smallest prime is smaller than

$$64 \cdot m \cdot \log_2 m \cdot (\log_2 \log_2 m)^2$$

for infinitely many m. This is equivalent to saying that there are at least m primes among the first $64 \cdot m \cdot \log_2 m \cdot (\log_2 \log_2 m)^2$ positive integers, i.e.,

$$Prim\left(64 \cdot m \cdot \log_2 m \cdot (\log_2 \log_2 m)^2\right) \geq m \tag{2.7}$$

for infinitely many m.

We substitute $k = 64 \cdot m \cdot \log_2 m \cdot (\log_2 \log_2 m)^2$ in the inequality (2.7). Since $k \geq m$, we obtain

$$\begin{aligned} Prim\,(k) \geq m &= \frac{k}{64 \cdot \log_2 m \cdot (\log_2 \log_2 m)^2} \\ &\geq \frac{k}{64 \cdot \log_2 k \cdot (\log_2 \log_2 k)^2}. \end{aligned}$$

□

Exercise 2.73. Consider $\overline{Bin}(\lceil \log_2 m \rceil) \cdot Bin(m) \cdot Bin(n/p_m)$ as the compressed representation of $n = p_m(n/p_m)$. Which lower bound on $Prim\,(k)$ would be derived if one takes this representation of n in the proof of Theorem 2.67?

Exercise 2.74. Which assertion could be made if one takes

$$\begin{aligned} &\overline{Bin}(\lceil \log_2 \lceil \log_2 \lceil \log_2 (m+1) \rceil \rceil \rceil) Bin(\lceil \log_2 \lceil \log_2 (m+1) \rceil \rceil) \\ &Bin(\lceil \log_2 (m+1) \rceil) Bin(m) Bin(n/p_m) \end{aligned}$$

as the representation of $n = p_m \cdot (n/p_m)$ in the proof of Theorem 2.67?

Exercise 2.75.* What is the tightest possible lower bound on $Prim\,(k)$ that is achievable by the argument used in the proof of Theorem 2.67?

Exercise 2.76.* Let $p_1, p_2, p_3, \ldots$ be the increasing sequence of all primes. Apply the Prime Number Theorem to prove that there exists a constant c such that

$$K(p_m) \leq \lceil \log_2 p_m \rceil - \lceil log_2 \log_2 p_m \rceil + c$$

for all positive integers m.

2.5 Summary and Outlook

In this chapter we have introduced the fundamental terms such as alphabet, word and language. An alphabet is an arbitrary nonempty finite set of symbols equivalent to the script of a natural language. The term word over an alphabet corresponds to any text that is comprised of symbols of this alphabet. Any set of words (texts) over the same alphabet is called a language.

These basic terms are used in all areas of data processing. We begin with the specification of algorithmic tasks, where inputs and outputs are always represented as words. A decision problem is in fact specified by a language L. The task is to decide whether a given word is in L or not. An optimization problem has a more complex, structured definition. Any input word codes a set of constraints and determines costs of objects (feasible solutions) that satisfy these constraints. The constraints are typically satisfied by many feasible solutions and the objective is to find a solution that is optimal in terms of the cost.

Kolmogorov complexity is a reasonable concept for measuring the size of the information content (the amount of information) of words over the alphabet $\Sigma_{\text{bool}} = \{0, 1\}$. The Kolmogorov complexity of a word x is the binary length of the shortest Pascal program that generates x. We can consider a program that generates a word x as a compression (a compressed representation) of x. Hence, the Kolmogorov complexity of x is the length of the shortest compressed representation of x over $\{0, 1\}$. Incompressible words are very irregular words and such words are considered to be random. There are infinitely many random words over $\{0, 1\}$, at least one for every length. The Kolmogorov complexity helps not only to measure the size of information content of a word and the degree of randomness of a word. It is also used as a powerful, transparent instrument for proving results in many areas of computer science and mathematics. We have applied the Kolmogorov complexity argument not only to show that there are many primes, but also to explain why there must be so many. A small density of primes among positive integers would imply the existence of compressed representations for all words, which contradicts the proven existence of random (incompressible) words.

The study of alphabets, words, languages, and their representations is the topic of the formal language theory. This theory is one of the oldest and most classical areas of computer science. Since key computer science objects such as information, data, programs, theorems, communication messages, memory contents, proofs, and computations are represented as words, the formal language theory provides a basis for other fundamental areas of computer science such as computability, compiler design, etc. In this introductory material we reduced the presentation of the knowledge of the formal language theory to a minimum necessary and sufficient for following the main topics of this book presented in the next chapters. One reason for this is that there are already many excellent textbooks on formal language theory and we do not see any reason for writing the '101-st' book on this topic. We warmly recommend "Introduction to Automata Theory, Language, and Computation" by Hopcroft and Ullman [28] as one of the most successful classical textbooks in theoretical computer science to the reader interested in fundamentals of formal language theory. Note that there is also a new edition of this book [27] that is extended by some new current topics. Another wonderful classical textbook specialized on formal languages is that by Salomaa [60]. The main feature of the textbooks we mentioned above is the excellent, transparent way of introducing new concepts and explaining the arguments at the intuitive level as well as rigorously in detail. It is precisely this feature that made these books milestones in the development of computer science education. Among the more recent German textbooks we would recommend Erk and Priese [18], Schöning [62], and Wegener [71].

The idea of introducing Kolmogorov complexity as a measure of the information content of words was discovered in the 1960s independently by Kolmogorov [39, 40] and Chaitin [9, 10, 11]. An exhaustive survey on this topic is provided by the monograph [43] written by Li and Vitányi. Unfortunately this book is primarily written for researchers and so it is not easily readable for beginners. Transparent and exciting examples of Kolmogorov complexity applications can be found in Schöning [62].

The proof of a high education
is the ability
to speak about complex matters
as simple as possible.

R. Emerson

3

Finite Automata

3.1 Objectives

A finite automaton is the simplest computing model considered in computer science. In fact, finite automata correspond to special restricted programs that solve decision problems without the use of any variable. Finite automata work in real time in the sense that the input word is read only once from the left to the right and the output (result) is fixed immediately after the last symbol of the input has been read.

The main reason for introducing finite automata here is not to present the fundamentals of automata theory. Instead, this chapter serves a didactic purpose – finite automata provide a very transparent and simple way for explaining how to model computations. Introducing the basic notions of theoretical computer science such as configuration, computation step, computation, simulation, determinism, and nondeterminism for the finite automaton model gives one a reasonable intuition about the general meaning of these fundamental terms and also some experience in working with them. This intuition helps us later in simplifying the way of explaining the exact meaning of these terms in the framework of general models of algorithmic computations.

Thus, in this chapter we will learn how to rigorously but transparently investigate and model a subclass of algorithms. Besides the first encounter with the above-mentioned fundamental terms of computer science, we also learn what proving the unsolvability of a concrete task in a restricted subclass of algorithms means.

3.2 Different Representations of Finite Automata

When defining a computing model, one should answer the following questions:

1. Which elementary operations are allowed, i.e., what instructions are available for building a program?

2. How large is the available memory and how may one work with (access) it?
3. How do inputs enter the computer (computing model)?
4. How is the output (result) determined?

A finite automaton does not have any memory, except the memory where the program is saved and the pointer showing the currently executed row (instruction) of the program. This means that the program may not use any variable. At first glance this may seem to be surprising because one can ask how can a program compute without the use of variables. The idea is that the content of the pointer, i.e., the number of the actual (executed) program row is the only changing information and the program computes with this pseudovariable.

If $\Sigma = \{a_1, a_2, \ldots, a_k\}$ is the alphabet over which the inputs are represented, then a finite automaton may only use the following instructions:

$$\texttt{select } \mathit{input} = a_1 \texttt{ goto } i_1$$

$$\mathit{input} = a_2 \texttt{ goto } i_2$$

$$\vdots$$

$$\mathit{input} = a_k \texttt{ goto } i_k$$

The meaning of this instruction (elementary operation) `select` is that the program reads the next input symbol and compares it with all $a_1, a_2, \ldots, a_k$. If the input symbol read is equal to a_j, then the program continues to work in row i_j (i.e., the next instruction executed is the instruction written in row i_j). The execution of this instruction is automatically coupled with the deletion of the symbol read, allowing the finite automaton to read the next input symbol in row i_j. Every row of the program contains exactly one instruction of the above type. The rows of the program are numbered by $0, 1, 2, 3, \ldots$ and the work (computation) of the program always starts in row 0. When Σ consists of two symbols only (for instance, 0 and 1), one could use the following simpler instruction `if` ... `then` ... `else` instead of `select`.

$$\texttt{if } \mathit{input} = 1 \texttt{ then goto } i \texttt{ else goto } j.$$

Such programs are used for solving decision problems. The answer is determined by the row numbers. If a program consists of m rows $\{0, 1, 2, \ldots, m-1\}$, then one could choose a subset $F \subset \{0, 1, 2, \ldots, m-1\}$. We can interpret F as the accepting set. Upon reading the last symbol of the input word, and the program reaches a row $j \in F$, then the program accepts this word (i.e., the program gives the answer 'yes' for this input). If $j \in \{0, 1, 2, \ldots, m-1\} - F$, then the program does not accept the input (i.e., the answer is 'no'). The set

L of all words accepted by such a program (finite automaton) is the language recognized by this program. In other words, we say that a finite automaton solves a decision problem (Σ, L) if it recognizes the language L.

To illustrate the above description of a finite automaton, consider the following program (finite automaton) A that works over the alphabet Σ_{bool}.

0: if *input* = 1 then goto 1 else goto 2

1: if *input* = 1 then goto 0 else goto 3

2: if *input* = 0 then goto 0 else goto 3

3: if *input* = 0 then goto 1 else goto 2

Let us choose $F = \{0, 3\}$. The program A works on the input 1011 as follows. It starts in row 0 and moves to row 1 after reading the first 1 of the input. After that it reads the symbol 0 in row 1 and moves to row 3. In row 3 it reads 1 and goes to row 2. Finally after reading the last symbol 1 of the input, it moves back to row 3. The computation is over and since $3 \in F$, the word 1011 is accepted.

Figure 3.2 presents a schema that is often used to introduce a finite automaton as a computing model. Here, we see the three main components of this model – the **program** saved in the memory, the **tape** (as the input medium) that contains the input word, and the **reading head** that may move on the tape from left to right only.[1] The tape, also called the input tape, can be viewed as a linear memory for the input. The tape consists of a sequence of squares (cells). A square is considered to be a basic memory unit that may contain exactly one symbol of the alphabet. The actual tape always has exactly as many squares as the number of symbols of the input word.

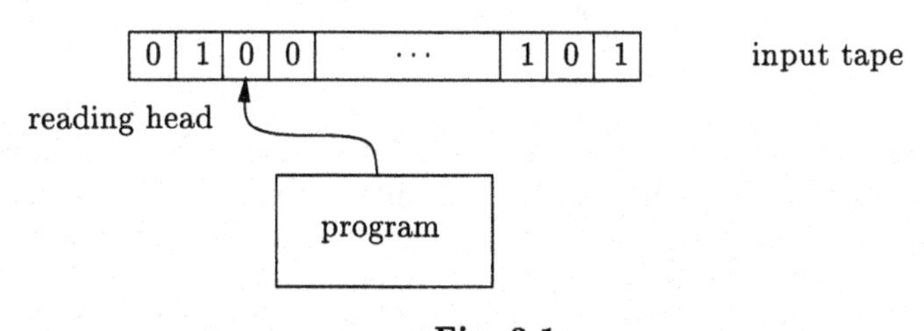

Fig. 3.1.

[1] The schema of components of general computing models involves additionally a memory and the description of possibilities for accessing the memory (for reading, writing, deleting, and inserting data). A general model may contain an output medium, too.

The class of programs described above is seldom used to define finite automata, because these programs use the goto instruction and hence they usually lack a nice structure.[2] Therefore, this way of modeling finite automata is not very transparent and is unwieldy for most purposes. The idea of a user-friendly definition of a finite automaton is based on the following visualization of our special programs. To any program A we assign a directed labeled graph $G(A)$. $G(A)$ has exactly as many vertices as the numbers of rows of A. Each vertex of $G(A)$ corresponds to exactly one row of A, and the order of this row is the name (label) of this vertex. If program A moves from the row i to the row j upon reading an input symbol b, then $G(A)$ contains the directed edge (i, j) that is labeled by b. Because our special programs without variables have the instruction goto for each symbol $a \in \Sigma$ in every row,[3] each vertex of $G(A)$ has the outdegree[4] $|\Sigma|$. Figure 3.2 shows the graph $G(A)$ that corresponds to the above-presented program A of 4 rows. The vertices corresponding to the rows 0 and 3 in F are drawn as double rings in order to distinguish them from the others. The vertex that corresponds to the row 0 is distinguished by an additional, special pointer (arc) entering it (Figure 3.2).

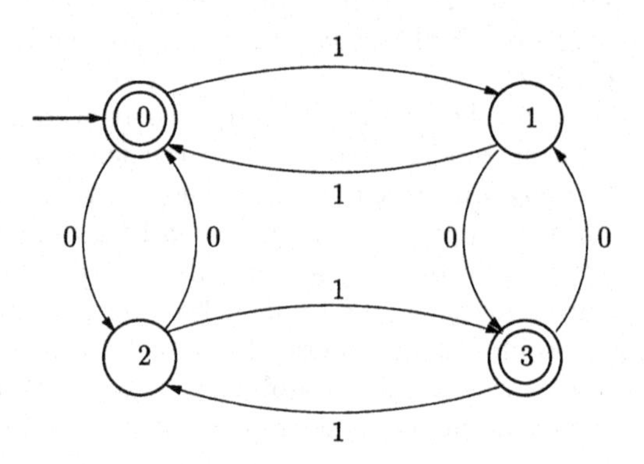

Fig. 3.2.

Starting with this graphic representation of programs without variables we will develop a standard formal definition of finite automata. But we continue to use this graphic representation since it provides a very transparent and unambiguous way of describing a finite automaton. However, the following definition is more convenient for studying the properties of finite automata and for proving results about them. For this purpose we shall slightly modify the standard terminology employed in automata theory. The terms "row of

[2] In terms of program schemas

[3] Every row is a select instruction over all symbols of the alphabet.

[4] The outdegree of a vertex of a directed graph G is the number of directed edges (arcs) leaving this vertex, i.e., the outdegree of a vertex v is $|\{(v, u) \mid (v, u) \text{ is an edge in G}\}|$.

a program" or equivalently "vertex of the corresponding graph" will be exchanged for (renamed by) the notion state of a finite automaton. The directed edges of the graph corresponding to the goto instructions of the program will be described by the so-called transition function that assigns a new state for each current state and the symbol read.

We bring attention to the important fact that the following definition of a finite automaton follows a general schema that can be applied for defining all computing models. First, one defines a structure that provides an exact description of every object of the modeled class of algorithms. Then, the semantic of this structure is fixed and explained. This is done in the following order. First, one defines the term configuration. A configuration is a complete description of the current situation (of the current general state) of our machine. Then one defines a step as a movement from one configuration to another determined by the execution of an actual instruction (elementary operation) of the machine (program). A computation is viewed as a sequence of steps. When the term of computation has been defined, one can assign the result of the computation of the machine on any input u to this input.

Definition 3.1. *A (deterministic)* **finite automaton (FA)** *is a 5-tuple* $M = (Q, \Sigma, \delta, q_0, F)$, *where*

(i) Q *is a finite nonempty set of* **states**
  {formerly, the set of all rows of a program without variables},
(ii) Σ *is an alphabet, called the* **input alphabet** *of* M
  {i.e., feasible inputs are all words over Σ *},*
(iii) $q_0 \in Q$ *is the initial state*
  {previously, the row 0 of a program without variables},
(iv) $F \subseteq Q$ *is the* **set of all accepting states,**[5] *and*
(v) δ *is a function from* $Q \times \Sigma$ *to* Q, *called the* **transition function** *of* M
  {$\delta(q, a) = p$ *states that* M *moves from the state* q *to the state* p *when the symbol* a *has been read (Figure 3.3)}.*

A **configuration** *of* M *is any element from* $Q \times \Sigma^*$.

{When M *has reached a configuration* $(p, w) \in Q \times \Sigma^*$, *which means that* M *is in the state* p *and* M *has still to read the suffix* w *of an input word (i.e., the reading head is adjusted on the first symbol of* w *as depicted in Figure 3.3.}*

A configuration $(q_0, x) \in \{q_0\} \times \Sigma^*$ *is called the* **initial configuration of** M **on** x.

[5] Some authors also use the term final state instead of the term accepting state. But this can be misleading. First, a computation can finish in any state and then the adjective final is not very appropriate. Second and mainly, the meaning of the terms final state and accepting state do not overlap in more general computation models such as Turing machines. Thus, we discourage the use of the term final state when dealing with automata.

{The work (computation) of M on x must start in the initial configuration (q_0, x)}.
Any configuration from $Q \times \{\lambda\}$ is called a **final configuration**.

A step of M is a relation (on configurations)

$$\vdash_{\overline{M}} \subseteq (Q \times \Sigma^*) \times (Q \times \Sigma^*),$$

defined by

$$(q, w) \vdash_{\overline{M}} (p, x) \Leftrightarrow w = ax, \ a \in \Sigma \text{ and } \delta(q, a) = p.$$

{A step corresponds to an application of the transition function on the actual configuration, in which M is in a state q and reads an input symbol a on the tape.}

A **computation** *C of M is a finite sequence $C = C_0, C_1, \ldots, C_n$ of configurations such that $C_i \vdash_{\overline{M}} C_{i+1}$ for all i, $0 \leq i \leq n - 1$.*

{Thus, one can also view a computation as a sequence of steps of M. Sometimes we may even prefer the description $C_0 \vdash_{\overline{M}} C_1 \vdash_{\overline{M}} \ldots \vdash_{\overline{M}} C_n$ instead of $C_0, C_1, \ldots, C_n$.}

C is the **computation of M on an input $x \in \Sigma^*$**, *if $C_0 = (q_0, x)$ and $C_n \in Q \times \{\lambda\}$.*

{The computation on x has to start in the initial configuration (q_0, x) of M on x and may terminate only when all input symbols have been read.}

If $C_n \in F \times \{\lambda\}$, we would say that C is an **accepting computation** *of M on x and equivalently that* **M accepts x**.

If $C_n \in (Q - F) \times \{\lambda\}$, we would say that C is a **rejecting computation** *of M on x and equivalently that* **M rejects x**.

{Note, that M has exactly one computation for any input $x \in \Sigma^*$.}

The **language $L(M)$ accepted by M** *is defined as*

$$\begin{aligned} L(M) &:= \{w \in \Sigma^* \mid \text{the computation of } M \text{ on } w \text{ finishes in} \\ &\qquad\qquad \text{a final configuration } (q, \lambda) \text{ with } q \in F\} \\ &= \{w \in \Sigma^* \mid M \text{ accepts } w\}. \end{aligned}$$

The **class of regular languages**

$$\mathcal{L}(\mathbf{FA}) = \{L(M) \mid M \text{ is an FA}\}$$

is the class of all languages that are accepted by finite automata. Any language L from $\mathcal{L}$(FA) is **regular**.

Let us once again use the program A to illustrate the above definition of a finite automaton. The formal description of the corresponding finite automaton is $M = (Q, \Sigma, \delta, q_0, F)$, where

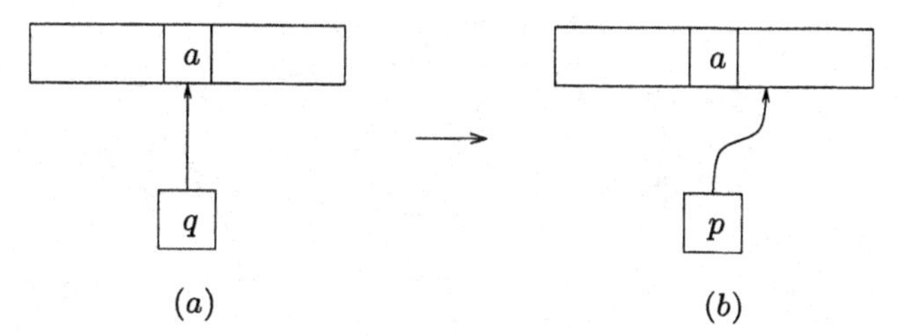

Fig. 3.3.

$Q = \{q_0, q_1, q_2, q_3\}$,
$\Sigma = \{0, 1\}$,
$F = \{q_0, q_3\}$, and
$\delta(q_0, 0) = q_2$, $\delta(q_0, 1) = q_1$, $\delta(q_1, 0) = q_3$, $\delta(q_1, 1) = q_0$,
$\delta(q_2, 0) = q_0$, $\delta(q_2, 1) = q_3$, $\delta(q_3, 0) = q_1$, $\delta(q_3, 1) = q_2$.

A transparent description of the above-described transition function δ is presented in Table 3.1.

Table 3.1.

State	Input	
	0	1
q_0	q_2	q_1
q_1	q_3	q_0
q_2	q_0	q_3
q_3	q_1	q_2

The graphic representation $G(A)$ of A in Figure 3.2 is, as previously mentioned, a transparent representation of A. In what follows we shall use the slightly modified[6] representation of A in Figure 3.4 as the standard representation of the corresponding finite automaton. The computation of M on the input 1011 is

$$(q_0, 1011) \vdash_{\overline{M}} (q_1, 011) \vdash_{\overline{M}} (q_3, 11) \vdash_{\overline{M}} (q_2, 1) \vdash_{\overline{M}} (q_3, \lambda).$$

Since $q_3 \in F$, we have $1011 \in L(M)$, i.e., M accepts 1011.

The following definition introduces some notation that is useful for the formal work with finite automata, especially for giving a short, elegant presentation of some proofs.

[6] The only modification is labeling vertices using the names of the states instead of natural numbers corresponding to the order of program rows. This kind of labeling can sometimes increase the degree of transparency (Figure 3.6), since suitably chosen names of the states may provide more information than the orders of any linear ordering of the states.

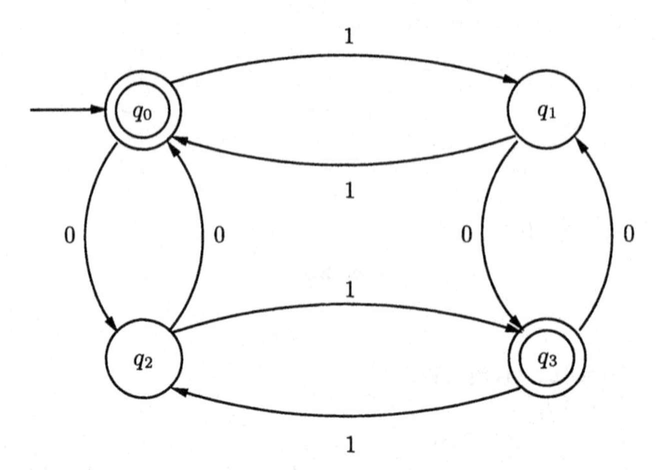

Fig. 3.4.

Definition 3.2. *Let $M = (Q, \Sigma, \delta, q_0, F)$ be a finite automaton. We define $\vdash_M^*$ as the reflexive and transitive closure of the step relation $\vdash_M$ of M, i.e.,*

$$(q, w) \vdash_M^* (p, u) \Leftrightarrow (q = p \text{ and } w = u)$$

or $\exists k \in \mathbb{N} - \{0\}$, such that

(i) $w = a_1 a_2 \ldots a_k u$, $a_i \in \Sigma$ for $i = 1, 2, \ldots, k$ and
(ii) $\exists r_1, r_2, \ldots, r_{k-1} \in Q$, such that
$(q, w) \vdash_M (r_1, a_2 \ldots a_k u) \vdash_M (r_2, a_3 \ldots a_k u) \vdash_M \cdots (r_{k-1}, a_k u) \vdash_M (p, u)$.

We define $\hat{\delta} : Q \times \Sigma^* \to Q$ by

(i) $\hat{\delta}(q, \lambda) = q$ for all $q \in Q$, and
(ii) $\hat{\delta}(q, wa) = \delta(\hat{\delta}(q, w), a)$ for all $a \in \Sigma$, $w \in \Sigma^*$, $q \in Q$.

The meaning of

$$(q, w) \vdash_M^* (p, u)$$

is that a computation of M starting in the configuration (q, w) reaches the configuration (p, u). The equality

$$\hat{\delta}(q, w) = p$$

means that if M has started to read the word w in the state q, then M finishes its work in the state p, i.e.,

$$\hat{\delta}(q, w) = p \text{ is equivalent to } (q, w) \vdash_M^* (p, \lambda).$$

Hence, we obtain the following equivalent descriptions of $L(M)$:

$$\begin{aligned} L(M) &= \{w \in \Sigma^* \mid (q_0, w) \vdash_M^* (p, \lambda) \text{ with } p \in F\} \\ &= \{w \in \Sigma^* \mid \hat{\delta}(q_0, w) \in F\}. \end{aligned}$$

Now, we will try to recognize which language is accepted by the finite automaton M in Figure 3.4, i.e., to determine $L(M)$. We observe that the computation of M on words with even [odd] number of 1s finishes either in q_0 or in q_2 [either in q_1 or in q_3]. Similarly, if the numbers of 0s in an input word x is even [odd], then $\hat{\delta}(q_0, x) \in \{q_0, q_1\}$ [$\hat{\delta}(q_0, x) \in \{q_2, q_3\}$]. This observation leads to the following assertion.

Lemma 3.3. $L(M) = \{w \in \{0,1\}^* \mid |w|_0 + |w|_1 \equiv 0 \pmod 2\}$.

Proof. First we observe that every FA partitions Σ^* in $|Q|$ classes

$$\begin{aligned} \mathbf{Kl[p]} &= \{w \in \Sigma^* \mid \hat{\delta}(q_0, w) = p\}, \\ &= \{w \in \Sigma^* \mid (q_0, w) \vdash_{\overline{M}} (p, x)\} \end{aligned}$$

and it is obvious that for all $p, q \in Q$, $p \neq q$,

$$\bigcup_{p \in Q} \mathrm{Kl}[p] = \Sigma^* \text{ and } \mathrm{Kl}[p] \cap \mathrm{Kl}[q] = \emptyset.$$

Using this terminology

$$L(M) = \bigcup_{p \in F} \mathrm{Kl}[p].$$

In other words, the following relation

$$x R_\delta y \Leftrightarrow \hat{\delta}(q_0, x) = \hat{\delta}(q_0, y)$$

is an equivalence relation on Σ^*, which determines finitely many classes Kl[p] (Figure 3.5).

Thus, a sure way of successfully determining $L(M)$ is to determine the sets Kl[q_0], Kl[q_1], Kl[q_2], and Kl[q_3] for our FA M. To do so we propose the following induction hypothesis,

$$\begin{aligned} \mathrm{Kl}[q_0] &= \{w \in \{0,1\}^* \mid |w|_0 \text{ and } |w|_1 \text{ are even}\}, \\ \mathrm{Kl}[q_1] &= \{w \in \{0,1\}^* \mid |w|_0 \text{ is even, } |w|_1 \text{ is odd}\}, \\ \mathrm{Kl}[q_2] &= \{w \in \{0,1\}^* \mid |w|_0 \text{ is odd, } |w|_1 \text{ is even}\}, \text{ and} \\ \mathrm{Kl}[q_3] &= \{w \in \{0,1\}^* \mid |w|_0 \text{ and } |w|_1 \text{ are odd}\}. \end{aligned}$$

Since

$$\mathrm{Kl}[q_0] \cup \mathrm{Kl}[q_3] = \{w \in \{0,1\}^* \mid |w|_0 + |w|_1 \equiv 0 \pmod 2\},$$

the claim of Lemma 3.3 is a direct consequence of the above hypothesis. Hence, to complete the proof of Lemma 3.3 it is sufficient to proof this induction hypothesis. We shall do it by induction over the input length.

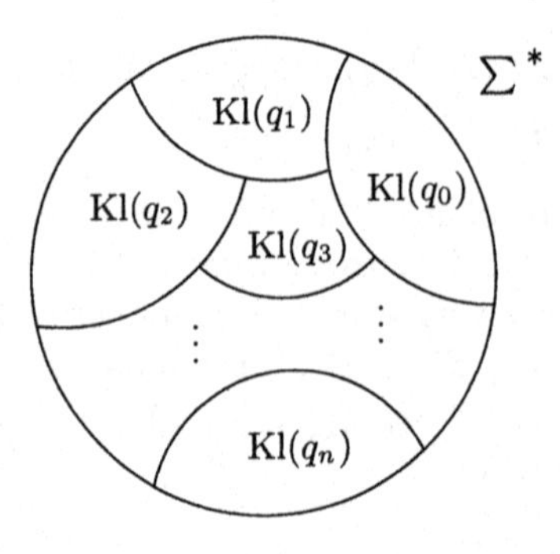

Fig. 3.5.

1. *Induction basis.*
   We prove the induction hypothesis for all words of at most length 2.
   $\hat{\delta}(q_0, \lambda) = q_0 \Rightarrow \lambda \in \mathrm{Kl}[q_0]$.
   $\hat{\delta}(q_0, 1) = q_1 \Rightarrow 1 \in \mathrm{Kl}[q_1]$.
   $\hat{\delta}(q_0, 0) = q_2$ and so $0 \in \mathrm{Kl}[q_2]$.
   $(q_0, 00) \underset{M}{\vdash} (q_2, 0) \underset{M}{\vdash} (q_0, \lambda) \Rightarrow 00 \in \mathrm{Kl}[q_0]$.
   $(q_0, 01) \underset{M}{\vdash} (q_2, 1) \underset{M}{\vdash} (q_3, \lambda) \Rightarrow 01 \in \mathrm{Kl}[q_3]$.
   $(q_0, 10) \underset{M}{\vdash} (q_1, 0) \underset{M}{\vdash} (q_3, \lambda) \Rightarrow 10 \in \mathrm{Kl}[q_3]$.
   $(q_0, 11) \underset{M}{\vdash} (q_1, 1) \underset{M}{\vdash} (q_0, \lambda) \Rightarrow 11 \in \mathrm{Kl}[q_0]$.

   Hence, the induction hypothesis is true for the words of length 0, 1, and 2.
2. *Induction step.*
   Let $i \geq 2$ be an integer. Assume that the induction hypothesis holds for all $x \in \{0,1\}^*$, $|x| \leq i$. Our aim is to show that it also holds for the words of length $i+1$. If we prove the induction step for all $i \geq 2$, then the induction hypothesis would hold for all words from $(\Sigma_{\text{bool}})^*$. Let w be an arbitrary word from $(\Sigma_{\text{bool}})^{i+1}$. Then, $w = za$, where $z \in \Sigma^i$ and $a \in \Sigma$. We distinguish four possibilities with respect to the parities of $|z|_0$ and $|z|_1$.
   (a) $|z|_0$ and $|z|_1$ are both even.
   (b) $|z|_0$ and $|z|_1$ are both odd.
   (c) $|z|_0$ is even and $|z|_1$ is odd.
   (d) $|z|_0$ is odd and $|z|_1$ is even.
   We now deal with each possibility.
   (a) Let both $|z|_0$ and $|z|_1$ be even. Following the induction hypothesis for z (note that $|z| = i$), we obtain $\hat{\delta}(q_0, z) = q_0$, i.e., $z \in \mathrm{Kl}[q_0]$. Therefore

$$\hat{\delta}(q_0, za) = \delta(\hat{\delta}(q_0, z), a) \underset{ind.}{=} \delta(q_0, a) = \begin{cases} q_1, \text{ if } a = 1, \\ q_2, \text{ if } a = 0. \end{cases}$$

Since $|z1|_0$ is even and $|z1|_1$ is odd, the result $\hat{\delta}(q_0, z1) = q_1$ corresponds to the induction hypothesis $z1 \in \mathrm{Kl}[q_1]$.
Since $|z0|_0$ is odd and $|z0|_1$ is even, the above result $\hat{\delta}(q_0, z0) = q_2$ agrees with the induction hypothesis $z0 \in \mathrm{Kl}[q_2]$.

(b) Let both $|z|_0$ and $|z|_1$ be odd. Applying the induction hypothesis for z we obtain $\hat{\delta}(q_0, z) = q_3$, i.e., $z \in \mathrm{Kl}[q_3]$.

$$\hat{\delta}(q_0, za) = \delta(\hat{\delta}(q_0, z), a) \underset{ind.}{=} \delta(q_3, a) = \begin{cases} q_2, \text{ if } a = 1, \\ q_1, \text{ if } a = 0. \end{cases}$$

This result agrees with the induction hypothesis that claims $z0 \in \mathrm{Kl}[q_1]$ and $z1 \in \mathrm{Kl}[q_2]$.

The cases (c) and (d) are similar and we leave it to the reader to complete the proof. □

Exercise 3.4. Complete the proof of the validity of the induction hypothesis in Lemma 3.3, i.e., solve the cases (c) and (d).

Exercise 3.5. Rewrite the proof of Lemma 3.3 using the notation $\vdash_M^*$ and $\vdash_M$ instead of $\hat{\delta}$ (i.e., the notation $\hat{\delta}$ is forbidden).

Exercise 3.6. Let $L = \{w \in (\Sigma_{\text{bool}})^* \mid |w|_0 \text{ is odd}\}$. Design an FA M with $L(M) = L$ and prove $L(M) = L$.

Exercise 3.7. Design finite automata for the languages $\emptyset, \Sigma^*$ and Σ^+ for an arbitrary alphabet Σ. Give the formal 5-tuple representations as well as the corresponding graphic representations.

If a transparent and detailed description of an FA A is present, then one can also determine $L(A)$ without proving it formally. Usually one dispenses with the kind of formal proofs[7] as given in Lemma 3.3. But reading the proof of Lemma 3.3 we have learned something very important for the design of finite automata. A good design strategy is to partition the set Σ^* of all inputs into subclasses with respect to some properties of the words and to determine the transitions between these classes in the context of concatenating a symbol from Σ to the words in these subclasses. Consider this strategy for the design of an FA for the language

$$U = \{w \in (\Sigma_{\text{bool}})^* \mid |w|_0 = 3 \text{ and } (|w|_1 \geq 2 \text{ or } |w|_1 = 0)\}.$$

To check if $|w|_0 = 3$, every FA B with $L(B) = U$ has to be able to distinguish the cases

[7] Note, that this situation is similar to designing programs. Because of an enormous amount of work one usually dispenses with proving the formal correctness of one's program.

$$|w|_0 = 0,\ |w|_0 = 1,\ |w|_0 = 2,\ |w|_0 = 3 \text{ and } |w|_0 \geq 4$$

for any word w (i.e., B has to count the number of 0s (up to four) in all prefixes read). Simultaneously, A has to start to count the number of the occurrences of the symbol 1 in order to be able to distinguish the following three cases

$$|w|_1 = 0,\ |w|_1 = 1 \text{ and } |w|_1 \geq 2.$$

The result of the above consideration is the idea to take the following set of states (the following partition of $\{0,1\}^*$ into subclasses):

$$Q = \{q_{i,j} \mid i \in \{0,1,2,3,4\}, j \in \{0,1,2\}\}.$$

The meaning of these states is as follows:
For all $i \in \{0,1,2,3\}$ and all $j \in \{0,1\}$,

$$\mathrm{Kl}[q_{i,j}] = \{w \in (\Sigma_{\mathrm{bool}})^* \mid |w|_0 = i \text{ and } |w|_1 = j\},$$

$$\mathrm{Kl}[q_{i,2}] = \{w \in (\Sigma_{\mathrm{bool}})^* \mid |w|_0 = i \text{ and } |w|_1 \geq 2\},$$

$$\mathrm{Kl}[q_{4,j}] = \{w \in (\Sigma_{\mathrm{bool}})^* \mid |w|_0 \geq 4 \text{ and } |w|_1 = j\}, \text{ and}$$

$$\mathrm{Kl}[q_{4,2}] = \{w \in (\Sigma_{\mathrm{bool}})^* \mid |w|_0 \geq 4 \text{ and } |w|_1 \geq 2\}.$$

Clearly, $q_{0,0}$ is the initial state of B. The transition function of B can be directly determined from the meaning[8] of the states $q_{i,j}$ as we have done in Figure 3.6. One easily observes that

$$U = \mathrm{Kl}[q_{3,0}] \cup \mathrm{Kl}[q_{3,2}]$$

and fixes $F = \{q_{3,0}, q_{3,2}\}$.

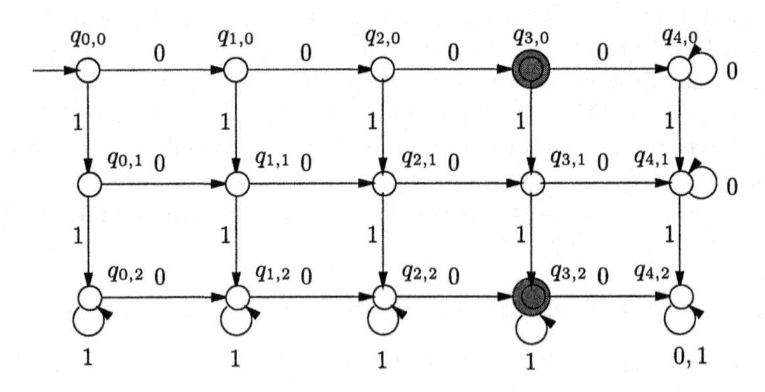

Fig. 3.6.

[8] More precisely, from the definition of the classes $Kl[q_{i,j}]$

Exercise 3.8. (a) Prove that $L(B) = U$ for the finite automaton B in Figure 3.6.
(b) Design an FA A such that $L(A) = U$ and A has a smaller number of states than B. Determine the class $\text{Kl}[q]$ for every state q of your FA A.

Exercise 3.9. Design an FA for each of the following regular languages:

(a) $\{w \in \{0,1,2\}^* \mid w = 002122x,\ x \in (\Sigma_{\text{bool}})^*\}$,
(b) $\{w \in \{a,b,c\}^* \mid w = xabcabc,\ x \in \{a,b,c\}^*\}$,
(c) $\{w \in \{a,b,c\}^* \mid w = xaabby,\ x, y \in \{a,b,c\}^*$,
(d) $\{w \in \{0,1\}^* \mid |w|_0 \equiv 1 \pmod 3$ and $w = x111y$ for $x, y \in \{0,1\}^*\}$,
(e) $\{abbxb^3y \mid x, y \in \{a,b\}^*\}$, and
(f) $\{w \in \{a,b\}^* \mid w = abbz$ for a $z \in \{a,b\}^*$ and $w = ub^3v$ for $u, v \in \{a,b\}^*\}$.

It is sufficient to give a graphic representations of these automata. Give the class $\text{Kl}[q]$ for every state q of each designed FA.

3.3 Simulations

Simulation is one of the most frequent terms in computer science. Despite its importance nobody has tried to provide a formal definition of this term. The reason for this is that this notion has different interpretations within different frameworks. The most narrow definition of the simulation of a computation requires for every (elementary) step of the simulated computation to be mimicked by exactly one step of the simulating computation. A slightly weaker requirement is that a step of the simulated computation may be mimicked by several steps of the simulating computation. One can also simulate without mimicking every step and only require that the counterparts to some important configurations of the simulated computation are reached. The most general definition only requires the same input-output behavior and it does not matter how the outputs to given inputs are computed.

In this section we show a simulation in the narrow sense. We do this by constructing an FA that can simultaneously simulate the computations of two other finite automata in a step-by-step manner.

Lemma 3.10. *Σ be an alphabet and let $M_1 = (Q_1, \Sigma, \delta_1, q_{01}, F_1)$ and $M_2 = (Q_2, \Sigma, \delta_2, q_{02}, F_2)$ be finite automata. For each set operation $\odot \in \{\cup, \cap, -\}$ there exists an* FA *M, such that*

$$L(M) = L(M_1) \odot L(M_2).$$

Proof. The idea of the proof is to construct the FA M in such a way that M can simultaneously simulate the work of both finite automata M_1 and M_2 on the same word.[9] The simulation idea is simple. The states of M are pairs

[9] In fact, there is no other possibility, because M may read the input only once and thus M cannot simulate first M_1 and then M_2.

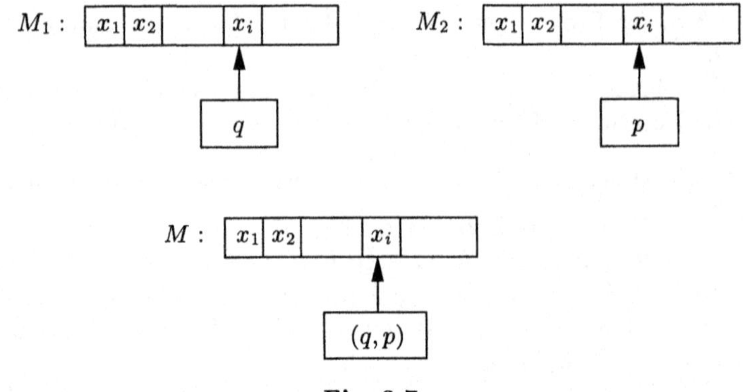

Fig. 3.7.

(q, p), where q is a state of M_1 and p is a state of M_2. The first element of (q, p) has to be q exactly when M_1 has reached the state q. Analogously, the second element of the state of M should have to be p when M_2 is in the state p. (Figure 3.7).

We present the formal proof[10] in two steps. First, we give the formal construction of the FA M and then we prove that M simulates both M_1 and M_2.

The construction of M.

Let $M = (Q, \Sigma, \delta, q_0, F_\odot)$, where

(i) $Q = Q_1 \times Q_2$,
(ii) $q_0 = (q_{01}, q_{02})$,
(iii) for all $q \in Q_1$, $p \in Q_2$ and $a \in \Sigma$, $\delta((q,p),a) = (\delta_1(q,a), \delta_2(p,a))$,
(iv) if $\odot = \cup$, then $F = F_1 \times Q_2 \cup Q_1 \times F_2$
{At least one of M_1 and M_2 stops in an accepting state.},
if $\odot = \cap$, then $F = F_1 \times F_2$
{Both M_1 and M_2 must accept}, and
if $\odot = -$, then $F = F_1 \times (Q_2 - F_2)$
{M_1 must accept and M_2 may not accept.}.

Proof of the claim $L(M) = L(M_1) \odot L(M_2)$.

To prove the claim for any $\odot \in \{\cup, \cap, -\}$, it is sufficient to prove that for all $x \in \Sigma^*$,

$$\hat{\delta}((q_{01}, q_{02}), x) = (\hat{\delta}_1(q_{01}, x), \hat{\delta}_2(q_{02}, x)). \tag{3.1}$$

We prove (3.1) by induction according to $|x|$.

1. *Induction basis.*
   If $x = \lambda$, then it is obvious that (3.1) holds.

[10] As usual in such cases

2. *Induction step.*
   For each positive integer i, we prove that the validity of (3.1) for all $x \in \Sigma^*$ with $|x| \leq i$ implies the validity of (3.1) for all $w \in \Sigma^{i+1}$.

   Let w be an arbitrary word from Σ^{i+1}. We can express w as

   $$w = za \text{ where } z \in \Sigma^i \text{ and } a \in \Sigma.$$

   Following the definition of $\hat{\delta}$, we obtain

   $$\begin{aligned}
   \hat{\delta}((q_{01}, q_{02}), w) &= \hat{\delta}((q_{01}, q_{02}), za) \\
   &= \delta(\hat{\delta}((q_{01}, q_{02}), z), a) \\
   &\underset{(3.1)}{=} \delta((\hat{\delta}_1(q_{01}, z), \hat{\delta}_2(q_{02}, z)), a) \\
   &\underset{\text{Def.}\delta}{=} (\delta_1(\hat{\delta}_1(q_{01}, z), a), \delta_2(\hat{\delta}_2(q_{02}, z), a)) \\
   &= (\hat{\delta}(q_{01}, za), \hat{\delta}(q_{02}, za)) \\
   &= (\hat{\delta}(q_{01}, w), \hat{\delta}(q_{02}, w)).
   \end{aligned}$$

□

Exercise 3.11. Let $L \subseteq \Sigma^*$ be a regular language. Prove that $L^{\mathsf{C}} = \Sigma^* - L$ is also a regular language.

3.4 Proofs of Nonexistence

To prove that a concrete language L is not regular, it is sufficient to show that there does not exist any finite automaton that accepts L. Usually, showing that a given problem is not solvable in a specific class of algorithms (i.e., that no program from this class solves the problem) belongs to the hardest tasks in theoretical computer science. The proofs of such assertions are called nonexistence proofs. In contrast to constructive proofs, where one proves the existence of an object with required properties by simply constructing such an object (for instance, one constructs an FA M with four states for a given regular language), proving the nonexistence of an object with required properties in an infinite set of candidates (for instance, all finite automata) cannot be done by checking whether each candidate fulfills the required property. To prove the nonexistence of an object with given properties in an infinite class of candidates, one typically has to uncover a deep understanding of this candidate class that contradicts the required properties.

Since the class of finite automata is a class of very strongly restricted programs, the nonexistence proofs of the kind "there does not exist any FA that accepts a given language L" are relatively easy. We use this fact to present a simple and transparent introduction to the methodology of creating nonexistence proofs.

We know that finite automata are called finite because their only possible memory content[11] is the actual state (the order of the actual row of the program). The consequence is that if an FA A has reached the same state when reading two different words x and y (i.e., $\hat{\delta}(q_0, x) = \hat{\delta}(q_0, y)$), then A can no longer distinguish x from y. Formally, this means that

$$\hat{\delta}(q_0, x) = \hat{\delta}(q_0, y)$$

implies that, for all $z \in \Sigma^*$,

$$\hat{\delta}_A(q_0, xz) = \hat{\delta}_A(q_0, yz).$$

We formulate this important property of finite automata in the following lemma.

Lemma 3.12. *Let $A = (Q, \Sigma, \delta_A, q_0, F)$ be an* FA. *Let, for some $x, y \in \Sigma^*$, $x \neq y$,*

$$(q_0, x) \vdash_A^* (p, \lambda) \text{ and } (q_0, y) \vdash_A^* (p, \lambda)$$

for a $p \in Q$ (i.e., $\hat{\delta}_A(q_0, x) = \hat{\delta}_A(q_0, y) = p$ $(x, y \in \mathrm{Kl}[p])$). Then for any $z \in \Sigma^*$ there exists an $r \in Q$, such that xz and $yz \in \mathrm{Kl}[r]$ and consequently

$$xz \in L(M) \Leftrightarrow yz \in L(M).$$

Proof. The existence of the computations

$$(q_0, x) \vdash_A^* (p, \lambda) \text{ and } (q_0, y) \vdash_A^* (p, \lambda)$$

of A on x and y implies the existence of the following computations on xz and yz

$$(q_0, xz) \vdash_A^* (p, z) \quad \text{and} \quad (q_0, yz) \vdash_A^* (p, z)$$

for all $z \in \Sigma^*$. When $r = \hat{\delta}_A(p, z)$ (i.e., when $(p, z) \vdash_A^* (r, \lambda)$ is the computation of A on z from the state p), then the computation of A on xz is

$$(q_0, xz) \vdash_A^* (p, z) \vdash_A^* (r, \lambda)$$

and the computation of A on yz is

$$(q_0, yz) \vdash_A^* (p, z) \vdash_A^* (r, \lambda).$$

If $r \in F$, then both words xz and yz belong to $L(A)$. If $r \notin F$, then both xz and yz do not belong to $L(A)$. □

Lemma 3.12 is a special property of all deterministic computing models. If a deterministic machine (algorithm) has reached the same configuration[12]

[11] Only possibility of saving information

[12] This is true only if a configuration is considered as a complete description of the general state of the machine that also includes the still accessible part of the input.

in computations on two different inputs, then the remaining parts of both computations are identical. In the case of decision problems, the consequence is that either both inputs are accepted or both inputs are rejected.

One can easily apply Lemma 3.12 to show that some languages are not regular. Let us illustrate this for the language

$$L = \{0^n 1^n \mid n \in \mathbb{N}\}.$$

Intuitively, L should be too hard for any finite automaton because to compare the number of zeros with the number of ones it is necessary first to memorize (count) the number of zeros. But the number of zeros in the prefix 0^n can be arbitrarily large and since every FA has a fixed size (a fixed number of states), no FA can count the number of zeros in input words. One only needs to give a formal argument why this counting is necessary in order to accept L. We show by contradiction that $L \notin \mathcal{L}(\text{FA})$.

Assume $L \in \mathcal{L}(\text{FA})$. Let $A = (Q, \Sigma_{\text{bool}}, \delta_A, q_0, F)$ be an FA with $L(A) = L$. Consider the words

$$0^1, 0^2, 0^3, \ldots, 0^{|Q|+1}.$$

Since the number of these words is $|Q|+1$, there exist $i, j \in \{1, 2, \ldots, |Q|+1\}$, $i < j$, such that

$$\hat{\delta}_A(q_0, 0^i) = \hat{\delta}_A(q_0, 0^j).$$

Following Lemma 3.12

$$0^i z \in L \Leftrightarrow 0^j z \in L$$

for all $z \in (\Sigma_{\text{bool}})^*$. But this is not true because for $z = 1^i$,

$$0^i 1^i \in L \text{ and } 0^j 1^i \notin L.$$

Exercise 3.13. Prove by Lemma 3.12 that the following languages are not in $\mathcal{L}(\text{FA})$:

(a) $\{w \in \{a, b\}^* \mid |w|_a = |w|_b\}$,
(b) $\{a^n b^m c^n \mid n, m \in \mathbb{N}\}$,
(c) $\{w \in \{0, 1, \#\}^* \mid w = x\#x \text{ for an } x \in \{0, 1\}^*\}$,
(d) $\{x1y \in \{0, 1\}^* \mid |x| = |y|\}$.

To give a simple and transparent method for proving nonregularity of concrete languages, one can search for easily verifiable properties [conditions] that every regular language has [satisfies]. If a language does not have this property, one can directly conclude that L is not regular. In what follows we present two methods for proving claims of the kind $L \notin \mathcal{L}(\text{FA})$. The first method is called "pumping". It is based on the following idea. If

$$(p, x) \vdash_A^* (p, \lambda),$$

for a state p and a word x, then

$$(p, x^i) \vdash_A^* (p, \lambda)$$

for all natural numbers i (Figure 3.8). This means that A cannot memorize how often it has read the word x and hence A cannot distinguish between words with different numbers of repetitions of x. Therefore, if $\hat{\delta}_A(q_0, y) = p$ for a $y \in \Sigma^*$ and $\hat{\delta}_A(p, z) = r$ for an $z \in \Sigma^*$ (Figure 3.8), then

$$(q_0, yx^iz) \vdash_A^* (p, x^iz) \vdash_A^* (p, z) \vdash_A^* (r, \lambda)$$

is the computation of A on yx^iz for all $i \in \mathbb{N}$, i.e.,

$$\{yx^iz \mid i \in \mathbb{N}\} \subseteq \mathrm{Kl}[r]$$

for an $r \in Q$. This means that A either accepts all words yx^iz for any $i \in \mathbb{N}$ (if $r \in F$), or A does not accept any word from $\{yx^iz \mid i \in \mathbb{N}\}$ (if $r \in Q - F$).

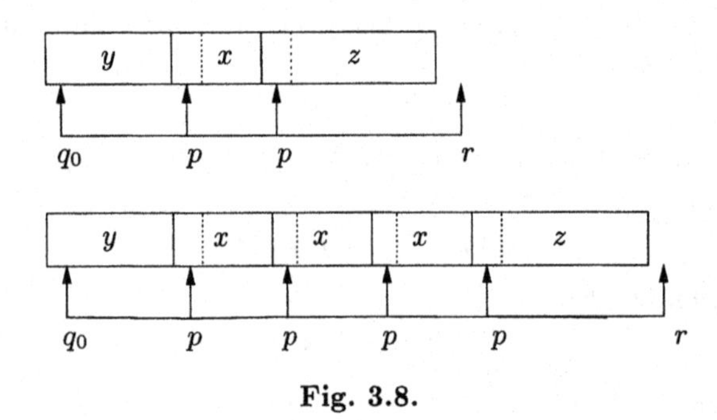

Fig. 3.8.

Lemma 3.14. The pumping lemma for regular languages

For every regular language L, there exists a constant $n_0 \in \mathbb{N}$ such that every word $w \in \Sigma^*$ with $|w| \geq n_0$ can be expressed as

$$w = yxz,$$

where

(i) $|yx| \leq n_0$,
(ii) $|x| \geq 1$, *and*
(iii) *either* $\{yx^kz \mid k \in \mathbb{N}\} \subseteq L$ *or* $\{yx^kz \mid k \in \mathbb{N}\} \cap L = \emptyset$.

Proof. Let $L \subseteq \Sigma^*$ be a regular language. Then there exists an FA $A = (Q, \Sigma, \delta_A, q_0, F)$, such that $L(A) = L$. We set

$$n_0 = |Q|.$$

Let w be an arbitrary word over Σ with $|w| \geq n_0$. Then w can be expressed as

$$w = w_1 w_2 \ldots w_{n_0} u,$$

where $w_i \in \Sigma$ for $i = 1, \ldots, n_0$ and $u \in \Sigma^*$. Consider the computation

$$(q_0, w_1 w_2 w_3 \ldots w_{n_0}) \vdash_A (q_1, w_2 w_3 \ldots w_{n_0}) \vdash_A (q_2, w_3 \ldots w_{n_0}) \vdash_A \cdots \vdash_A (q_{n_0-1}, w_{n_0}) \vdash_A (q_{n_0}, \lambda) \quad (3.2)$$

of A on $w_1 w_2 \ldots w_{n_0}$. We have $n_0 + 1$ states $q_0, q_1, q_2, \ldots, q_{n_0}$ in the $n_0 + 1$ configurations of this computation. Since $|Q| = n_0$,

there exist $i, j \in \{0, 1, \ldots, n_0\}$, $i < j$, such that $q_i \equiv q_j$.

Hence computation (3.2) can be written as

$$(q_0, w_1 \ldots w_{n_0}) \vdash_A^* (q_i, w_{i+1} \ldots w_{n_0}) \vdash_A^* (q_i, w_{j+1} \ldots w_{n_0}) \vdash_A^* (q_{n_0}, \lambda). \quad (3.3)$$

Now we set

$$y = w_1 \ldots w_i, \ x = w_{i+1} \ldots w_j \text{ and } z = w_{j+1} \ldots w_{n_0} u.$$

It is obvious that $w = yxz$. We check that properties (i), (ii), and (iii) are fulfilled for this partition of w into the three subwords y, x, and z.

(i) $yx = w_1 \ldots w_i w_{i+1} \ldots w_j$ and so $|yx| = j \leq n_0$.
(ii) Since $i < j$ and $|x| = j - i$, we obtain $x \neq \lambda$ ($|x| \geq 1$).
(iii) If one exchanges the notation y for $w_1 \ldots w_i$ and the notation x for $w_{i+1} \ldots w_j$ in computation (3.3), then the computation of A on yx could be written as follows:

$$(q_0, yx) \vdash_A^* (q_i, x) \vdash_A^* (q_i, \lambda). \quad (3.4)$$

Hence, computation (3.3) implies that for all $k \in \mathbb{N}$,

$$(q_i, x^k) \vdash_A^* (q_i, \lambda).$$

Then

$$(q_0, yx^k z) \vdash_A^* (q_i, x^k z) \vdash_A^* (q_i, z) \vdash_A^* (\hat{\delta}_A(q_i, z), \lambda)$$

is the computation of A on $yx^k z$ for any $k \in \mathbb{N}$. We see that, for all $k \in \mathbb{N}$, the computation ends in the same state $\hat{\delta}_A(q_i, z)$.
If $\hat{\delta}_A(q_i, z) \in F$, then A accepts all words from $\{yx^k z \mid k \in \mathbb{N}\}$.
If $\hat{\delta}_A(q_i, z) \notin F$, then A does not accept any word from $\{yx^k z \mid k \in \mathbb{N}\}$.

This completes the proof of Lemma 3.14. □

How do we apply Lemma 3.14 to show that a specific language is not regular? Let us again use the language $L = \{0^n 1^n \mid n \in \mathbb{N}\}$ to illustrate the application of this method. Our aim is to show that L does not have the property of regular languages formulated in the pumping lemma. We do this by contradiction. Assume L is regular. Then there exists a constant n_0 with the properties formulated in Lemma 3.14. This means that for every word of length at least n_0, a partition of w with the properties (i), (ii), and (iii) must exist. Therefore, to prove that $L \notin \mathcal{L}(\text{FA})$ it is sufficient to find a word w with $|w| \geq n_0$ such that none of its partitions satisfies all three properties (i), (ii), and (iii). For instance, choose the word

$$w = 0^{n_0}1^{n_0}.$$

Obviously, $|w| = 2n_0 \geq n_0$ and so y, x, and z must exist such that $w = yxz$ satisfy properties (i), (ii), and (iii). Following (i) we have $|yx| \leq n_0$ and hence $y = 0^l$ and $x = 0^m$ for some $l, m \in \mathbb{N}$. Following (ii), $m \neq 0$. Since

$$w = 0^{n_0}1^{n_0} = yxz \in L \text{ and } yxz \in \{yx^kz \mid k \in \mathbb{N}\},$$

the property (iii) implies

$$\{yx^kz \mid k \in \mathbb{N}\} = \{0^{n_0-m+km}1^{n_0} \mid k \in \mathbb{N}\} \subseteq L.$$

This is a contradiction since

$$yx^0z = yz = 0^{n_0-m}1^{n_0} \notin L$$

(it is even so that $0^{n_0}1^{n_0}$ is the only word in $\{yx^kz \mid k \in \mathbb{N}\}$ that belongs to L) and hence L is not regular.

The crucial point in the application of the pumping lemma is that we have the choice of a sufficiently large word w, because the pumping lemma holds for all sufficiently large words. This choice is especially important for the following two reasons. First, one can make a bad choice that is not helpful in proving $L \notin \mathcal{L}(\text{FA})$. For the language $L = \{0^n1^n \mid n \in \mathbb{N}\}$, an example of a bad choice is the word

$$w = 0^{n_0} \notin L.$$

One can partition w as

$$w = yxz \text{ with } y = 0,\ x = 0 \text{ and } z = 0^{n_0-2}.$$

Obviously, the pumping lemma holds for such w because all words in

$$\{yx^kz \mid k \in \mathbb{N}\} = \{0^{n_0-1+k} \mid k \in \mathbb{N}\}$$

do not belong to L and so all three properties (i), (ii), and (iii) are satisfied for $w = yxz$.

Secondly, a choice may help to prove $L \notin \mathcal{L}(\text{FA})$, but which is not as good as others in the sense that one has to do more work to prove that every partition of the chosen word does not satisfy all the properties (i), (ii), and (iii) than for a convenient choice. For instance, consider the word

$$w = 0^{\lceil n_0/2 \rceil}1^{\lceil n_0/2 \rceil}.$$

This word can be used to prove that $L = \{0^n1^n \mid n \in \mathbb{N}\}$ is not regular, but considering all possible partitions yxz of w one has to consider at least three cases:[13]

(a) $y = 0^i$, $x = 0^m$, $z = 0^{\lceil n_0/2\rceil - m - i}1^{\lceil n_0/2\rceil}$ for any $i \in \mathbb{N}$ and any $m \in \mathbb{N} - \{0\}$, i.e., x consists of zeros only.
{In this case property (iii) is not satisfied using a similar argument as above for the choice $w = 0^{n_0}1^{n_0}$.}

(b) $y = 0^{\lceil n_0/2\rceil - m}$, $x = 0^m1^j$, $z = 1^{\lceil n_0/2\rceil - j}$ for any positive integers m and j, i.e., x contains at least one 0 and at least one 1.
{In this case property (iii) is not satisfied because $x = yxz \in L$ and $yx^2z \notin L$ because yx^2z does not have the form a^*b^*.}

(c) $y = 0^{\lceil n_0/2\rceil}1^i$, $x = 1^m$, $z = 1^{\lceil n_0/2\rceil - i - m}$ for any $i \in \mathbb{N}$ and any $m \in \mathbb{N} - \{0\}$.
{In this case one can pump 1s without increasing the number of 0s and analogously to the case (a) property (iii) is not satisfied.}

Thus, we see that the use of the word $0^{\lceil n_0/2\rceil}1^{\lceil n_0/2\rceil}$ for proving $L \notin \mathcal{L}(\text{FA})$ by the pumping lemma requires more work than the use of the word $0^{n_0}1^{n_0}$.

Exercise 3.15. Prove by the pumping lemma that the following languages are not regular.

(i) $\{ww \mid w \in \{0,1\}^*\}$
(ii) $\{a^nb^nc^n \mid n \in \mathbb{N}\}$
(iii) $\{w \in \{0,1\}^* \mid |w|_0 = |w|_1\}$
(iv) $\{a^{n^2} \mid n \in \mathbb{N}\}$
(v) $\{a^{2^n} \mid n \in \mathbb{N}\}$
(vi) $\{w \in \{0,1\}^* \mid |w|_0 = 2|w|_1\}$
(vii) $\{x1y \mid x, y \in \{0,1\}^*, |x| = |y|\}$

Exercise 3.16. For every language over Σ_{bool} from Exercise 3.15, find a word u such that every partition of u into yxz satisfies all three conditions (i), (ii), and (iii) of the pumping lemma.

Exercise 3.17. Prove the following version of the pumping lemma.

For every regular language $L \subseteq \Sigma^*$, there exists a constant $n_0 \in \mathbb{N}$, such that every word $w \in \Sigma^*$ with $|w| \geq n_0$ can be expressed as $w = yxz$, where

(i) $|xz| \leq n_0$,
(ii) $|x| \geq 1$, and
(iii) either $\{yx^kz \mid k \in \mathbb{N}\} \subseteq L$ or $\{yx^kz \mid k \in \mathbb{N}\} \cap L = \emptyset$.

Exercise 3.18.* Formulate and prove a general form of the pumping lemma that involves Lemma 3.14 and the pumping lemma from Exercise 3.17 as special cases.

---

[13] For the word $0^{n_0}1^{n_0}$, property (i) ensures that x consists of 0s only. However, for the word $w = 0^{\lceil n_0/2\rceil}1^{\lceil n_0/2\rceil}$, x may be any subword of w.

The next method for proving $L \notin \mathcal{L}(\mathrm{FA})$ for a given language L is based on the Kolmogorov complexity argument. In some sense this method shows with the aid of a different combinatorial argument that finite automata cannot count arbitrarily far (or cannot save too much information of the word that has been read). This method is based on the following theorem that shows that all suffixes of words of a regular language have a small Kolmogorov complexity. This can be viewed as a refinement of Theorem 3.19 that claims that all words of a recursive language with a small density have a small Kolmogorov complexity.

Theorem 3.19.* *Let $L \subseteq (\Sigma_{\mathrm{bool}})^*$ be a regular language. For each $x \in \Sigma^*$, let*

$$L_x = \{y \in \Sigma^* \mid xy \in L\}.$$

Then there exists a constant const, such that, for all $x, y \in \Sigma^*$,

$$K(y) \leq \lceil \log_2(n+1) \rceil + const$$

if y is the n-th word in the language L_x.

Proof. Since L is regular, there exists an FA M with $L(M) = L$. The idea of the proof is somewhat (though not entirely) similar to the idea of the proof of Theorem 2.65, but not the same. If one follows the proof of Theorem 2.65 exactly, one would generate all words z from Σ^* in the canonical order, and simulate the computation of M on every xz in order to determine whether $xz \in L = L(M)$ (i.e., whether $z \in L_x$). Finding the n-th word xz with the fixed prefix x corresponds to the fact that z is the n-th word in L_x and so one has a program for generating z. The drawback of this approach is that the resulting program that generates the n-th word y in L_x needs to obtain not only n and M, but also x. But the word x can have an arbitrarily large Kolmogorov complexity $K(x)$ in comparison with $K(y)$.

The crucial point is that in fact the program generating y does not need to obtain the full information about x. It is sufficient to include the state $\hat{\delta}(q_0, x)$ into the program and then to start the simulation of the computation of M on the generated words z always from the state $\hat{\delta}(q_0, x)$ (instead of simulating the work of M on xz from the initial state q_0 for every generated z).

Let y be the n-th word in L_x for an $x \in \Sigma^*$. The program $A_{x,y}$ generating y can be described as follows:

```
A_{x,y}:  begin
            z := λ; i := 0;
            while i < n do begin
              Simulate the work of M from the state δ̂(q_0, x) on z;
              if δ̂(δ̂(q_0, x), z) ∈ F then
                begin i := i + 1;
                  y := z
```

```
        end;
        z := the canonical successor of z.
      end;
      write(y);
    end
```

For all $x, y \in \Sigma^*$, the program $A_{x,y}$ is the same except n and the state $\hat{\delta}(q_0, x)$. The special role of $\hat{\delta}(q_0, x)$ can be marked within the description of the FA M by a special pointer to $\hat{\delta}(q_0, x)$. Since there are only $|Q|$ possibilities for the pointer, there exists a constant $const_M$ that is an upper bound on the binary length of the representation of M with the pointer to a special state (it does not matter to which one). The length of the binary representation of $A_{x,y}$ except the parameters n, M, and $\hat{\delta}(q_0, x)$ is a constant d independent of x and y. Since the positive integer n can be binary coded by $\lceil \log_2(n+1) \rceil$ bits, one obtains

$$K(y) \leq \lceil \log_2(n+1) \rceil + const_M + d.$$

This completes the proof of Theorem 3.19. □

We apply Theorem 3.19 in order to show once again that

$$L = \{0^n 1^n \mid n \in \mathbb{N}\} \notin \mathcal{L}(\text{FA}).$$

The proof is done (as usual) by contradiction. Assume, L is regular. For every $m \in \mathbb{N}$, the word 1^m is the first word in the language

$$L_{0^m} = \{y \mid 0^m y \in L\} = \{0^j 1^{m+j} \mid j \in \mathbb{N}\}.$$

Following Theorem 3.19 there exists a constant c independent of m (i.e., independent of the word 1^m), such that

$$K(1^m) \leq K(1) + c.$$

Since $K(1)$ is constant too,

$$K(1^m) \leq d \tag{3.5}$$

for the constant $d = K(1) + c$ and all $m \in \mathbb{N} - \{0\}$. But this leads to the contradiction between the following facts:

(i) the number of all programs whose binary length is at most d is finite[14] , and
(ii) the set $\{1^m \mid m \in \mathbb{N}\}$ is infinite.

Instead of using the argument that finitely many programs cannot generate infinitely many different words, one can alternatively apply the assertion in

---

[14] More precisely, at most 2^d

Exercise 2.29. This assertion guarantees the existence of infinitely many positive integers m with

$$K(m) \geq \lceil \log_2(m+1) \rceil - 1. \tag{3.6}$$

Since there exists a $b \in \mathbb{N}$ such that $|K(1^m) - K(m)| \leq b$ for all $m \in \mathbb{N}$, the inequality (3.6) contradicts inequality (3.5) for infinitely many $m \in \mathbb{N}$.

Exercise 3.20. Prove that there exists a constant b such that, for all $m \in \mathbb{N}$,

$$|K(1^m) - K(m)| \leq b.$$

Exercise 3.21. Apply Theorem 3.19 to prove that the following languages are not regular.

(i) $\{0^{n^2} \mid n \in \mathbb{N}\}$
(ii) $\{0^{2^{2n}} \mid n \in \mathbb{N}\}$
(iii) $\{w \in \{0,1\}^* \mid |w|_0 = 2 \cdot |w|_1\}$
(iv) $\{w \in \{0,1\}^* \mid w = xx \text{ for an } x \in \{0,1\}^*\}$

3.5 Nondeterminism

The usual programs as well as the introduced finite automata are models of deterministic computations. Determinism means that in every configuration it is unambiguously determined what will happen in the next computation step. Therefore a (deterministic) program and its input determine unambiguously the computation of A on x. In contrast to determinism, nondeterminism allows in each configuration a choice from several (finitely many) possible actions.[15] The consequence is that a nondeterministic program can have many different computations on the same input. The only requirement is that at least one of these possibilities yields the correct result. This may seem like an artificial rule because it corresponds to the assumption that a nondeterministic program always chooses the right computation. This choice from several possibilities is called a **nondeterministic decision**. For a decision problem (Σ, L), it means that a nondeterministic program (a nondeterministic FA) A accepts the language L if, for every $x \in L$, there is at least one accepting computation of A on x, and, for every $y \in \Sigma^* - L$, none of the computations of A on y is an accepting one. Although a nondeterministic program does not seem to be useful for practical purposes,[16] the study of nondeterminism and nondeterministic computations has essentially contributed to our understanding of deterministic computations and especially to the investigation of

[15] That is, to have several possibilities how to continue in the computation
[16] This is because we do not have any oracle that would help us to take the right nondeterministic decisions.

the limits for solving problems by algorithms. As we will see later, nondeterminism has become a very powerful instrument for investigating quantitative laws of computing and this is one of the main reasons for introducing it and learning to work with it here.

The goal of this section is to introduce and to study nondeterminism for the model of finite automata. Our main interest is devoted to answering the following questions. Can one simulate nondeterministic computations using deterministic ones? If yes, what is the price[17] of being able to do so? One can introduce nondeterminism for programs by simply allowing the additional instruction

"choose goto i or goto j".

For finite automata, this simply corresponds to allowing several transitions from a state by reading the same input symbol (i.e., by allowing several directed edges labeled by the same alphabet symbol from a state (Figure 3.9)).

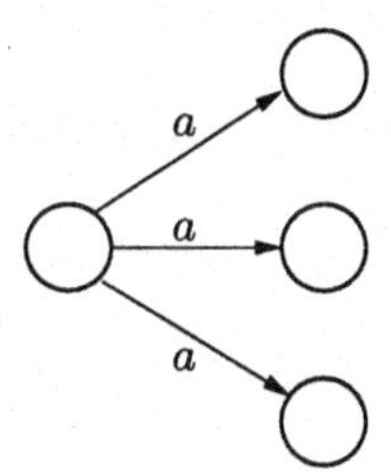

Fig. 3.9.

Definition 3.22. *A* **nondeterministic finite automaton** *(***NFA***) is a* 5-*tuple* $M = (Q, \Sigma, \delta, q_0, F)$, *where*

(i) Q *is a finite set, called the* **set of states,**
(ii) Σ *is an alphabet, called* **input alphabet,**
(iii) $q_0 \in Q$ *is the* **initial state,**
(iv) $F \subseteq Q$ *is the set of* **accepting states,** *and*
(v) δ *is a function*[18] *from* $Q \times \Sigma$ *to* $\mathcal{P}(Q)$, *called the* **transition function.**

{We note, that Q, Σ, q_0, *and* F *have the same meaning as for a (deterministic)* FA. *But following (v), for a state* q *and a symbol* b *read, an* NFA *can have several successor states or none, i.e.,* $\delta(q, b)$ *is a set of states (possible actions) in contrast to having exactly one state (action) in the case of an* FA.*}*

A **configuration** *of* M *is an element from* $Q \times \Sigma^*$. *The configuration* (q_0, x) *is the* **initial configuration of the input word** x.

[17] The size of the additional computational effort (resources)
[18] Alternatively, one can define δ as a relation on $(Q \times \Sigma) \times Q$.

A **step of** M *is a relation*

$$\vdash_{\overline{M}} \subseteq (Q \times \Sigma^*) \times (Q \times \Sigma^*)$$

defined by

$$(q, w) \vdash_{\overline{M}} (p, x) \Leftrightarrow w = ax \text{ for an } a \in \Sigma \text{ and } p \in \delta(q, a).$$

A **computation of** M *is a finite sequence* $D_1, D_2, \ldots, D_k$ *of configurations, such that* $D_i \vdash_{\overline{M}} D_{i+1}$ *for* $i = 1, \ldots, k-1$.

A **computation on** x *is a computation* $C_0, C_1, \ldots, C_m$ *of* M, *where*

(i) $C_0 = (q_0, x)$, *i.e.,* C_0 *is the initial configuration of* M *on* x, *and*
(ii) either $C_m \in Q \times \{\lambda\}$
or $C_m = (q, ax)$ *for an* $a \in \Sigma$, *and a* $q \in Q$ *such that* $\delta(q, a) = \emptyset$.

{A computation of M *on* x *can stop when the whole input has been read, or (in contrast to* FA*s) if there is no possibility of continuing the computation for some argument* (q, a).*}*

$C_0, C_1, \ldots, C_m$ *is an* **accepting computation of** M **on** x, *when* $C_m = (p, \lambda)$ *for a* $p \in F$.

{M *accepts* x *in a computation* C *only if the whole word* x *has been read in* C *and the computation finishes in an accepting state.}*

If there exists an accepting computation of M *on* x, *then one says that* M **accepts the word** x.

The relation $\vdash^*_{\overline{M}}$ *denotes the reflexive and transitive closure*[19] *of the relation* $\vdash_{\overline{M}}$.

The **language accepted by** M *is*

$$L(M) = \{w \in \Sigma^* \mid (q_0, w) \vdash^*_{\overline{M}} (p, \lambda) \text{ for a } p \in F\}.$$

For the transition function δ, *we define the function* $\hat{\delta}$ *from* $Q \times \Sigma^*$ *to* $\mathcal{P}(Q)$ *for all* $q \in Q, a \in \Sigma$ *and* $w \in \Sigma^*$:

(i) $\hat{\delta}(q, \lambda) = \{q\}$,
(ii) $\hat{\delta}(q, wa) = \{p \mid \text{there exists an } r \in \hat{\delta}(q, w), \text{ such that } p \in \delta(r, a)\}$
$= \bigcup_{r \in \hat{\delta}(q,w)} \delta(r, a)$.

We see that a word $x \in L(M)$ if M has at least one accepting computation on x. For any accepting computation, one requires that all symbols of the input are read and that M reaches an accepting state immediately after reading the last symbol.[20] In contrast to (deterministic) finite automata, a nonaccepting computation can finish without reading the whole input. This would happen when there is no transition for the given argument, i.e., if $\delta(q, a) = \emptyset$ for the current state q and symbol a read.

[19] Exactly as in the definition of finite automata

[20] This requirement is the same as for (deterministic) finite automata.

Following the definition $\hat{\delta}$, we see that $\hat{\delta}(q_0, w)$ is the set of all states from Q, which can be reached after reading the whole word w from the state q_0, i.e.,

$$\hat{\delta}(q_0, w) = \{p \in Q \mid (q_0, w) \underset{M}{\overset{*}{\vdash}} (p, \lambda)\}.$$

Therefore

$$L(M) = \{w \in \Sigma^* \mid \hat{\delta}(q_0, w) \cap F \neq \emptyset\}$$

is an equivalent definition of the language accepted by an NFA M.

Consider the following example of an NFA. Let $M = (Q, \Sigma, \delta, q_0, F)$, where

$Q = \{q_0, q_1, q_2\}$,
$\Sigma = \{0, 1\}$,
$F = \{q_2\}$ and
$\delta(q_0, 0) = \{q_0\}$, $\delta(q_0, 1) = \{q_0, q_1\}$,
$\delta(q_1, 0) = \emptyset$, $\delta(q_1, 1) = \{q_2\}$,
$\delta(q_2, 0) = \{q_2\}$, $\delta(q_2, 1) = \{q_2\}$.

Using the same procedure as for (deterministic) finite automata, one can derive a graphic representation of nondeterministic finite automata. The above-presented NFA M is depicted in Figure 3.10.

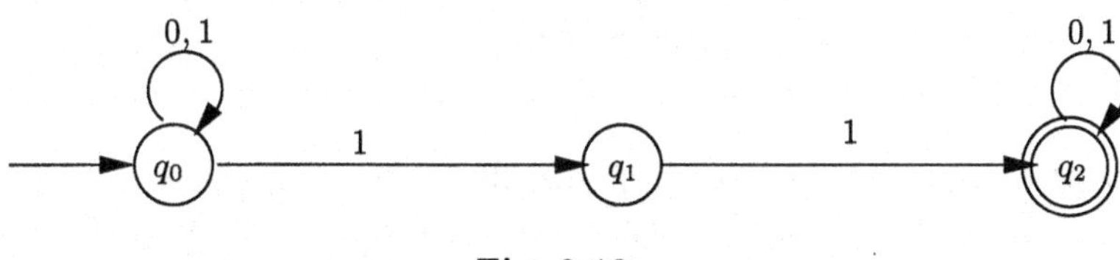

Fig. 3.10.

The word 10110 is in $L(M)$, because

$$(q_0, 10110) \underset{M}{\vdash} (q_0, 0110) \underset{M}{\vdash} (q_0, 110) \underset{M}{\vdash} (q_1, 10) \underset{M}{\vdash} (q_2, 0) \underset{M}{\vdash} (q_2, \lambda)$$

is an accepting computation of M on 10110.

To decide whether an NFA M accepts a word x one has to follow all computations of M on x. A transparent representation of all computations of M on x can be given by the so-called **computation tree $\mathcal{B}_M(x)$ of M on x**. The vertices of this tree are configurations of M. The root of $\mathcal{B}_M(x)$ is the initial configuration of M on x. The sons of a vertex (q, α) are all configurations that are reachable in one computation step from (q, α), i.e., all configurations (p, β) such that $(q, \alpha) \underset{M}{\vdash} (p, \beta)$. A leaf of $\mathcal{B}_M(x)$ is either a final configuration (r, λ) or a configuration $(s, a\beta)$ with an $a \in \Sigma$ and $\delta(s, a) = \emptyset$. Hence the leaves are configurations from which no further computation step is possible. In this representation, any path of $\mathcal{B}_M(x)$ from the root to a leaf corresponds to a computation of M on x, and vice versa. Therefore the number of leaves of $\mathcal{B}_M(x)$ equals the number of different computations of M on x.

Exercise 3.23. Design an NFA that has exactly $2^{|x|}$ distinct computations on any input $x \in (\Sigma_{\text{bool}})^*$.

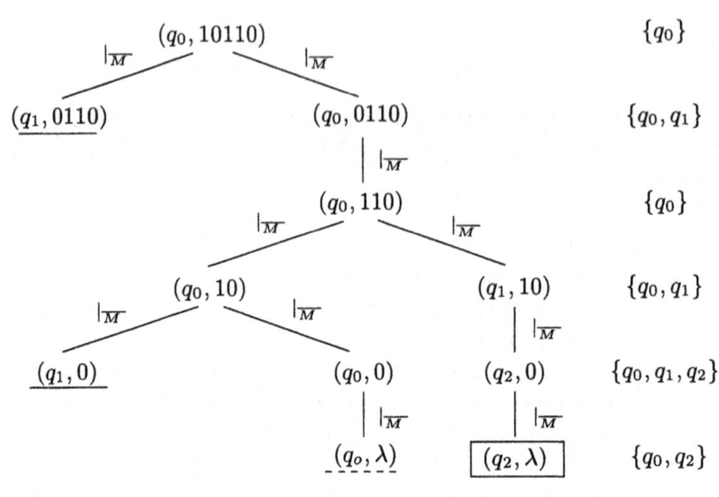

Fig. 3.11.

Let M be the NFA depicted in Figure 3.10. The computation tree $\mathcal{B}_M(x)$ of M on the input $x = 10110$ is depicted in Figure 3.11. The tree $B_M(10110)$ has four leaves. Two leaves $(q_1, 0110)$ and $(q_1, 0)$ correspond to computations, for which M fails to read the whole input word (because $\delta(q_1, 0) = \emptyset$). Hence, these two computations are not accepting. The leaves (q_0, λ) and (q_2, λ) correspond to two computations, in which the 10110 has been entirely read. Since $q_2 \in F$, the corresponding[21] computation is an accepting computation, and hence $10110 \in L(M)$.

Since q_2 is the only accepting state of M and the only possibility of reaching q_2 from q_0 is to read the subword 11, one can guess that $L(M)$ is the set of all words of the form $x11y$, where x, y are arbitrary words in $(\Sigma_{\text{bool}})^*$. The following lemma confirms our hypothesis.

Lemma 3.24. *Let M be the* NFA *from Figure 3.10. Then*

$$L(M) = \{x11y \mid x, y \in (\Sigma_{\text{bool}})^*\}.$$

Proof. We prove the equality of these two sets by proving the corresponding two inclusions.

(i) First, we prove $\{x11y \mid x, y \in (\Sigma_{\text{bool}})^*\} \subseteq L(M)$.

[21] The computation that finishes in the configuration (q_2, λ)

Let $w \in \{x11y \mid x, y \in (\Sigma_{\text{bool}})^*\}$, i.e., $w = x11y$ for some $x, y \in (\Sigma_{\text{bool}})^*$. It is sufficient to prove the existence of an accepting computation of M on w.

Since $q_0 \in \delta(q_0, 0) \cap \delta(q_0, 1)$, we have the following computation of M on every $x \in (\Sigma_{\text{bool}})^*$:

$$(q_0, x) \mid\!\frac{*}{M}\, (q_0, \lambda). \tag{3.7}$$

Since $q_2 \in \delta(q_2, 0) \cap \delta(q_2, 1)$, there exists the following computation of M on every $y \in (\Sigma_{\text{bool}})^*$:

$$(q_2, y) \mid\!\frac{*}{M}\, (q_2, \lambda). \tag{3.8}$$

Hence, the computation

$$(q_0, x11y) \mid\!\frac{*}{M}\, (q_0, 11y) \mid\!\frac{}{M}\, (q_1, 1y) \mid\!\frac{}{M}\, (q_2, y) \mid\!\frac{*}{M}\, (q_2, \lambda)$$

is an accepting computation of M on $x11y$.

(ii) We prove $L(M) \subseteq \{x11y \mid x, y \in (\Sigma_{\text{bool}})^*\}$.

Let $w \in L(M)$. Hence, there exists an accepting computation C of M on w. Since this accepting computation C on w must start in the initial state q_0 and finish in the only accepting state q_2, and the only path from q_0 to q_2 goes via q_1, the computation C has the following form:

$$(q_0, w) \mid\!\frac{*}{M}\, (q_1, z) \mid\!\frac{*}{M}\, (q_2, \lambda). \tag{3.9}$$

Every computation of M may contain at most one configuration with the state q_1 because

(a) $q_1 \notin \delta(q_1, a)$ for any $a \in \Sigma_{\text{bool}}$, and

(b) if M has left q_1, then M can return to q_1 no longer.

As such, one can express the accepting computation C as follows:

$$(q_0, w) \mid\!\frac{*}{M}\, (q_0, abu) \mid\!\frac{}{M}\, (q_1, bu) \mid\!\frac{}{M}\, (q_2, u) \mid\!\frac{*}{M}\, (q_2, \lambda), \tag{3.10}$$

where $a, b \in \Sigma_{\text{bool}}$ and $u \in \Sigma_{\text{bool}}^*$. The only possibility to reach q_1 is to apply the transition $q_1 \in \delta(q_0, 1)$, i.e., by reading 1 in the state q_0. This implies $a = 1$ in computation (3.10). Since $(q_1, 0) = \emptyset$ and $\delta(q_1, 1) = \{q_2\}$, the only possibility to execute a computation step from q_1 is to read 1, and hence $b = 1$. Rewriting[22] computation (3.10), we obtain that C must have the following form:

$$(q_0, w) \mid\!\frac{*}{M}\, (q_0, 11u) \mid\!\frac{}{M}\, (q_1, 1u) \mid\!\frac{}{M}\, (q_2, u) \mid\!\frac{*}{M}\, (q_2, \lambda).$$

The consequence is that w must contain the subword 11 and hence

$$w \in \{x11y \mid x, y \in (\Sigma_{\text{bool}})^*\}.$$

□

[22] Inserting 1 for a and b in computation (3.10)

Exercise 3.25. Design nondeterministic finite automata for the following languages.

(a) $\{1011x00y \mid x, y \in (\Sigma_{\text{bool}})^*\}$
(b) $\{01, 101\}^*$
(c) $\{x \in (\Sigma_{\text{bool}})^* \mid x$ contains the subwords 01011 and 01100$\}$
(d) $\{x \in (\Sigma_{10})^* \mid \mathit{Number}_{10}(x)$ is divisible by 3$\}$

Try to construct as simple NFAs as possible, i.e., minimize the number of states (vertices) and the number of transitions (edges).

Let

$$\mathcal{L}(\text{NFA}) = \{L(M) \mid M \text{ is an NFA}\}.$$

The central question of this section is whether $\mathcal{L}(\text{NFA}) = \mathcal{L}(\text{FA})$, more precisely, whether finite automata can simulate the work of nondeterministic finite automata. This question is also crucial for general computing models. From widespread scientific experience, we know that the simulation of nondeterminism by determinism is executable only if there is a possibility of mimicking all computations of a given nondeterministic model on an input by one deterministic computation. This is also the case for finite automata.

The idea of the simulation of an NFA M by an FA A is based on the breadth-first search in the computation trees of M. The first important assumption for this simulation is that all configurations of a computation tree at the same distance i from the root have the same second element, because they are reached after reading the first i symbols of the input word. Thus, these configurations at the same distance from the root may differ only in states. Although the number of configurations at a distance i from the root can be exponential in i, it does not mean that one needs to simulate exponentially many different computations. The NFA M has only finitely many states and so there are only finitely many different configurations at a distance i from the root for any $i \in \mathbb{N}$. If two different vertices u and v of the computation tree are labeled by the same configurations then the subtrees rooted by u and v are identical and hence it is sufficient to search for an accepting configuration in only one of the trees (Figure 3.12). This means that for a simulation of the work of the NFA M on an input x, it suffices to determine the set of all states reachable after i steps of M on x for any $i \in \{0, 1, \ldots, |x|\}$. For any i, this set is none other than $\hat{\delta}(q_0, z)$, where z is the prefix of x of length i. From the right border of Figure 3.11 one sees the sets $\hat{\delta}(q_0, 10) = \{q_0\}$, $\hat{\delta}(q_0, 101) = \{q_0, q_1\}$, $\hat{\delta}(q_0, 1011) = \{q_0, q_1, q_2\}$, and $\hat{\delta}(q_0, 10110) = \{q_0, q_2\}$, which correspond to the sets of reachable states after $i = 0, 1, \ldots, 5$ steps (i.e., after reading the prefix of length i).

This observation triggers the idea to take subsets of the set of states Q of a given NFA $M = (Q, \Sigma, \delta, q_0, F)$ as states of the simulating (deterministic) FA A. Therefore, the following construction of the FA A is called the powerset construction in the automata theory. A state $\langle P \rangle$ of A for a set $P \subseteq Q$ has the

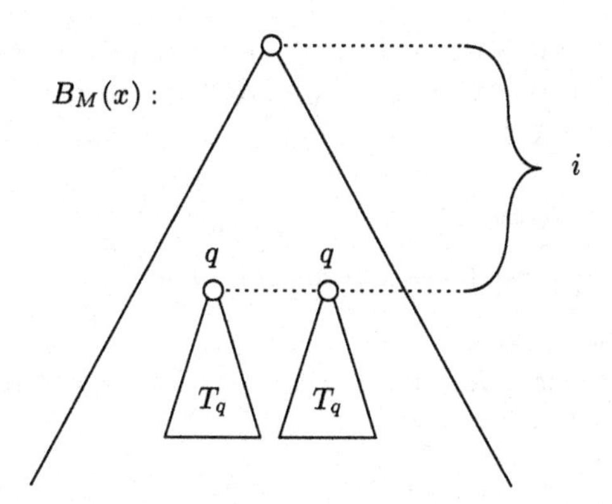

Fig. 3.12.

interpretation that exactly the states of P are reachable[23] in the computations of M on an input z, i.e., that $P = \hat{\delta}(q_0, z)$. A computation step of A from a state $\langle P \rangle$ by reading a symbol a is determined by

$$\bigcup_{p \in P} \delta(p, a),$$

i.e., by the set of all states that can be reached when M reads the symbol a in a state from P. A formalization of this idea is given by the next theorem.

Theorem 3.26. *For any* NFA M*, there exists an* FA A*, such that*

$$L(M) = L(A).$$

Proof. Let $M = (Q, \Sigma, \delta_M, q_0, F)$ be an NFA. We construct an equivalent FA $A = (Q_A, \Sigma_A, \delta_A, q_{0A}, F_A)$ as follows:

(i) $Q_A = \{\langle P \rangle \mid P \subseteq Q\}$,
(ii) $\Sigma_A = \Sigma$,
(iii) $q_{0A} = \langle \{q_0\} \rangle$,
(iv) $F_A = \{\langle P \rangle \mid P \subseteq Q \text{ and } P \cap F \neq \emptyset\}$,
(v) δ_A is a function from $Q_A \times \Sigma_A$ to Q_A, such that for any $\langle P \rangle \in Q_A$ and $a \in \Sigma_A$,

$$\begin{aligned}\delta_A(\langle P \rangle, a) &= \left\langle \bigcup_{p \in P} \delta_M(p, a) \right\rangle \\ &= \langle \{q \in Q \mid \exists p \in P, \text{ such that } q \in \delta_M(p, a)\} \rangle .\end{aligned}$$

[23] We use the notation $\langle P \rangle$ instead of P to clearly distinguish between a state $\langle P \rangle$ of the FA A that corresponds to the set P of reachable states of M, and a set P of states of M.

Clearly, A is an FA. Figure 3.13 shows the FA A that is created by the powerset construction[24] from the NFA M in Figure 3.10. To prove that the FA A is equivalent to the NFA M, it is sufficient to prove that for all $x \in \Sigma^*$,

$$\hat{\delta}_M(q_0, x) = P \iff \hat{\delta}_A(q_{0A}, x) = \langle P \rangle . \tag{3.11}$$

We prove (3.11) by induction according to $|x|$.

(i) *Induction basis.*
Let $|x| = 0$, i.e., $x = \lambda$. Since $\hat{\delta}_M(q_0, \lambda) = \{q_0\}$ and $q_{0A} = \langle\{q_0\}\rangle$, equivalence (3.11) is true for $x = \lambda$.

(ii) *Induction step.*
Let equivalence (3.11) hold for all $z \in \Sigma^*$ with $|z| \leq m$ where $m \in \mathbb{N}$. We prove that equivalence (3.11) also holds for all words from Σ^{m+1}.
Let y be an arbitrary word from Σ^{m+1}. Then, $y = xa$ for an $x \in \Sigma^m$ and a symbol $a \in \Sigma$. Following the definition of the function $\hat{\delta}_A$, we have

$$\hat{\delta}_A(q_{0A}, xa) = \delta_A(\hat{\delta}_A(q_{0A}, x), a). \tag{3.12}$$

Applying the induction hypothesis (3.11) for x, we obtain

$$\hat{\delta}_A(q_{0A}, x) = \langle R \rangle \Leftrightarrow \hat{\delta}_M(q_0, x) = R,$$

and thus

$$\hat{\delta}_A(q_{0A}, x) = \left\langle \hat{\delta}_M(q_0, x) \right\rangle . \tag{3.13}$$

According to (v) of the construction of A, we obtain for all $R \subseteq Q$ and $a \in \Sigma$, that

$$\delta_A(\langle R \rangle , a) = \left\langle \bigcup_{p \in R} \delta_M(p, a) \right\rangle . \tag{3.14}$$

Summarizing,

$$\begin{aligned}
\hat{\delta}_A(q_{0A}, xa) &\underset{(3.12)}{=} \delta_A(\hat{\delta}_A(q_{0A}, x), a) \\
&\underset{(3.13)}{=} \delta_A(\left\langle \hat{\delta}_M(q_0, x) \right\rangle , a) \\
&\underset{(3.14)}{=} \left\langle \bigcup_{p \in \hat{\delta}_M(q_0, x)} \delta_M(p, a) \right\rangle \\
&= \left\langle \hat{\delta}_M(q_0, xa) \right\rangle .
\end{aligned}$$

This completes the proof of the equivalence (3.11) and hence the proof of Theorem 3.26. □

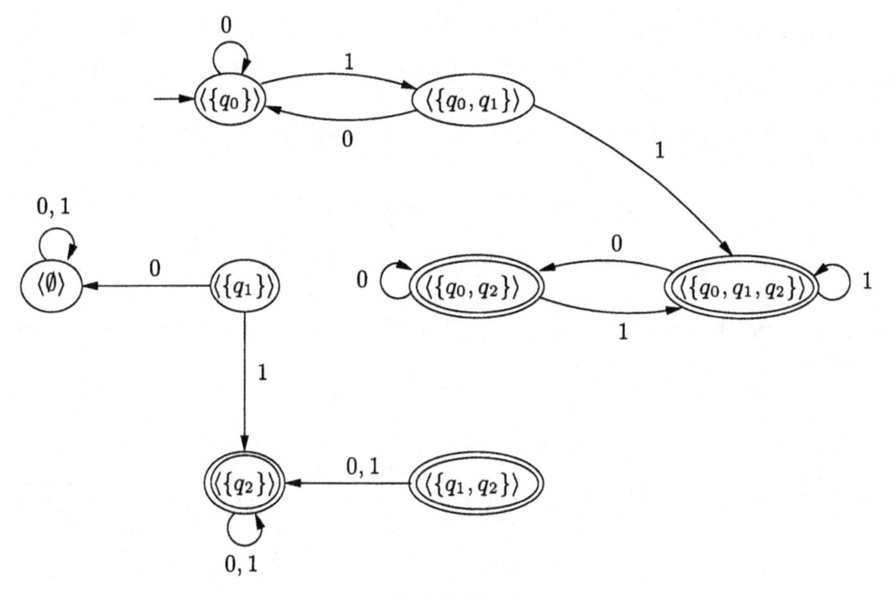

Fig. 3.13.

In what follows we say that two automata A and B are **equivalent** if $L(A) = L(B)$.

Exercise 3.27. Apply the powerset construction of Theorem 3.26 in order to build an FA that is equivalent to the NFA in Figure 3.14.

Exercise 3.28. Applying the powerset construction, construct finite automata that are equivalent to the nondeterministic finite automata you have designed for Exercise 3.25 (b) and (d).

A consequence of Theorem 3.26 is that

$$\mathcal{L}(\text{FA}) = \mathcal{L}(\text{NFA}),$$

i.e., finite automata are as powerful as nondeterministic finite automata with respect to language recognition. But we observe that the finite automata constructed by the powerset construction are essentially (exponentially) larger than the corresponding nondeterministic finite automata. We now deal with the following questions. Is there an alternative simulation of NFAs by FAs that guarantees the existence of an FA that is small with respect to the size

[24] Note, that the states $\langle\emptyset\rangle$, $\langle\{q_1\}\rangle$, $\langle\{q_2\}\rangle$, and $\langle\{q_1, q_2\}\rangle$ of A in Figure 3.10 are not reachable from the initial state $\langle\{q_0\}\rangle$, i.e., there is no word whose handling by A would finish in any of these states. Hence, the removal of these states does not change the language accepted by A.

of its nondeterministic counterpart? Or, are there any regular languages for which the simulation of nondeterminism by determinism inevitably results in an exponential growth in the number of states? We show that one cannot improve on the powerset construction. To do this, we consider the following regular language

$$L_k = \{x1y \mid x \in (\Sigma_{\text{bool}})^*, y \in (\Sigma_{\text{bool}})^{k-1}\}$$

for any finite integer k. The NFA A_k in Figure 3.14 accepts L_k by guessing in the state q_0, for every symbol 1 of the input, whether this symbol is the k-th symbol from the end of the input word. A_k then checks in a deterministic way whether this guess was correct.

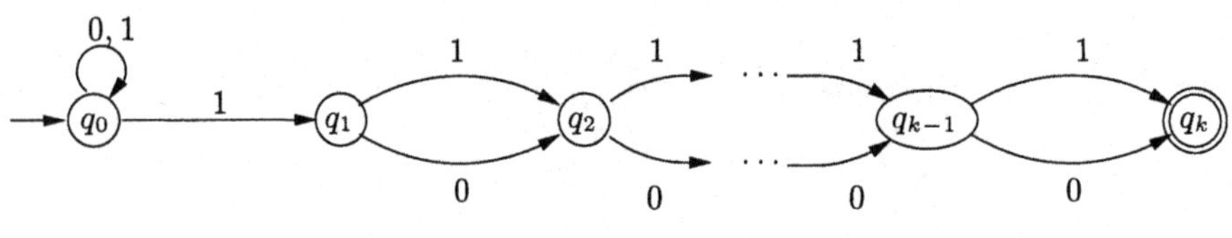

Fig. 3.14.

Exercise 3.29. Give a formal description of the NFA A_k in Figure 3.14 and prove that A_k accepts the language L_k. Construct, for every positive integer k, an FA B_k such that $L_k = L(B_k)$.

The NFA A_k has $k + 1$ states. We prove that every FA accepting L_k has a number of states that is exponential in the size of A_k.

Lemma 3.30. *For any positive integer k, every finite automaton accepting L_k has at least 2^k states.*

Proof. Let $B_k = (Q_k, \Sigma_{\text{bool}}, \delta_k, q_{0k}, F_k)$ be an FA with $L(B_k) = L_k$. To show that B_k has at least 2^k states, we use the same proof technique as in Section 3.4 for proving the nonexistence[25] of finite automata accepting some specific languages. If

$$\hat{\delta}_k(q_{0k}, x) = \hat{\delta}_k(q_{0k}, y)$$

for some words x and y over Σ_{bool}, then for all $z \in (\Sigma_{\text{bool}})^*$,

$$xz \in L(B_k) \iff yz \in L(B_k). \tag{3.15}$$

The idea of this proof is to find a large set S_k of words, such that for any two different words $x, y \in S_k$, the equality $\hat{\delta}_k(q_{0k}, x) = \hat{\delta}_k(q_{0k}, y)$ does not hold, because there exists a word z such that

[25] This is not surprising because we are creating a nonexistence proof. More precisely, we are proving that there does not exist any FA that accepts L_k with fewer than 2^k states.

$$xz \in L(B_k) \text{ and } yz \notin L(B_k).$$

Then, the FA B_k has to have at least $|S_k|$ different states.[26]

We choose

$$S_k = (\Sigma_{\text{bool}})^k$$

and show, that $\hat{\delta}_k(q_{0k}, u)$ must be pair-wise different states of B_k for all $u \in S_k$. We prove this by contradiction. Let

$$x = x_1 x_2 \dots x_k \text{ and } y = y_1 y_2 \dots y_k,$$

be two different words from $S_k = (\Sigma_{\text{bool}})^k$, where x_i, $y_i \in \Sigma_{\text{bool}}$ for $i = 1, \dots, k$. Assume

$$\hat{\delta}_k(q_{0k}, x) = \hat{\delta}_k(q_{0k}, y).$$

Since $x \neq y$, there exists a $j \in \{1, \dots, k\}$, such that $x_j \neq y_j$. Without loss of generality we may assume

$$x_j = 1 \text{ and } y_j = 0.$$

Now consider the word $z = 0^{j-1}$. Then

$$xz = x_1 \dots x_{j-1} 1 x_{j+1} \dots x_k 0^{j-1} \text{ and } yz = y_1 \dots y_{j-1} 0 y_{j+1} \dots y_k 0^{j-1}$$

and thus

$$xz \in L_k \text{ and } yz \notin L_k.$$

This contradicts equality (3.15), i.e., $\hat{\delta}_k(q_{0k}, x)$ and $\hat{\delta}_k(q_{0k}, y)$ must be different. Hence, B_k has at least $|S_k| = 2^k$ states. □

Lemma 3.30 presents a simple technique for proving lower bounds on the sizes of finite automata for recognizing concrete regular languages. To practice this technique in a transparent way, we prove now a lower bound for the simple regular language

$$L = \{x11y \mid x, y \in \{0,1\}^*\}$$

used to illustrate the powerset construction in Figure 3.10. To prove that every finite automaton accepting L has at least three states, we choose the following three words:

$$\lambda,\ 1,\ 11.$$

For any of the three pairs (x, y) of different words from $S = \{\lambda, 1, 11\}$ we have to show that there exists a word z such that exactly one of the words xz and yz is in L (i.e., that x and y do not satisfy the equivalence (3.15)).

For $x = \lambda$ and $y = 1$ we choose $z = 1$. Then

$$xz = 1 \notin L \text{ and } yz = 11 \in L$$

[26] If $|S_k|$ is infinite, then we would prove the nonexistence of finite automata recognizing the language considered.

and thus $\hat{\delta}(q_0, x) \neq \hat{\delta}(q_0, y)$ for any finite automaton accepting L.

For $x = \lambda$ and $y = 11$ we choose $z = 0$. Then

$$xz = 0 \notin L \text{ and } yz = 110 \in L.$$

For $x = 1$ and $y = 11$ we choose $z = \lambda$. Then

$$xz = 1 \notin L \text{ and } yz = 11 \in L.$$

Hence, the states

$$\hat{\delta}(q_0, \lambda), \hat{\delta}(q_0, 1) \text{ and } \hat{\delta}(q_0, 11)$$

have to be pair-wise different for any $A = (Q, \Sigma_{\text{bool}}, \delta, q_0, F)$ that accepts L.

Exercise 3.31. Consider the language $L = \{x11y \mid x, y \in \{0,1\}^*\}$ and the three words $\lambda, 1, 11$. For any pair x, y of different words from $S = \{\lambda, 1, 11\}$, estimate the set $Z(x, y) \subseteq (\Sigma_{\text{bool}})^*$ such that for any $z \in Z(x, y)$

$$(xz \notin L \text{ and } yz \in L) \text{ or } (xz \in L \text{ and } yz \notin L).$$

Exercise 3.32. Prove that any FA accepting $L = \{x11y \mid x, y \in \{0,1\}^*\}$ has at least three states by choosing three words different from $\lambda, 1$, and 11.

Exercise 3.33. Let $L = \{x011y \mid x, y \in \{0,1\}^*\}$.

(i) Construct an NFA M with 4 states for L and prove $L = L(M)$.
(ii) Prove that there is no FA M with 3 states and $L = L(M)$.
(iii) Apply the powerset construction to construct an FA that accepts L.

Exercise 3.34.* An FA A is called **minimal** for the regular language $L(A)$, if there is no smaller (with respect to the number of states) finite automaton that accepts $L(A)$. Construct minimal finite automata for the languages from Exercise 3.25 and prove their minimality.

Exercise 3.35. Design an NFA with at most six states, such that $L(M) = \{0x \mid x \in \{0,1\}^*$ and x contains at least one of the subwords 11 and 100$\}$.

3.6 Summary

In this chapter we have introduced finite automata as a model of simple computations that does not use any variable and hence does not have any memory. The main goal was not the study of finite automata, but a transparent introduction to modeling of computers (algorithms). A standard definition of a computing model starts with the description of the components of the model and the fixing of the instructions (elementary actions) of the model. Then the term configuration, which is a complete description of the general state of the computing model in a time unit, is defined. The execution of a (computation) step corresponds to the application of an instruction (elementary operation of the model) on the current configuration. Hence, a step is a movement from a configuration C to a configuration D. The difference between C and D can be achieved by the execution of an elementary operation. A computation on an input x starts in an initial configuration that includes x as the input. The end configuration determines the output (result).

Finite automata correspond to a subclass of simple algorithms for language recognition (for solving decision problems). A finite automaton does not use any variable (memory). A finite automaton moves only among its finitely many states (rows of the program) and accepts the input word if it finishes in an accepting state after reading the whole input.

To prove that a problem is not solvable by any algorithm from a special subclass of algorithms, one needs to perform a nonexistence proof. This usually requires a deep understanding of the nature of the considered subclass of algorithms. In the case of finite automata, the proof of $L \notin \mathcal{L}(\mathrm{FA})$ is based on showing that the structure of L is too complex to be described by a finite equivalence relation on Σ^*. Differently phrased, finitely many states are not sufficient to save all important characteristics of the read prefix of an input. This argumentation can also be used for proving a lower bounds on the number of states of any finite automaton that recognizes a given regular language L.

In contrast to their deterministic counterparts, nondeterministic computing models (algorithms) allow a choice from finitely many possible actions in any computation step. In this way a nondeterministic algorithm can have exponentially many (with respect to the input length) computations on an input. The interpretation of nondeterminism is an optimistic one. We assume that a nondeterministic algorithm would always makes the right choice, if such a choice exists. This means that a nondeterministic algorithm is considered to be successful in solving a problem if there exists a computation on every problem instance x that outputs the correct result. For a decision problem (Σ, L), it means that the nondeterministic algorithm A solving (Σ, L) has to have an accepting.[27] computation for every $x \in L$ and, for any $y \notin L$, all computations of A on y are not accepting.[28]

In general, we do not have any more efficient simulation of a nondeterministic algorithm A by a deterministic algorithm B, other than to let B simulate all computations of A for any input. This is also the case for finite automata, where B performs the breadth-first search in the computation tree of A on the given input. Since B on reading z saves the set of all states of A reachable in computations of A in which z has been read, the construction of the FA B for a given NFA A is called the powerset construction.

This chapter is devoted to some elementary aspects of automata theory only and should not be viewed as an exhaustive introduction to this area. Es-

[27] A computation with the output $x \in L$

[28] Computations with the output $y \notin L$

pecially when it comes to regular languages, we have provided a very selective introduction. Regular languages can be described and represented not only by finite automata, but by several further formal concepts that are not based on computing models. The most important examples of such concepts are the regular grammars as generation mechanisms and regular expressions as an algebraic representation. Further reading on the class of regular languages can be found in the textbook by Hopcroft, Motwani and Ullman [27].

Only those who have the patience
to do simple things perfectly
will acquire the skill
to do difficult things easily.

F. Schiller

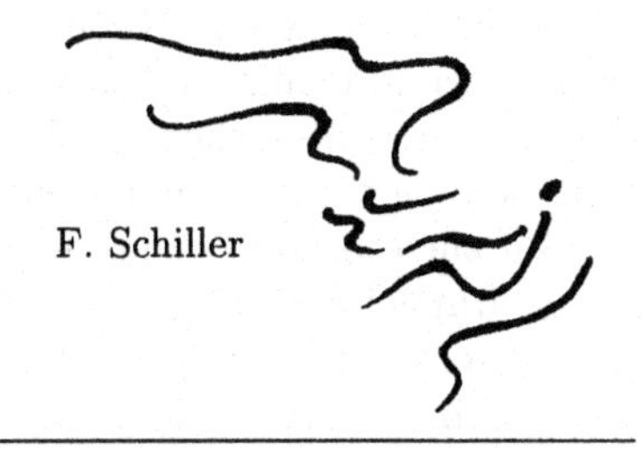

4

Turing Machines

4.1 Objectives

A long time ago, if one had had to explain an approach for solving some specific problems in mathematics, one had to describe it as a mathematical method in a formal way. A painstaking formal description of a method has the advantage that the user does not need to understand why the method works and can nevertheless successfully apply this method to solving her/his specific problem instance. The only assumption for a successful application is understanding the formal language, in which the method is represented. The development of computers led to the description of methods for solving problems by programs. The mathematical formalism[1] is given here by the programming language used. But the crucial feature of the original description remains. A computer does not posses any intellect and hence understands neither the problem nor the methods for solving it. Despite this, the computer can execute the program on the given input, solving the corresponding problem instance. Because of this, one can speak of automatic solvability or algorithmic solvability of problems. To show that a problem is automatically solvable it suffices to find a method solving the problem and to express this method in the form of a program (algorithm). Therefore one does not need to fix any formal definition of the term algorithm (program) when formulating positive assertions about algorithmic (automatic) solvability of problems. It suffices to give a rough, partially informal description of a method (algorithm) and the method can obviously be implemented as a program. Hence it is no surprise that the mathematics has connected the solvability of mathematical problems with the existence of general solution methods[2] in the sense of automatic solvability long before the discovery of computers.

The necessity for giving exact formal definition of the notion "algorithm" as a method for solving problems did not arise until mathematicians started

[1] The formal language

[2] Today we would say algorithms

to consider proving unsolvability of some specific problems. With this in mind, several formal definitions have been proposed and developed, and all reasonable ones have been shown to be equivalent to each other. Also, any programming language is a feasible formalization of automatic (algorithmic) solvability. But such formalizations are not very convenient for proving nonexistence of algorithms for concrete problems, because due to their user friendliness they contain instructions that are too complex. For this purpose one needs models of algorithms that are very simple and use only elementary instructions, but yet have the full computational power of programs of any high programming language. The Turing machine is such a model that has became the standard model of the computability theory. The aim of this chapter is to introduce this model and hence to build the basis for the presentation of the theory of algorithmic solvability[3] and the complexity theory in the next chapters.

This chapter is organized as follows. Section 4.2 introduces the formal model of Turing machines and practices working with them. Section 4.3 is devoted to the multitape version of Turing machines which is the fundamental computing model of the complexity theory. In this section we also discuss the equivalence between Turing machine models and programming languages. Section 4.4 introduces nondeterministic Turing machines and investigates possible simulations of nondeterministic Turing machines by deterministic ones. A coding of Turing machines over the alphabet Σ_{bool} is presented in Section 4.5.

4.2 The Turing Machine Model

A Turing machine can be viewed as a generalization of a finite automaton. It consists of (Figure 4.1):

(i) a finite control that contains the program,
(ii) a tape that serves as an input tape and simultaneously as a memory, and
(iii) a read/write head that may move in both directions.

The similarity to an FA lies in the control by a finite set of states (a program) and in the tape that contains an input word at the beginning of a computation. The main difference between a Turing machine and a finite automaton is the use of the tape. While a finite automaton may only read from the tape, a Turing machine may use the tape as a memory and write symbols on the tape. The tape is considered to be infinite in the sense that a Turing machine can use an arbitrary large finite part[4] of the tape in any computation. Technically, this difference is achieved by exchanging the reading head of an FA by a read/write head and additionally allowing the head to also move to

[3] Called the computability theory

[4] This means that a Turing machine can use as many tape squares (memory cells) as it needs. This is similar to real programs, where one can use as many variables as necessary and the number of variables may grow with the input length.

the left. An instruction (an elementary operation) of a Turing machine can be described as the following action. The arguments are

(i) the current state of the Turing machine, and
(ii) the symbol of the square (cell) of the tape, on which the read/write head is adjusted.

Depending on the values of these arguments, the Turing machine does the following activities:

(i) it changes its state (we can also say, it moves to a new state),
(ii) it writes a symbol in the square of the tape where the read/write head is adjusted (this can also be viewed as an exchange of the symbol read by a new one), and
(iii) it moves the read/write head one square to the right or the left, or it does not move the head.

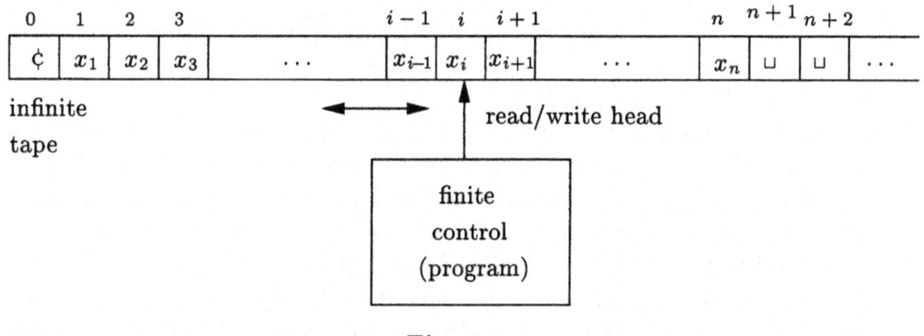

Fig. 4.1.

Some important technicalities are as follows:

1. The leftmost symbol of the tape is the so-called **left endmarker** ¢. This symbol may not be exchanged for another one by the Turing machine and the Turing machine may never go to the left when it has read ¢. The use of the left endmarker ¢ enables the enumeration of the tape squares from the left to the right starting with 0 on the square tape that contains ¢.
2. The tape is infinite in the direction to the right.[5]

Now, we give a formal definition of a Turing machine in the same way as we have done for finite automata. First, we describe the main components

[5] Note that within a finite time, a Turing machine can visit at most finitely many cells of the tape and thus the size of the actual memory (the number of the squares that contain symbols different from ⊔) is always finite. The sense of taking an unbounded-sized memory (an infinite tape) is to always have as large a memory as needed.

and the set of instructions (elementary operations). Then, we choose a representation of configurations and define the computation step as a relation on configurations. Finally, the definitions of a computation and of the language accepted by the given Turing machine follow.

Definition 4.1. *A* **Turing machine** (**TM** *for short*) *is a 7-tuple* $M = (Q, \Sigma, \Gamma, \delta, q_0, q_{\text{accept}}, q_{\text{reject}})$, *where*

(i) Q *is a finite set, called the* **set of states** *of* M,

(ii) Σ *is the* **input alphabet**, *where* ¢ *and the blank symbol* ␣ *are not in* Σ {Σ *is used for the representation of input words in the same way as for finite automata*},

(iii) Γ *is an alphabet, called* **working alphabet**, *where* $\Sigma \subseteq \Gamma$, *and* ¢, ␣ $\in \Gamma$ {Γ *contains all symbols that may occur in the squares of the tape, i.e., all symbols used as variable values in the memory.*},

(iv) $\delta : (Q - \{q_{\text{accept}}, q_{\text{reject}}\}) \times \Gamma \to Q \times \Gamma \times \{\text{L}, \text{R}, \text{N}\}$ *is a mapping, called the* **transition function of** M, *with the property*

$$\delta(q, ¢) \in Q \times \{¢\} \times \{\text{R}, \text{N}\}$$

for all $q \in Q - \{q_{\text{accept}}, q_{\text{reject}}\}$
{δ *determines the instructions of* M. M *may perform an instruction* $(q, X, Z) \in Q \times \Gamma \times \{\text{L}, \text{R}, \text{N}\}$, *if* M *is in a state* p, *reads a symbol* $Y \in \Gamma$ *and* $\delta(p, Y) = (q, X, Z)$. *Execution of this instruction implies moving from state* p *to state* q, *replacing the symbol* Y *by* X, *and moving the read/write head according to* Z. *If* $Z = \text{L}$, *then* M *should move one square to the left, if* $Z = \text{R}$, *then* M *should move one square to the right, and* $Z = \text{N}$ *means that the head remains stationary. The property* $\delta(q, ¢) \in Q \times \{¢\} \times \{\text{R}, \text{N}\}$ *assures that* M *neither rewrites the left endmarker* ¢ *nor moves outside the tape to the left.*},

(v) $q_0 \in Q$ *is the* **initial state**,

(vi) $q_{\text{accept}} \in Q$ *is the* **accepting state**
{M *has exactly one accepting state. If* M *enters the accepting state* q_{accept}, *then it would accept the input regardless of the position of its read/write head on the tape. Once* q_{accept} *is reached, there is no possibility of continuing the computation.*[6]},

(vii) $q_{\text{reject}} \in Q - \{q_{\text{accept}}\}$ *is the* **rejecting state**
{*If* M *has reached* q_{reject}, *then* M *stops the computation and rejects (does not accept) the input.*}.

A **configuration** C *of* M *is an element from*

$$\textbf{conf}(M) = \{¢\} \cdot \Gamma^* \cdot Q \cdot \Gamma^+ \cup Q \cdot \{¢\} \cdot \Gamma^*.$$

[6] Among others, it means that in contrast to finite automata M does not need to read the whole input before deciding about its acceptance or its rejection.

{A configuration w_1qaw_2, $w_1 \in \{¢\}\Gamma^*$, $w_2 \in \Gamma^*$, $a \in \Gamma$, $q \in Q$ (Figure 4.2), is a complete description of the following situation. M is in the state q, the content of the tape is $¢w_1aw_2\sqcup\sqcup\sqcup\dots$, and the head is adjusted on the $(|w_1|+1)$-th square of the tape and reads the symbol a. A configuration[7] $p¢w$, $p \in Q$, $w \in \Gamma^*$ describes the situation in which the content of the tape is $¢w\sqcup\sqcup\sqcup\dots$ and the head is positioned on the 0-th position of the tape thus reading the left endmarker ¢.}

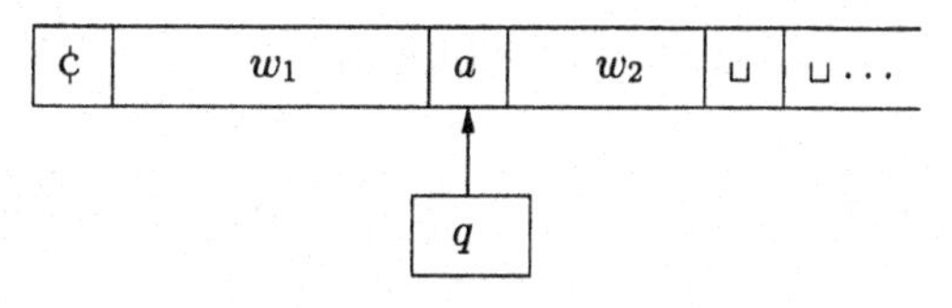

Fig. 4.2.

The **initial configuration** *of M on an input word x is $q_0¢x$.*

A **step of M** *is a relation $\vdash_M$ on the set of configurations, i.e., $\vdash_M \subseteq \text{conf}(M) \times \text{conf}(M)$, defined by*

(i) $x_1x_2\dots x_{i-1}qx_ix_{i+1}\dots x_n \vdash_M x_1x_2\dots x_{i-1}pyx_{i+1}\dots x_n$,
if $\delta(q, x_i) = (p, y, \mathrm{N})$ *(Figure 4.3a),*

(ii) $x_1x_2\dots x_{i-1}qx_ix_{i+1}\dots x_n \vdash_M x_1x_2\dots x_{i-2}px_{i-1}yx_{i+1}\dots x_n$,
if $\delta(q, x_i) = (p, y, \mathrm{L})$ *(Figure 4.3b),*

(iii) $x_1x_2\dots x_{i-1}qx_ix_{i+1}\dots x_n \vdash_M x_1x_2\dots x_{i-1}ypx_{i+1}\dots x_n$,
if $\delta(q, x_i) = (p, y, \mathrm{R})$ *for $i < n$ (Figure 4.3c) and*
$x_1x_2\dots x_{n-1}qx_n \vdash_M x_1x_2\dots x_{n-1}yp\sqcup$
if $\delta(q, x_n) = (p, y, \mathrm{R})$ *(Figure 4.3d).*

A **computation of M** *is a potentially infinite sequence of configurations $C_0, C_1, C_2, \dots$, such that $C_i \vdash_M C_{i+1}$ for all $i = 0, 1, 2, \dots$. If*

$$C_0 \vdash_M C_1 \vdash_M \cdots \vdash_M C_i$$

for an $i \in \mathbb{N}$, then we write

$$C_0 \vdash_M^* C_i.$$

The **computation of M on an input** *x is a computation that starts with the initial configuration $C_0 = q_0¢x$ and it is either infinite or stops in a configuration w_1qw_2, where $q \in \{q_{\text{accept}}, q_{\text{reject}}\}$.*

---

[7] One has a choice between several appropriate ways of representing configurations of a Turing machine. For instance, one can choose the representation $(q, w, i) \in Q \times \Gamma^* \times \mathbb{N}$ for describing the general state, where M is in the (internal) state q, $w\sqcup\sqcup\sqcup\dots$ is the content of the tape, and the head is adjusted at the i-th squares of the tape.

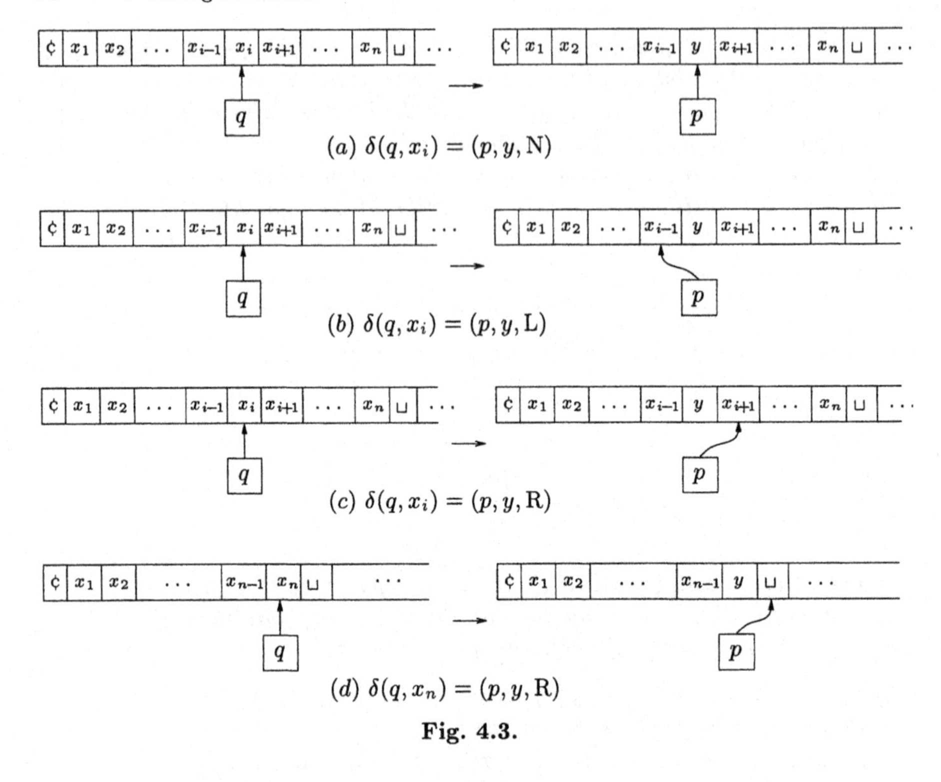

(a) $\delta(q, x_i) = (p, y, \mathrm{N})$

(b) $\delta(q, x_i) = (p, y, \mathrm{L})$

(c) $\delta(q, x_i) = (p, y, \mathrm{R})$

(d) $\delta(q, x_n) = (p, y, \mathrm{R})$

Fig. 4.3.

The computation of M on x is called **accepting** *if it finishes in an accepting configuration $w_1 q_{\text{accept}} w_2$. The computation of M on x is called* **rejecting**, *if it ends in a rejecting configuration $w_1 q_{\text{reject}} w_2$. If the computation of M on x is accepting [rejecting] we say that M* **accepts** [**rejects**] *x. If the computation of M on x is rejecting or infinite then we say that M* **does not accept**[8] *x.*

The language $L(M)$ accepted by M is defined by

$$\begin{aligned} \boldsymbol{L(M)} &= \{w \in \Sigma^* \mid q_0 ¢ w \mathrel{\vert\!\frac{*}{M}} y q_{\text{accept}} z, \text{ for some } y, z \in \Gamma^*\} \\ &= \{w \in \Sigma^* \mid \ M \text{ accepts } w\ \}. \end{aligned}$$

We say that M computes a function $F : \Sigma^* \to \Gamma^*$, if

$$\text{for all } x \in \Sigma^* : q_0 ¢ x \mathrel{\vert\!\frac{*}{M}} q_{\text{accept}} ¢ F(x).$$

A language is called **recursively enumerable**, *if there exists a* TM *M such that $L = L(M)$.*

The set

[8] Note that we distinguish between rejecting and not accepting. More precisely, the rejection is a special subcase of nonacceptance.

$$\mathcal{L}_{\text{RE}} = \{L(M) \mid M \textit{ is a } \text{TM}\}$$

is called the **class of recursively enumerable languages.**

A language $L \subseteq \Sigma^*$ *is called* **recursive** *or the decision problem* (Σ, L) *is called* **decidable,** *if* $L = L(M)$ *for a* TM M*, such that, for all* $x \in \Sigma^*$,

(i) $q_0 ¢x \mid\frac{*}{M}\, yq_{\text{accept}}z$, $y, z \in \Gamma^*$, *if* $x \in L$, *and*
(ii) $q_0 ¢x \mid\frac{*}{M}\, uq_{\text{reject}}v$, $u, v \in \Gamma^*$, *if* $x \notin L$.

{*This means that* M *does not have any infinite computations.*}

If (i) and (ii) hold, we would say that M **halts on every input** *or that* M **always halts.**

{*A Turing machine that always halts is a formal model of the notion "algorithm".*}

The set

$$\mathcal{L}_{\text{R}} = \{L(M) \mid M \textit{ is a } \text{TM}, \textit{ that always halts}\}$$

is the **class of recursive (algorithmically recognizable) languages.**

A function $F : \Sigma_1^* \to \Sigma_2^*$ *for two alphabets* Σ_1, Σ_2 *is called* **computable,** *if there exists a* TM M *that computes* [9] F.

Turing machines, that always halt, represent algorithms, i.e., programs that always terminate and output the right answers. So, recursive languages (decidable decision problems) are exactly languages (decision problems) that are algorithmically recognizable (decidable).

Exercise 4.2. Reformalize the definition of a TM using the triple $(q, ¢w, i) \in Q \times \{¢\}\Gamma^* \times \mathbb{N}$ to represent the configurations. A triple $(q, ¢w, i)$ describes the situation in which the Turing machine is in the state q, the content of the tape is $¢w\sqcup\sqcup\sqcup\ldots$, and the head is adjusted at the i-th square of the tape. Give the definition of a (computation) step and of a computation using this representation.

In what follows we present a few concrete Turing machines and, similarly as for finite automata, we develop a transparent graphic representation of Turing machines. Let

$$L_{\text{middle}} = \{w \in (\Sigma_{bool})^* \mid w = x1y, \text{ where } |x| = |y|\}.$$

Hence, L_{middle} contains all words of an odd length with the symbol 1 in the middle.

Exercise 4.3. Prove that $L_{\text{middle}} \notin \mathcal{L}(\text{FA})$.

[9] Note that M always halts.

Now we shall describe a TM M such that $L(M) = L_{\text{middle}}$. The idea is to first check whether the input length is odd and then estimate the middle position of the input and check the symbol at this position. Let $M = (Q, \Sigma, \Gamma, \delta, q_0, q_{\text{accept}}, q_{\text{reject}})$, where

$Q = \{q_0, q_{\text{even}}, q_{\text{odd}}, q_{\text{accept}}, q_{\text{reject}}, q_A, q_B, q_1, q_{\text{left}}, q_{\text{right}}, q_{\text{middle}}\}$,
$\Sigma = \{0, 1\}$,
$\Gamma = \Sigma \cup \{¢, \sqcup\} \cup (\Sigma \times \{A, B\}) = \{0, 1, ¢, \sqcup, \binom{0}{A}, \binom{0}{B}, \binom{1}{A}, \binom{1}{B}\}$ and

$$\begin{aligned}
\delta(q_0, ¢) &= (q_{\text{even}}, ¢, \mathrm{R}),\\
\delta(q_0, a) &= (q_{\text{reject}}, a, \mathrm{N}) \text{ for all } a \in \{0, 1, \sqcup\},\\
\delta(q_{\text{even}}, b) &= (q_{\text{odd}}, b, \mathrm{R}) \text{ for all } b \in \Sigma,\\
\delta(q_{\text{even}}, \sqcup) &= (q_{\text{reject}}, \sqcup, \mathrm{N}),\\
\delta(q_{\text{odd}}, b) &= (q_{\text{even}}, b, \mathrm{R}) \text{ for all } b \in \Sigma,\\
\delta(q_{\text{odd}}, \sqcup) &= (q_B, \sqcup, \mathrm{L}).
\end{aligned}$$

Applying the above transitions we see that, after reading a prefix of an even [odd] length, M enters the state q_{even} [q_{odd}]. Hence, if M reads the symbol $\sqcup$ in the state q_{even}, then the input word is of an even length and must be rejected. If M reads the symbol $\sqcup$ in the state q_{odd}, then M moves to the state q_B in which the second phase of the computation starts.

In the second phase M determines the middle of the input and by alternatively replacing (rewriting) the leftmost symbol $a \in \{0, 1\}$ for $\binom{a}{B}$ and the rightmost symbol $b \in \Sigma$ with $\binom{b}{A}$. This can be executed by applying the following transitions:

$$\begin{aligned}
\delta(q_B, a) &= (q_1, \tbinom{a}{B}, \mathrm{L}) \text{ for all } a \in \{0, 1\},\\
\delta(q_1, a) &= (q_{\text{left}}, a, \mathrm{L}) \text{ for all } a \in \{0, 1\},\\
\delta(q_1, c) &= (q_{\text{middle}}, c, R) \text{ for all } c \in \{¢, \tbinom{0}{A}, \tbinom{1}{A}\},\\
\delta(q_{\text{middle}}, \tbinom{0}{B}) &= (q_{\text{reject}}, 0, \mathrm{N}),\\
\delta(q_{\text{middle}}, \tbinom{1}{B}) &= (q_{\text{accept}}, 1, \mathrm{N}),\\
\delta(q_{\text{left}}, a) &= (q_{\text{left}}, a, \mathrm{L}) \text{ for all } a \in \{0, 1\},\\
\delta(q_{\text{left}}, c) &= (q_A, c, \mathrm{R}) \text{ for all } c \in \{\tbinom{0}{A}, \tbinom{1}{A}, ¢\},\\
\delta(q_A, b) &= (q_{\text{right}}, \tbinom{b}{A}, \mathrm{R}) \text{ for all } b \in \{0, 1\},\\
\delta(q_{\text{right}}, b) &= (q_{\text{right}}, b, \mathrm{R}) \text{ for all } b \in \{0, 1\},\\
\delta(q_{\text{right}}, d) &= (q_B, d, \mathrm{L}) \text{ for all } d \in \{\tbinom{0}{B}, \tbinom{1}{B}\}.
\end{aligned}$$

The missing arguments such as the pair $(q_{\text{right}}, ¢)$ cannot occur in any computation and hence one can complete the formal definition by assigning these missing arguments to the state q_{reject}.

If one takes the graphic representation of Figure 4.4 of the instruction

$$\delta(q, a) = (p, b, X)$$

for any $q, p \in Q$, $a, b \in \Sigma$ and $X \in \{\mathrm{L}, \mathrm{R}, \mathrm{N}\}$, then one would obtain the graphic representation of M depicted in Figure 4.5.

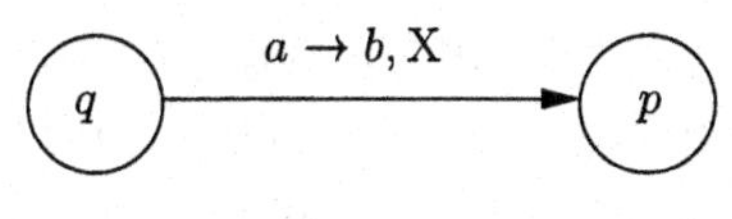

Fig. 4.4.

The graphic representation of Turing machines is similar to that of finite automata. The difference lies only in the labeling of the edges because now we have to include for each symbol a read, both the new symbol b and the direction X of the head movement on the tape.

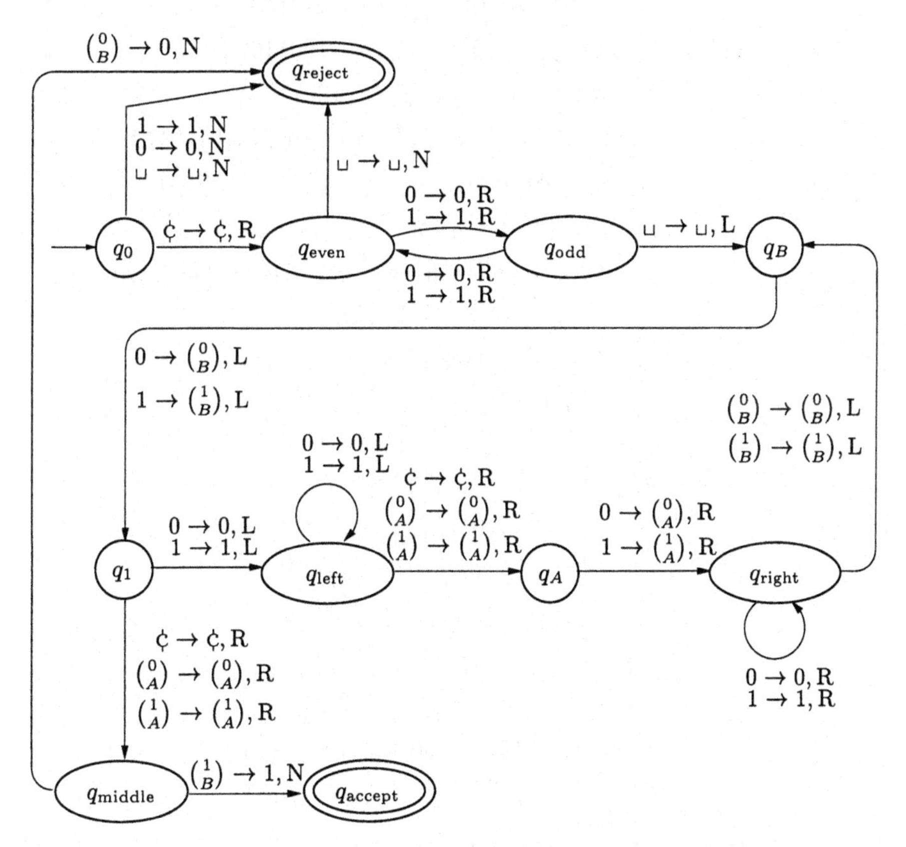

Fig. 4.5.

Let us consider the work of M on the input word $x = 1001101$. The first phase of the computation on x is as follows:

$$
\begin{aligned}
q_0\text{¢}1001101 &\vdash_M \text{¢}q_{\text{even}}1001101 \vdash_M \text{¢}1q_{\text{odd}}001101 \vdash_M \text{¢}10q_{\text{even}}01101\\
&\vdash_M \text{¢}100q_{\text{odd}}1101 \vdash_M \text{¢}1001q_{\text{even}}101 \vdash_M \text{¢}10011q_{\text{odd}}01\\
&\vdash_M 100110q_{\text{even}}1 \vdash_M \text{¢}1001101q_{\text{odd}}\sqcup \vdash_M \text{¢}100110q_B1
\end{aligned}
$$

Finishing in the state q_B has the meaning that x is of an odd length. Now, M alternatively exchanges symbols a on the right side (boundary) of the tape by $\binom{a}{B}$ and on the left side (boundary) of the tape by $\binom{a}{A}$.

$$
\begin{aligned}
\text{¢}100110q_B1 &\vdash_M \text{¢}10011q_10\tbinom{1}{B} \vdash_M \text{¢}1001q_{\text{left}}10\tbinom{1}{B}\\
&\vdash_M \text{¢}100q_{\text{left}}110\tbinom{1}{B} \vdash_M \text{¢}10q_{\text{left}}0110\tbinom{1}{B}\\
&\vdash_M \text{¢}1q_{\text{left}}00110\tbinom{1}{B} \vdash_M \text{¢}q_{\text{left}}100110\tbinom{1}{B}\\
&\vdash_M q_{\text{left}}\text{¢}100110\tbinom{1}{B} \vdash_M \text{¢}q_A100110\tbinom{1}{B}\\
&\vdash_M \text{¢}\tbinom{1}{A}q_{\text{right}}00110\tbinom{1}{B} \vdash_M \text{¢}\tbinom{1}{A}0q_{\text{right}}0110\tbinom{1}{B}\\
&\vdash_M^* \text{¢}\tbinom{1}{A}00110q_{\text{right}}\tbinom{1}{B} \vdash_M \text{¢}\tbinom{1}{A}0011q_B0\tbinom{1}{B}\\
&\vdash_M \text{¢}\tbinom{1}{A}001q_11\tbinom{0}{B}\tbinom{1}{B} \vdash_M \text{¢}\tbinom{1}{A}00q_{\text{left}}11\tbinom{0}{B}\tbinom{1}{B}\\
&\vdash_M^* \text{¢}q_{left}\tbinom{1}{A}0011\tbinom{0}{B}\tbinom{1}{B} \vdash_M \text{¢}\tbinom{1}{A}q_A0011\tbinom{0}{B}\tbinom{1}{B}\\
&\vdash_M \text{¢}\tbinom{1}{A}\tbinom{0}{A}q_{\text{right}}011\tbinom{0}{B}\tbinom{1}{B}\\
&\vdash_M^* \text{¢}\tbinom{1}{A}\tbinom{0}{A}011q_{\text{right}}\tbinom{0}{B}\tbinom{1}{B}\\
&\vdash_M \text{¢}\tbinom{1}{A}\tbinom{0}{A}01q_B1\tbinom{0}{B}\tbinom{1}{B}\\
&\vdash_M \text{¢}\tbinom{1}{A}\tbinom{0}{A}0q_11\tbinom{1}{B}\tbinom{0}{B}\tbinom{1}{B}\\
&\vdash_M^* \text{¢}\tbinom{1}{A}q_{\text{left}}\tbinom{0}{A}01\tbinom{1}{B}\tbinom{0}{B}\tbinom{1}{B}\\
&\vdash_M \text{¢}\tbinom{1}{A}\tbinom{0}{A}q_A01\tbinom{1}{B}\tbinom{0}{B}\tbinom{1}{B}\\
&\vdash_M \text{¢}\tbinom{1}{A}\tbinom{0}{A}\tbinom{0}{A}q_{\text{right}}1\tbinom{1}{B}\tbinom{0}{B}\tbinom{1}{B}\\
&\vdash_M \text{¢}\tbinom{1}{A}\tbinom{0}{A}\tbinom{0}{A}1q_{\text{right}}\tbinom{1}{B}\tbinom{0}{B}\tbinom{1}{B}\\
&\vdash_M \text{¢}\tbinom{1}{A}\tbinom{0}{A}\tbinom{0}{A}q_B1\tbinom{1}{B}\tbinom{0}{B}\tbinom{1}{B}\\
&\vdash_M \text{¢}\tbinom{1}{A}\tbinom{0}{A}q_1\tbinom{0}{A}\tbinom{1}{B}\tbinom{1}{B}\tbinom{0}{B}\tbinom{1}{B}\\
&\vdash_M \text{¢}\tbinom{1}{A}\tbinom{0}{A}\tbinom{0}{A}q_{\text{middle}}\tbinom{1}{B}\tbinom{1}{B}\tbinom{0}{B}\tbinom{1}{B}\\
&\vdash_M \text{¢}\tbinom{1}{A}\tbinom{0}{A}\tbinom{0}{A}q_{\text{accept}}1\tbinom{1}{B}\tbinom{0}{B}\tbinom{1}{B}.
\end{aligned}
$$

The TM finishes in q_{accept}, meaning that the middle position is occupied by the symbol 1, and accepts the word 1001101.

Exercise 4.4. Write the computations of the TM M in Figure 4.5 on the input words 010011 and 101.

Exercise 4.5. The TM M in Figure 4.5 works in such a way that the complete information about the original input word is saved. Use this property in order to extend M to a TM M' that accepts the language $L(M') = \{w \in \{0,1\}^* \mid w = x1x \text{ for a } x \in \{0,1\}^*\}$.

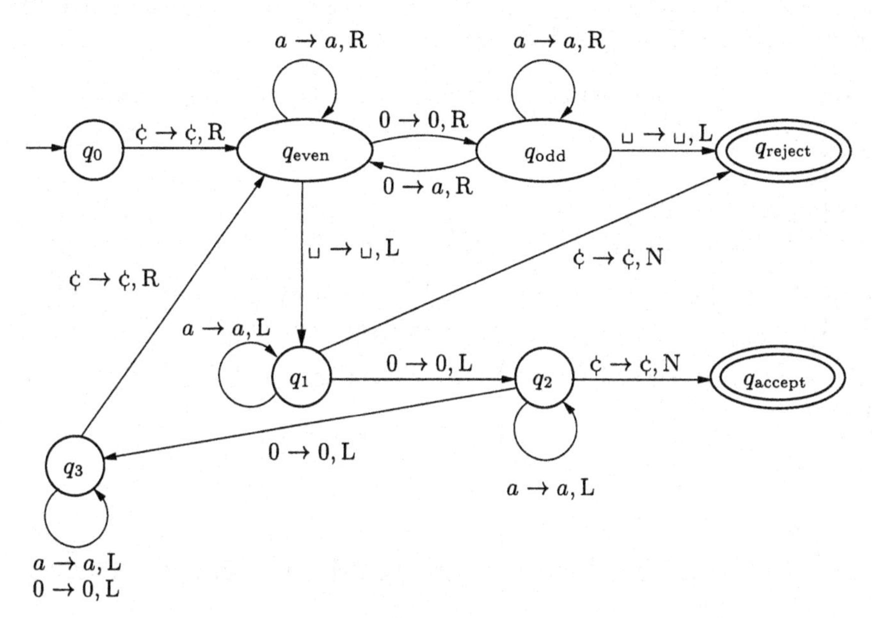

Fig. 4.6.

Consider now the language $L_{\mathcal{P}} = \{0^{2^n} \mid n \in \mathbb{N} - \{0\}\}$. A Turing machine accepting L can adopt the following strategy:

1. Run on the tape from the left endmarker ¢ to the first $\sqcup$ symbol on the right, deleting every second 0 on the tape (more precisely, exchanging every second 0 for the symbol a). If the number of 0s on the tape is odd, then stop in the state q_{reject}. Otherwise, continue with step 2.
2. In one run from the right (from the symbol $\sqcup$) to the left (to the left endmarker ¢) check whether the number of 0s on the tape is 1 or more.
   - If there is exactly one symbol 0 on the tape, then accept.
   - If there are at least 2 occurrences of 0s on the tape, then continue with step 1.

The idea of this strategy is that for any positive integer i, a number 2^i can be divided by 2 without remainder as long as one gets 1. A possible execution of this strategy in the formalism of a Turing machine is given by the TM $A = (\{q_0, q_{\text{even}}, q_{\text{odd}}, q_1, q_2, q_3, q_{\text{accept}}, q_{\text{reject}}\}, \{0\}, \{0, a, ¢, \sqcup\}, \delta_A, q_0, q_{\text{accept}}, q_{\text{reject}})$ shown in Figure 4.6.

Exercise 4.6. Another strategy for recognizing $L_{\mathcal{P}}$ is to rewrite an input 0^i by 0^j1^j if $i = 2j$ is an even natural number. Next, one checks whether j is even and if it is then one rewrites 0^j1^j by $0^{\frac{j}{2}}1^{\frac{j}{2}}1^j$. The input would be accepted only if this halving strategy results in the content 01^{i-1} of the tape. To implement this halving by a TM, one can use the above-described way of searching for the middle of an input. Give an explicit construction of a TM that accepts $L_{\mathcal{P}}$ using this strategy.

Exercise 4.7. Design Turing machines for the following languages:

(i) $\{a^nb^n \mid n \in \mathbb{N}\}$,
(ii) $\{0^{n^2} \mid n \in \mathbb{N}\}$,
(iii) $\{w\#w \mid w \in \{0,1\}^*\}$,
(iv) $\{x\#y \mid x, y \in \{0,1\}^*, \mathit{Number}(x) = \mathit{Number}(y) + 1\}$.

Exercise 4.8. Design Turing machines that, for every input word $x \in (\Sigma_{\text{bool}})^*$, finish in the state q_{accept} with the following tape content:

(i) $y \in (\Sigma_{\text{bool}})^*$, such that $\mathit{Number}(y) = \mathit{Number}(x) + 1$,
(ii) $x\#x$,
(iii) $z \in (\Sigma_{\text{bool}})^*$, such that $\mathit{Number}(z) = 2 \cdot \mathit{Number}(x)$,
(iv) $\#\#\#x$.

4.3 Multitape Turing Machines and the Church–Turing Thesis

Due to their simplicity, Turing machines represent the standard computing model in the computability theory, i.e., for the classification of problems with respect to their recursivity or recursive enumerability. But this model is not always convenient for the purpose of the complexity theory. The main drawback of the introduced Turing machine model is that it does not fit into the general framework of a computer model given by Von Neumann.[10] The Von Neumann computer model requires that all main components of a computer, namely memory where the program is saved, memory for data, CPU, and input medium are physically independent parts of the computer. Whereas in Turing machines, the input medium and the memory are the same thing – the tape. The second drawback of using Turing machines in the complexity theory is the linearity of the memory, i.e., the restricted access to the tape. If one wants to compare the contents of two different squares (cells) of the tape, one would need to execute as many operations (computation steps) as the distance[11] between these two squares.

[10] Called also a Von Neumann computer
[11] If the distance is large, too much work would be needed for the execution of a simple comparison.

The following model of multitape Turing machines overcomes the above-mentioned drawbacks of Turing machines to some extent and it is hence considered to be the fundamental computing model of the complexity theory.[12] For any positive integer k, a k-tape Turing machine has the following components (Figure 4.7):

- a finite control involving a program,
- a finite tape with a read-only head as an input medium, and
- k working tapes, each accompanied by a read/write head as memory for data.

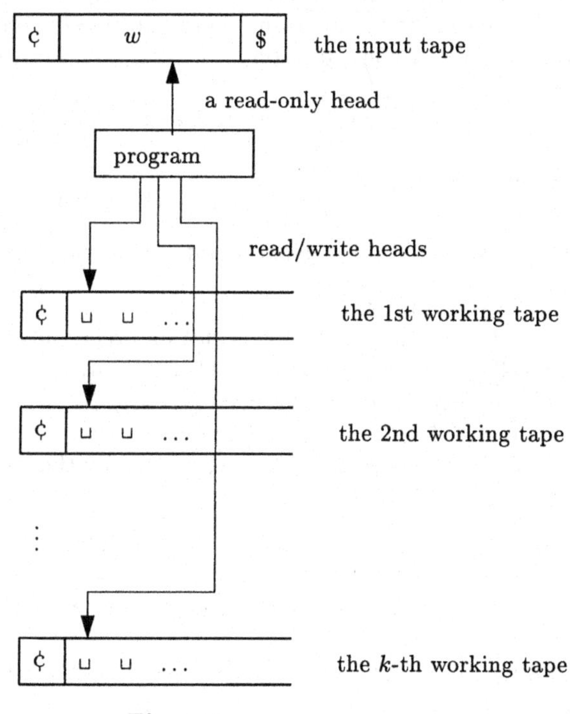

Fig. 4.7.

Before starting the computation on an input word w, a k-tape Turing machine is always in the following general state (initial configuration).

- The finite input tape contains $¢w\$$, where ¢ and \$ are the symbols that mark the left side and the right side of the input, respectively.
- The read-only head on the input tape is adjusted on the left endmarker ¢.

[12] More discussion about suitability of the multitape Turing machine model for the purposes of the complexity theory is given in Chapter 6.

- The content of each working tape is ¢⊔⊔⊔... and all read/write heads of the working tapes are adjusted on the square containing ¢.
- The finite control is in the initial state q_0.

During the computation, all heads may move to the left and to the right, except moving to the left from a left endmarker ¢ or moving to the right on the input tape from the right endmarker $. The read-only head of the input tape may not write, so the initial content ¢$w$$ of the input tape remains unchanged during the whole computation on w. Similar to Turing machines, the content of any cell (square) of the working tapes may be a symbol from the working alphabet Γ. The cells of all $k+1$ tapes are numbered from the left to the right, starting with 0 for the cell containing the left endmarker ¢. Hence, one can take the following representation of configurations of a k-tape Turing machine. A configuration

$$(q, w, i, u_1, i_1, u_2, i_2, \ldots, u_k, i_k)$$

is an element from

$$Q \times \Sigma^* \times \mathbb{N} \times (\Gamma^* \times \mathbb{N})^k.$$

It represents the following general state of the machine M:

- M is in the state q,
- the content of the input tape is ¢$w$$ and read-only head is adjusted on the i-th cell of the input tape (i.e., when $w = a_1 a_2 \ldots a_n$ for $a_i \in \Sigma$, then the read-only head is reading the symbol a_i), and
- for any $j \in \{1, 2, \ldots, k\}$, the content of the j-th tape is ¢u_j⊔⊔⊔..., and $i_j \leq |u_j|$ is the position of the cell visited by the head on the j-th working tape.

A (computation) step M can be described by a transition function

$$\delta : Q \times (\Sigma \cup \{¢, \$\}) \times \Gamma^k \to Q \times \{\mathrm{L}, \mathrm{R}, \mathrm{N}\} \times (\Gamma \times \{\mathrm{L}, \mathrm{R}, \mathrm{N}\})^k$$

The arguments $(q, a, b_1, \ldots, b_k) \in Q \times (\Sigma \cup \{¢, \$\}) \times \Gamma^k$ are:

- the actual state q,
- the symbol $a \in \Sigma \cup \{¢, \$\}$ read by the read-only head on the input tape,
- the k symbols $b_1, \ldots, b_k$ read on the k working tapes.

Following these arguments the execution of a step of M corresponds to the following actions.

- Any of the k symbols $b_1, \ldots, b_k$ read on the working tapes can be replaced for another symbol.
- M moves to a new state.
- Any of the $k+1$ heads can move one square to the left or to the right, except when endmarkers are read.

If the computation of M on w finishes in the state q_{accept}, then M accepts w. M does not accept w when the computation of M on w is infinite or finishes in the state q_{reject}. We continue using the term "M rejects w" if M reaches q_{reject} when computing on w.

We omit the presentation of the formal definition of a k-tape Turing machine. Given the above description, the specification of a formal definition for a k-tape Turing machine is a matter of routine, and thus we leave this to the reader as an exercise.

Exercise 4.9. Give an exact formal definition of a k-tape Turing machine, following the definition of a Turing machine step-by-step.

For any positive integer k, we use the abbreviation **k-tape-TM** to denote a k-tape Turing machine. For any $k \in \mathbb{N} - \{0\}$, any k-tape-TM is called a **multitape Turing machine**, MTM. Since the instructions of an MTM are slightly more complex than the instructions of a TM, one could expect that multitape Turing machines are able to solve some problems in a simpler or more efficient way than Turing machines. Consider the language

$$L_{\text{equal}} = \{w\#w \mid w \in (\Sigma_{\text{bool}})^*\}.$$

Let $x\#y$ be an input. A TM that compares x (the prefix before the first occurrence of the symbol #) with y (the suffix after the first occurrence of #) has to run many times the long distance to and fro on its tape.

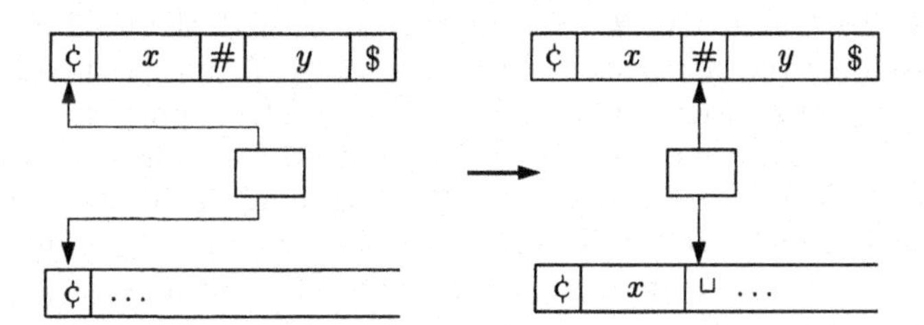

Fig. 4.8.

A 1-tape-TM A accepts L_{equal} using the following strategy.

1. A checks whether the input has the form[13] $x\#y$ with $x, y \in (\Sigma_{\text{bool}})^*$. If not, A rejects the input.
2. For the input $x\#y$, A makes a copy of x on the working tape, i.e., the working tape contains ¢x at the end of this step (Figure 4.8).
3. A adjusts the head of the working tape on the left endmarker ¢ while the head of the input tape is adjusted on #. Then, A simultaneously moves both heads to the right and compares x and y in this way. Should the

[13] This means that A checks whether the input contains exactly one symbol #.

heads read different symbols in a step, then A knows that $x \neq y$ and rejects the input. If all pairs of symbols are equal and both heads reach the symbol $\sqcup$ in the same step, then A accepts the input.

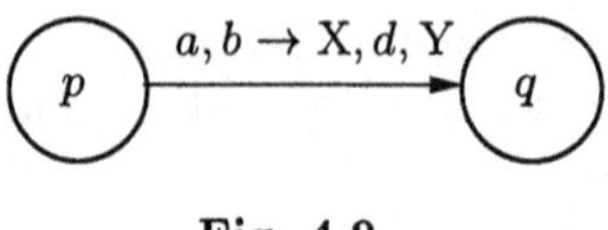

Fig. 4.9.

The graphic representation of the transition

$$\delta(p, a, b) = (q, \mathrm{X}, d, \mathrm{Y})$$

of a 1-tape-TM is depicted in Figure 4.9. The transition from state p to state q happens when a is read on the input tape and b is read on the working tape. The symbol b is replaced by the symbol d. $\mathrm{X} \in \{\mathrm{L}, \mathrm{R}, \mathrm{N}\}$ determines the movement of the head on the input tape, and $\mathrm{Y} \in \{\mathrm{L}, \mathrm{R}, \mathrm{N}\}$ determines the movement of the head on the working tape.

Following the graphic representation of an instruction (transition) shown in Figure 4.11, we can construct the graphic description Figure 4.12 of the 1-tape-TM presented above, which recognizes the language L_{equal}. The states q_0, q_1, q_2 and q_{reject} are used to execute the first phase of this strategy. If the input contains exactly one #, M reaches state q_2 with the head of the input tape adjusted on the last symbol of the input. Else, M finishes in q_{reject}. State q_2 is used to return the read-only head to the left endmarker ¢. With the help of q_{copy}, M copies the prefix of the input up to # on the working tape (Figure 4.10). Then state q_{adjust} is used for returning the head of the working tape to the left endmarker ¢. The comparison of x and y of the input $x\#y$ is executed by using the state q_{compare}. If $x = y$, M stops in q_{accept}. If x and y differ on a position or x and y have different lengths, the computation ends in q_{reject}.

Exercise 4.10. Describe 1-tape Turing machines informally as well as graphically for the following recursive languages:

1. $L = \{a^n b^n \mid n \in \mathbb{N}\}$,
2. $L = \{w \in (\Sigma_{\mathrm{bool}})^* \mid |w|_0 = |w|_1\}$,
3. $L = \{a^n b^n c^n \mid n \in \mathbb{N}\}$,
4. $L = \{www \mid w \in (\Sigma_{\mathrm{bool}})^*\}$,
5. $L = \{a^{n^2} \mid n \in \mathbb{N}\}$.

Now, we have two different computing models – the Turing machine and the multitape Turing machine. In order to use them interchangeably in the computability theory, one needs to prove their equivalence with respect to

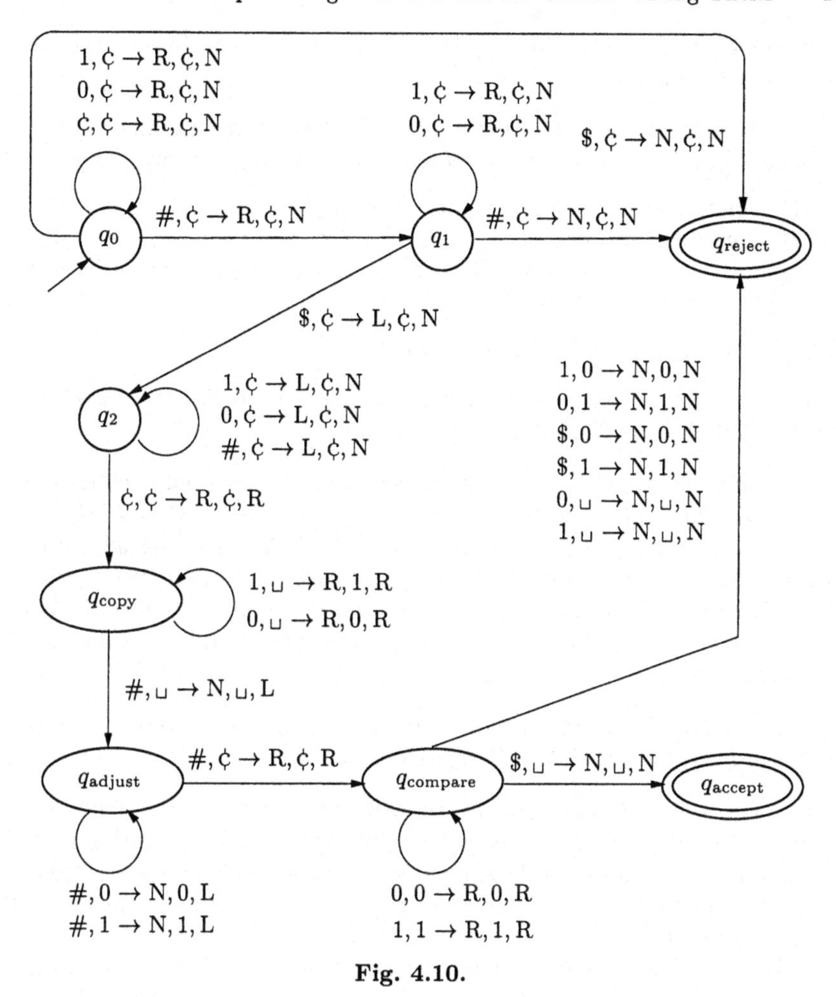

Fig. 4.10.

the classes of languages accepted by them. Let A and B be two machines (Turing machines, multitape Turing machines, or others) that work over the same input alphabet Σ. In what follows we say that A **is equivalent to** B if, for every input $x \in (\Sigma_{\text{bool}})^*$, the following conditions hold:

(i) A accepts x iff B accepts x,
(ii) A rejects x iff B rejects x, and
(iii) the computation of A on x is infinite iff the computation of B on x is infinite.

Clearly, the equivalence of A and B implies $L(A) = L(B)$, but $L(A) = L(B)$ alone does not imply that A and B are equivalent.

Lemma 4.11. *For any* TM A, *there exists a* 1-*tape*-TM B *such that* A *and* B *are equivalent.*

Proof. We describe a simulation of A by B without giving any formal construction[14] of B. The 1-tape-TM B works in the following two phases.

1. B copies the whole input w on the working tape.
2. B simulates the work of A on w on its working tape in a step-by-step manner. {This means that B does exactly the same work on its working tape as A does on its input tape.}

It is obvious that A and B are equivalent. □

Exercise 4.12. Give a formal construction of a 1-tape-TM B, which is equivalent to a TM $A = (Q, \Sigma, \Gamma, \delta, q_0, q_{\text{accept}}, q_{\text{reject}})$.

In what follows we will usually dispense with the formal constructions of Turing machines and with formal proofs of facts like $L(A) = L(B)$ for two machines A and B. The reason is similar to when describing algorithms or arguing for the correctness of programs. We save a lot of detailed work when doing away with formal proofs in situations where things are intuitively clear (for instance, that a TM accepts a given language or that a program executes the job required).

Lemma 4.13. *For every multitape Turing machine* A, *there exists a Turing machine* B, *such that* A *and* B *are equivalent.*

Proof. Let A be a k-tape-TM. We show how to construct a TM B that simulates A step-by-step. A convenient way of explaining a simulation is to first show how configurations of the simulated machine A can be represented by the configurations of the simulating machine B, and then explain the simulation of particular steps of A.

The idea of the representation of a configuration of A by a configuration of B is transparently described in Figure 4.11. B saves the contents of all $k+1$ tapes of A on its only tape. A visual interpretation is that B splits its tape into $2(k+1)$ tracks and uses these tracks to save the contents of particular tapes of A and its head positions. Technically, it can be done as follows. If Γ_A is the working alphabet of A, then

$$\Gamma_B = (\Sigma \cup \{¢, \$, \sqcup\}) \times \{\sqcup, \uparrow\} \times (\Gamma_A \times \{\sqcup, \uparrow\})^k \cup \Sigma_A \cup \{¢\}$$

is chosen to be the working alphabet of B. For a symbol

$$\alpha = (a_0, a_1, a_2, \ldots, a_{2k+1}) \in \Gamma_B$$

we say that the symbol a_i lies on the i-th track. So, the i-th elements of the symbols on the tape of B determines the content of the "hypothetical" i-th track.

[14] A formal construction requires a lot of routine work.

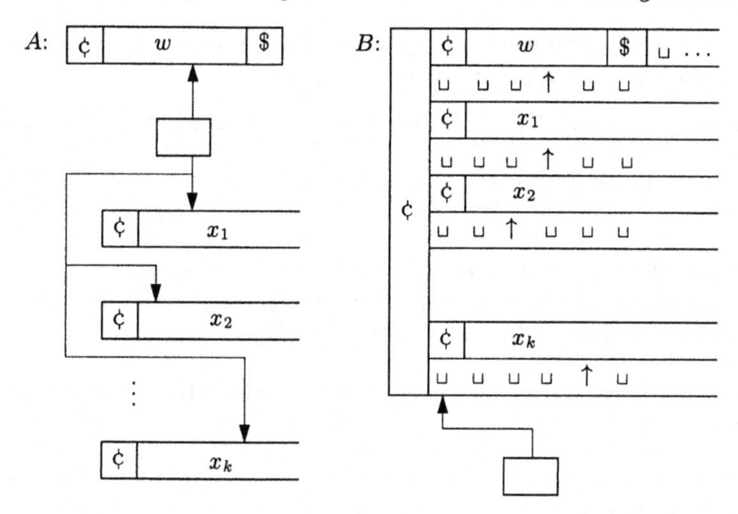

Fig. 4.11.

A configuration

$$(q, w, i, x_1, i_1, x_2, i_2, \ldots, x_k, i_k)$$

can be saved by B as follows. The state q of A is saved in the finite control of B too. Track 0 of B contains ¢w\$, i.e., the content of the input tape of A. For all $i \in \{1, \ldots, k\}$, the $(2i)$-th track of the tape B contains the word ¢x_i, i.e., the content of the i-th working tape of A. The position of the only symbol $\uparrow$ on the $(2i+1)$-th track determines the position of the head on the i-th tape of A.

A step of A can be simulated by the following procedure of B.

1. B reads the whole content of its tape from the left to the right. During this scan of the tape B saves all $k+1$ symbols,[15] read by the $k+1$ heads of A (these k symbols are exactly those symbols on even tracks, below them the symbols $\uparrow$ are positioned on the odd tracks).
2. After the first phase B knows the whole argument[16] of the transition function of A and is thus able to perform the corresponding activities (move the symbols $\uparrow$ with respect to the head movements of A, replace the symbol read and change the internal state) in one run over its tape from the right to the left.

Exercise 4.14. Give a formal construction of a TM that executes the first phase of the simulation of a step of A in the proof of Lemma 4.13.

[15] Formally, this means that the set of states of B contains the set $Q \times (\Sigma \cup \{¢, \$\}) \times \Gamma^k$, which is allowed, because this set is finite.

[16] Remember, that the state of A is saved in the state of B too.

Definition 4.15. *Two machine models (machine classes) $\mathcal{A}$ and $\mathcal{B}$ for solving decision problems are* **equivalent** *if*

(i) for every machine $A \in \mathcal{A}$, there exists a machine $B \in \mathcal{B}$ that is equivalent to A, and
(ii) for every machine $B \in \mathcal{B}$, there exists a machine $A \in \mathcal{A}$ that is equivalent to B.

Exercise 4.16. First give a formal definition of the equivalence of two machines that compute functions from Σ^* to Γ^*. Then give the definition of the equivalence of machine classes for computing functions.

Following Lemma 4.11 and Lemma 4.13 we directly obtain the following assertion.

Theorem 4.17. *The machine models of Turing machines and multitape Turing machines are equivalent.*

The knowledge that these two computing models are interchangeable with respect to the algorithmic solvability makes our next work easier. If one wants to prove that a language is recursive or recursively enumerable, it is sufficient to construct a multitape Turing machine, which is usually easier than to construct a Turing machine. But, if we aim to show that a language is not recursive (or recursively enumerable) then we will argue with the nonexistence of a Turing machine for this language. This strategy is similar to using a high-level programming language for showing the algorithmic solvability of a problem and an assembler or even machine code to prove the algorithmic unsolvability of a problem. This topic will be handled in the next section.

Therefore it is of interest to show the equivalence between Turing machines and a high-level programming language. A formal proof of this equivalence would require a lot of detailed, technical work which is too time consuming. Hence, we explain only the idea of how such an equivalence can be proved.

We hope that every reader, who has some experience in programming, believes that for any Turing machine M, we can write a program equivalent to M. We can go further by writing an interpreter C_{TM} for Turing machines. The interpreter C_{TM} obtains a description[17] of a TM M and a word w over the input alphabet of M as the input and then simulates the work of M on w.

How does one build a TM that is equivalent to a given program in a programming language with computer instructions? To answer this question let us look at the development of programming languages. At the very beginning, programs were written in an assembler or even in machine code. The only available instructions were a comparison of two integers and arithmetical operations. All complex instructions of high-level programming languages were created as small programs consisting of these elementary instructions in

[17] In a fixed formal representation

order to ease the work of a programmer. Therefore, we have no doubts that any assembler and any programming language are equivalent. Additionally, remember that compilers translate programs written in high-level programming languages into programs in machine codes.

Thus, it is sufficient to show the equivalence between an assembler and the Turing machine model. A possibility is to model an assembler by the so-called register machines and then show a step-by-step simulation of register machines by Turing machines. But this way is also too technical and so we will not use it. We simplify our task of simulating an assembler by a Turing machine. The operations multiplication and division can be executed by a program that consists of additions, subtractions, and integer comparisons only.

Exercise 4.18. Write a program that, for given integer variables I and J, computes the product $I \cdot J$. The program may only use the operations addition, subtraction, and the comparison of two integer in the framework of the `if ... then ... else` instruction.

Furthermore, we can omit the comparison of two integers by simulating it by a program that uses only the operations $+1$ ($I := I + 1$), -1 ($I := I - 1$), and the test on 0 (`if` $I = 0$ `then ... else ...`).

Exercise 4.19. Write a program that executes the instruction

$$\texttt{if } I \geq J \texttt{ then goto } 1 \texttt{ else goto } 2$$

by using the operations $+1$, -1, and the test on 0 only.

Finally, we can do away with addition and subtraction.

Exercise 4.20. Write programs that, for given two integer variables I and J, compute the subtraction $I - J$ and the addition $I + J$ by using only the operations $+1$, -1, and the test on 0.

The task of simulating programs consisting of instructions

- $I := I + 1$
- $I := I - 1$
- `if` $I = 0$ `then ... else ...`

by a multitape Turing machine does not take such great pains. The variables of such a program can be saved on the working tapes of an MTM in the form $x\#y$, where x is the binary code of the name of the variable I_x and y is the binary representation of the current value of I_x. The operations $+1$, -1, and the test $y = 0$ can be easily executed by an MTM. The only case when an MTM has to do more work is when the content of a working tape is $¢x\#y\#\#z\#u\#\# \ldots$ and the memory (the number of cells) for the value y of the variable I_x is too small for saving the current value y. When this happens,[18] the MTM has to move the suffix $\#\#z\#u\#\#\ldots$ of the tape content one square to the right in order to get one more bit for saving y.

In theoretical computer science hundreds of formal models (not only machine models) were designed for specifying the term of algorithmic solvability. All reasonable models are equivalent to the Turing machine model. This extensive experience led to the formulation of the famous Church–Turing thesis.

Church–Turing thesis

The Turing machines are the formalization of the notion "algorithm", i.e., the class of recursive languages (decidable decision problems) corresponds to the class of algorithmically (automatically) recognizable languages.

The Church–Turing thesis is not provable because it deals with the formalization of an intuitive term "algorithm", i.e., it is nothing more than a new axiom, which together with the axioms of mathematics build the fundamentals of theoretical computer science. An axiom cannot be proven because it formally fixes the interpretation of a fundamental notion and this is done with respect to our experience and belief only. Thus with this formalization of the notion algorithm, it is not possible to prove the nonexistence of another formal definition of the term algorithm such that

(i) it agrees with our intuition of the term algorithm, and
(ii) enables us to algorithmically solve decision problems that are not solvable by Turing machines.

The only thing that may happen is that such a stronger model of algorithms is found. Then, the fundamentals of theoretical computer science will have to be revised. However, the search for such a powerful model has yet been successful, and we even know that Turing machines are equivalent to the physical model of quantum computers.[19] So, there is a widespread belief that no model of algorithms that is more powerful than Turing machines exists.

Summarizing, the current state of theoretical computer science is similar to the situation in mathematics and physics. We accept the Church–Turing thesis because it agrees with our experience and we postulate it as an axiom. As already mentioned, the Church–Turing thesis has the characteristics of axioms of mathematics. It cannot be proven, but may be rebuttable, i.e., one cannot exclude the possibility that it might be disproved.[20] The Church–

[18] After adding 1 to I_x

[19] Quantum computers work on the principles of quantum mechanics.

[20] Disproving an axiom of a thesis should not be considered a disaster. Results of this kind are an unavoidable part of the development of science. The theory based on a disproved axiom need not be thrown out. Its results have to be only relativized because they are true assuming the axiom holds and there are certainly many frameworks where this axiom is true. Moreover, the revision of the theory by

Turing thesis is the only specific axiom of theoretical computer science. All other axioms stem from mathematics.

4.4 Nondeterministic Turing Machines

Nondeterminism can be introduced to Turing machines in the same way as we did for finite automata. For every argument, there exists the possibility of a choice from finitely many actions. On the formal level of transitions it means that transition function δ no longer maps from

$$Q \times \Gamma \text{ to } Q \times \Gamma \times \{\mathrm{L}, \mathrm{R}, \mathrm{N}\},$$

but rather, from

$$Q \times \Gamma \text{ to } \mathcal{P}(Q \times \Gamma \times \{\mathrm{L}, \mathrm{R}, \mathrm{N}\}).$$

Another possibility is to consider δ as a relation on

$$(Q \times \Gamma) \times (Q \times \Gamma \times \{\mathrm{L}, \mathrm{R}, \mathrm{N}\}).$$

A nondeterministic Turing machine M accepts an input word w if and only if there exists an accepting computation of M on w. This definition of acceptance corresponds to the optimistic point of view that M always makes a right choice. The formal definition of a nondeterministic Turing machine follows.

Definition 4.21. *A* **nondeterministic Turing machine (NTM)** *is a 7-tuple* $M = (Q, \Sigma, \Gamma, \delta, q_0, q_{\text{accept}}, q_{\text{reject}})$, *where*

(i) $Q, \Sigma, \Gamma, q_0, q_{\text{accept}}, q_{\text{reject}}$ *have the same meaning as in the definition of a (deterministic)* TM, *and*
(ii) $\delta : (Q - \{q_{\text{accept}}, q_{\text{reject}}\}) \times \Gamma \to \mathcal{P}(Q \times \Gamma \times \{\mathrm{L}, \mathrm{R}, \mathrm{N}\})$ *is the* **transition function** *of* M, *that satisfies*

$$\delta(p, ¢) \subseteq \{(q, ¢, X) \mid q \in Q, X \in \{\mathrm{R}, \mathrm{N}\}\}$$

for all $p \in Q - \{q_{\text{accept}}, q_{\text{reject}}\}$.
{The left endmarker may not be replaced (overwritten) by another symbol and the head may not move to the left when positioned on ¢.}

A **configuration** *of* M *is an element from*

$$\mathbf{conf}(M) = (\{¢\} \cdot \Gamma^* \cdot Q \cdot \Gamma^*) \cup (Q \cdot \{¢\} \cdot \Gamma^*).$$

{The meaning is the same as that for (deterministic) TM.*}*

The configuration $q_0¢w$ *is the* **initial configuration of M on w** *for any input word* $w \in \Sigma^*$. *A configuration is called* **accepting** *if it contains the state* q_{accept}. *A configuration is called* **rejecting,** *if it contains the state* q_{reject}.

using a new axiom usually belongs among the most important and big steps in the development of science.

A **step** *of M is a relation $\vdash_{\overline{M}}$ that is defined over the set of configurations (i.e., ($\vdash_{\overline{M}}\ \subseteq \mathrm{conf}(M) \times \mathrm{conf}(M)$) as follows. For all $p, q \in Q$ and all $x_1, x_2, \ldots, x_n, y \in \Gamma$,*

- $x_1x_2\ldots x_{i-1}qx_ix_{i+1}\ldots x_n \vdash_{\overline{M}} x_1x_2\ldots x_{i-1}pyx_{i+1}\ldots x_n$, *if* $(p, y, \mathrm{N}) \in \delta(q, x_i)$,
- $x_1x_2\ldots x_{i-2}, x_{i-1}qx_ix_{i+1}\ldots x_n \vdash_{\overline{M}} x_1x_2\ldots x_{i-2}px_{i-1}yx_{i+1}\ldots x_n$, *if* $(p, y, \mathrm{L}) \in \delta(q, x_i)$,
- $x_1x_2\ldots x_{i-1}qx_ix_{i+1}\ldots x_n \vdash_{\overline{M}} x_1x_2\ldots x_{i-1}ypx_{i+1}\ldots x_n$, *if* $(p, y, \mathrm{R}) \in \delta(q, x_i)$ *for* $i < n$ *and*
- $x_1x_2\ldots x_{n-1}qx_n \vdash_{\overline{M}} x_1x_2\ldots x_{n-1}yp\sqcup$, *if* $(p, y, \mathrm{R}) \in \delta(q, x_n)$.

The relation $\vdash^{*}_{\overline{M}}$ is the reflexive and transitive closure of $\vdash_{\overline{M}}$.

A **computation of M** *is a sequence $C_0, C_1, \ldots$ of configurations such that*

$$C_i \vdash_{\overline{M}} C_{i+1}$$

for $i = 0, 1, 2, \ldots$. {Sometimes we use the notation $C_0 \vdash_{\overline{M}} C_1 \vdash_{\overline{M}} C_2 \vdash_{\overline{M}} \cdots$ instead of the shorter $C_0, C_1, C_2 \ldots$ for a computation.}

A **computation of M on an input x** *is any computation that starts with the initial configuration $q_0¢x$ and is either infinite or halts in a configuration w_1qw_2 with $q \in \{q_{\mathrm{accept}}, q_{\mathrm{reject}}\}$. A computation of M on x is called* **accepting** *if it ends in an accepting configuration. A computation of M on x is called* **rejecting** *if it halts in a rejecting configuration.*

We say that M **accepts** *an input word w if there exists an accepting computation of M on w. Else, we say that M* **does not accept** *w (i.e., if all computations of M on w are rejecting or infinite).*

The **language $L(M)$ accepted by the NTM M** *is*

$$\begin{aligned} L(M) &= \{w \in \Sigma^* \mid q_0¢w \vdash^{*}_{\overline{M}} yq_{\mathrm{accept}}z \text{ for some } y, z \in \Gamma^*\} \\ &= \{w \in \Sigma^* \mid M \text{ accepts } w\}. \end{aligned}$$

Exercise 4.22. Describe both informally and formally a nondeterministic k-tape Turing machine.

Exercise 4.23. Let M be a nondeterministic multitape Turing machine. Describe the construction of an NTM M' such that $L(M) = L(M')$.

Similarly as for finite automata, nondeterminism can simplify the computation strategies of Turing machines. For instance, consider the language

$$L_{\mathrm{unequal}} = \{x\#y \mid x, y \in (\Sigma_{\mathrm{bool}})^*, x \neq \lambda, x \neq y\}.$$

A deterministic TM has to compare x and y symbol-by-symbol in order to determine a possible difference between x and y. Let $x = x_1x_2\ldots x_n$ and $y = y_1y_2\ldots y_m$ for $x_j, y_l \in \Sigma$ for $j = 1, \ldots n$, $l = 1, 2, \ldots m$. If x and y are different, an NTM can guess a position i at which x and y differ and then check its guess by comparing x_i with y_i. We describe a nondeterministic 1-tape-TM A that accepts L_{unequal} in what follows. A formal representation of A is given in Figure 4.12. For any input word w, A works in four phases (Figure 4.12):

1. A deterministically checks whether there is exactly one occurrence of the symbol # in w. This can be easily done by a single run on the input tape. (The states $q_0, q_1, q_{\text{reject}}$ are used for this purpose.) If w does not contain exactly one #, A rejects the input w. If $w = x\#y$ for some $x, y \in (\Sigma_{\text{bool}})^*$, then A continues on with phase 2 and enters state q_2.
2. A positions both heads on the left endmarkers ¢ of the input and working tapes. A is currently in state q_2.
3. A simultaneously moves both heads to the right and replaces the symbols ␣ on the working tape by a symbol $a \in \Gamma - \{0, 1, \#\}$. In every such step A nondeterministically guesses whether the current position in x is the position where x and y differ (the state q_{guess}). If A reads a symbol $b \in \{0, 1\}$ on its input tape and guesses that this is the position where x and y differ, then A saves b in its state and continues with the phase 4 to check the correctness of this nondeterministic decision (A moves to one of the states p_0 or p_1). If A reads # on the input tape, then it continues with phase 4 to check whether $|x| \neq |y|$.
4. The current situation assures that the distance between ¢ and the head position on the working tape is equal to the position i ($i \in N - \{0\}$) of the symbol $b \in \{0, 1, \#\}$ in $x\#$. Now, A moves its read-only head to the right until it reaches # without moving the head on the working tape (using states p_0 and p_1). After that A simultaneously moves the head of the input tape to the right and the head of the working tape to the left (using states s_0 and s_1). When the head on the working tape reaches ¢, then the head of the input tape is adjusted on the hypothesized position i in y. If the memorized symbol b of s_b differs from the currently read symbol of y on the input tape, then A accepts the input $w = x\#y$. A accepts w too, if $|x| < |y|$ ($(q_{\text{accept}}, \mathrm{N}, ¢, \mathrm{N}) \in \delta(p_{\#}, c, \mathrm{N})$ for all $c \in \{0, 1\}$) or if $|x| > |y|$ ($\delta(s_b, \$, d) = \{(q_{\text{accept}}, \mathrm{N}, d, \mathrm{N})\}$ for all $b \in \{0, 1\}, d \in \{a, ¢\}$).

The above-described strategy of nondeterministic guessing followed by its deterministic verification is typical[21] for nondeterministic computations. Another example is a nondeterministic 2-tape-TM B that accepts the language

$$L_{\text{quad}} = \{a^{n^2} \mid n \in \mathbb{N}\}.$$

For a given input w, B first guesses a positive integer n by adjusting the head of the first working tape on the position n. It then checks whether $|w| = n^2$ or not.

Exercise 4.24. Give a detailed description of the work of a nondeterministic 2-tape-TM B that accepts L_{quad} and present B graphically.

[21] In Chapter 6 we will learn how strongly nondeterminism can be characterized in this way.

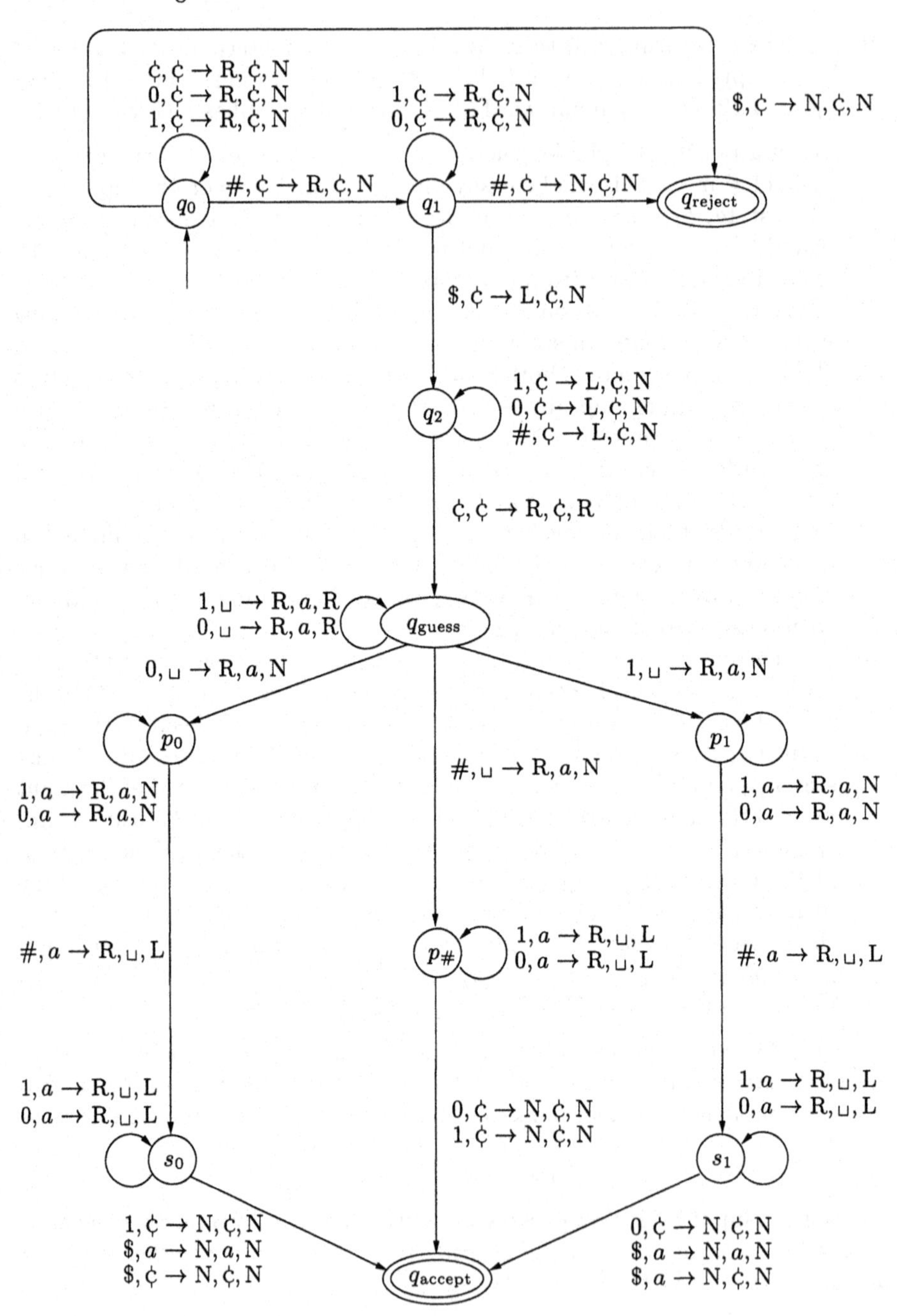
¢, ¢ → R, ¢, N
0, ¢ → R, ¢, N
1, ¢ → R, ¢, N
1, ¢ → R, ¢, N
0, ¢ → R, ¢, N
$, ¢ → N, ¢, N
#, ¢ → R, ¢, N
#, ¢ → N, ¢, N
q_0
q_1
q_{reject}
$, ¢ → L, ¢, N
q_2
1, ¢ → L, ¢, N
0, ¢ → L, ¢, N
#, ¢ → L, ¢, N
¢, ¢ → R, ¢, R
1, ⊔ → R, a, R
0, ⊔ → R, a, R
q_{guess}
0, ⊔ → R, a, N
1, ⊔ → R, a, N
p_0
p_1
#, ⊔ → R, a, N
1, a → R, a, N
0, a → R, a, N
1, a → R, a, N
0, a → R, a, N
#, a → R, ⊔, L
$p_\#$
1, a → R, ⊔, L
0, a → R, ⊔, L
#, a → R, ⊔, L
1, a → R, ⊔, L
0, a → R, ⊔, L
0, ¢ → N, ¢, N
1, ¢ → N, ¢, N
1, a → R, ⊔, L
0, a → R, ⊔, L
s_0
s_1
1, ¢ → N, ¢, N
$, a → N, a, N
$, ¢ → N, ¢, N
q_{accept}
0, ¢ → N, ¢, N
$, a → N, a, N
$, a → N, ¢, N

Fig. 4.12.

Now the principal question is whether nondeterministic Turing machines can accept a language that is not accepted by any (deterministic) Turing machine. Similarly as for finite automata, the answer is negative and the simulation strategy is based on the breadth-first search of the computation trees of the nondeterministic Turing machine.

Definition 4.25. *Let* $M = (Q, \Sigma, \Gamma, \delta, q_0, q_{\text{accept}}, q_{\text{reject}})$ *be an* NTM *and let* x *be a word over the input alphabet* Σ *of* M*. The* **computation tree** $\boldsymbol{T_{M,x}}$ **of** $\boldsymbol{M}$ **on** $\boldsymbol{x}$ *is a (possibly infinite) directed rooted tree defined as follows.*

(i) Every vertex of $T_{M,x}$ *is labeled by a configuration.*
(ii) The root is the only vertex of $T_{M,x}$ *with indegree* 0 *and it is labeled by the initial configuration* $q_0 ¢ x$ *of* M *on* x*.*
(iii) Every vertex of $T_{M,x}$ *labeled by a configuration* C *has exactly as many sons*[22] *as the number of successor configurations of* C *and these sons are labeled by the successor configurations of* C*.*

Clearly, the definition of a computation tree can be used for nondeterministic multitape Turing machines too.

Exercise 4.26. Draw the computation trees of the nondeterministic 1-tape-TM in Figure 4.12 on the inputs 01#01#1 and 01#0.

There are two essential differences between the computation trees of an NFA and an NTM. While the computation trees of nondeterministic finite automata are always finite, computation trees of nondeterministic Turing machines may be infinite. Secondly, the configurations with the same distance to the root of $T_{M,x}$ of an NTM M on x need not show any similarity and hence in contrast to finite automata they may have the head on different positions.

Theorem 4.27. *Let* M *be an* NTM*. There exists a* TM A*, such that*

(i) $L(M) = L(A)$*, and*
(ii) if M *does not have any infinite computation on a word from* $(L(M))^{\complement}$*, then* A *always halts.*

Proof. Following Lemma 4.13 it is sufficient to construct a 2-tape TM A with the properties (i) and (ii). We explain the work of A without giving a formal construction of A. The strategy of A is the breadth-first search in the computation trees of M.

Input: a word w
Phase 1. A copies the initial configuration $q_0 ¢ w$ of M on w on its first working tape.
Phase 2. A checks whether the first tape contains an accepting configuration[23] of M. If yes, A accepts w. Else, A continues with phase 3.

---

[22] Children nodes

[23] In fact, it suffices to look whether the accepting state q_{accept} is written on a cell of the first working tape.

Phase 3. A writes all successors of the configurations written on the first working tape on the second working tape. Note, that any configuration has only a finite number of successors (for any argument the set of possible actions of M is finite), so A can execute this in a finite time. If there is no successor for any configuration written on the first tape (i.e., the second working tape remains empty), then A halts in the state q_{reject}. Else, A continues with phase 4.

Phase 4. A erases the content of the first working tape and copies the content of the second working tape onto the first working tape. Then, A erases the content of the second working tape and continues with phase 2.

We see that after the i-th execution of phases 3 and 4 the first working tape of A contains all configurations of the computation tree $T_{M,w}$ with the distance i to the root (i.e., all configurations of M that are reachable after i steps of M).

If $w \in L(M)$, then there exists an accepting computation C of M on w. Since any accepting computation is finite, one may assume that C consists of j steps for a $j \in \mathbb{N} - \{0\}$. So, after j executions of phases 3 and 4, A finds an accepting configuration in phase 2 and accepts w.

If $w \notin L(M)$, it is obvious that A does not accept w. Moreover, since $T_{M,x}$ is finite, then A halts in the state q_{reject} (i.e., A rejects w). □

Exercise 4.28. Let A be the NTM in Figure 4.13.

(i) Give the first 6 levels (all configurations up to at most 5 steps of M) of the computation tree $T_A(x)$ for $x = 01$ and for $x = 0010$.
(ii) Determine the language $L(A)$.

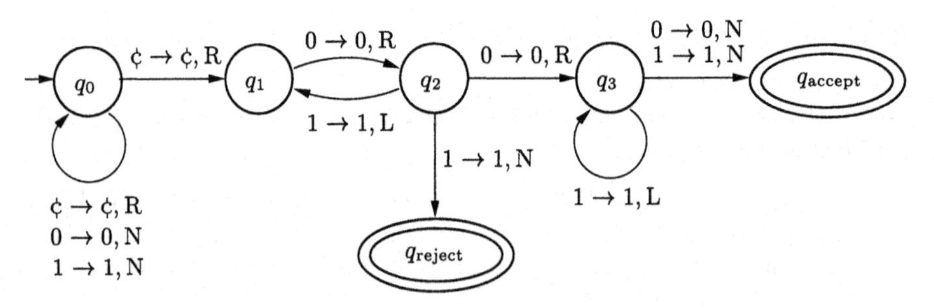

Fig. 4.13.

4.5 Coding of Turing Machines

Every program has a binary representation determined by its machine code. The job of translating a program from the alphabet Σ_{keyboard} into its machine code over Σ_{bool} is done by a computer. The goal of this chapter is to design a simple binary coding (representation) of Turing machines. We start by developing codes for Turing machines over the alphabet $\{0, 1, \#\}$.

Let $M = (Q, \Sigma, \Gamma, \delta, q_0, q_{\text{accept}}, q_{\text{reject}})$ be a TM, where

$$Q = \{q_0, q_1, \ldots, q_m, q_{\text{accept}}, q_{\text{reject}}\} \text{ and } \Gamma = \{A_1, A_2, \ldots, A_r\}.$$

First, we define the codes of specific symbols as follows:

$$\begin{aligned}
\text{Code}(q_i) &= 10^{i+1}1 \text{ for } i = 0, 1, \ldots, m,\\
\text{Code}(q_{\text{accept}}) &= 10^{m+2}1,\\
\text{Code}(q_{\text{reject}}) &= 10^{m+3}1,\\
\text{Code}(A_j) &= 110^j11 \text{ for } j = 1, \ldots, r,\\
\text{Code(N)} &= 1110111,\\
\text{Code(R)} &= 1110^2111,\\
\text{Code(L)} &= 1110^3111.
\end{aligned}$$

The codes of symbols are used for assigning the following codes to any particular transition.

$$\begin{aligned}
&\text{Code}(\delta(p, A_l) = (q, A_m, \alpha))\\
&= \#\text{Code}(p)\text{Code}(A_l)\text{Code}(q)\text{Code}(A_m)\text{Code}(\alpha)\#
\end{aligned}$$

for any transition $\delta(p, A_l) = (q, A_m, \alpha)$, where $p \in \{q_0, q_1, \ldots, q_m\}$, $q \in Q$, $l, m \in \{1, \ldots, r\}$, and $\alpha \in \{\text{N}, \text{L}, \text{R}\}$.

Our code of a Turing machine M begins with the global information, namely the number of states ($|Q|$) and the number of symbols of the working alphabet ($|\Gamma|$) of M and the list of all transitions follows. Hence,

$$\text{Code}(M) = \#0^{m+3}\#0^r\#\#\text{Code}(\mathit{Transition}_1)\#\text{Code}(\mathit{Transition}_2)\# \ldots.$$

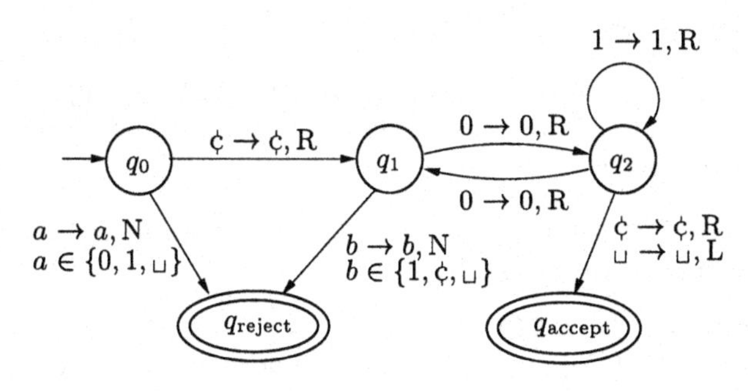

Fig. 4.14.

Exercise 4.29. Let M be the TM in Figure 4.14.

(i) Give Code(M) of M. Do so transparently by commenting any part of Code(M) between two # symbols.
(ii) Determine the language $L(M)$.
(iii) Is $L(M)$ regular? Prove your answer.

To get a code over Σ_{bool} one uses the following homomorphism

$$h : \{0, 1, \#\}^* \to (\Sigma_{\text{bool}})^*,$$

where

$$h(\#) = 01, \quad h(0) = 00, \quad h(1) = 11.$$

Definition 4.30. *For any Turing machine M,*

$$\mathbf{Kod}(\boldsymbol{M}) = h(\text{Code}(M))$$

is the **code of the TM M**.

$$\mathbf{KodTM} = \{\text{Kod}(M) \mid M \textit{ is a } \text{TM}\}$$

denotes the set of the codes of all Turing machines.

Clearly, the mapping assigning the code Kod(M) to any TM M is injective, so Kod(M) unambiguously describes the TM M.

Exercise 4.31. Design a program that, for any formal description of a TM M according to Definition 4.30, computes Code(M).

Exercise 4.32. Design and implement a program that, for an $x \in \{0,1\}^*$, decides whether $x \in$ KodTM for a TM M (i.e., whether x is the code of a Turing machine) or not.

In what follows $\boldsymbol{A_{\text{ver}}}$ denotes an algorithm (a TM) that decides the decision problem (Σ_{bool}, KodTM), i.e., that, for any given $x \in (\Sigma_{\text{bool}})^*$, decides whether x is the code of a TM.

An observation of crucial importance is that by fixing the binary representation (codes) of Turing machines one obtains a linear order on the set of all Turing machines.

Definition 4.33. *Let $x \in (\Sigma_{\text{bool}})^*$. For every positive integer i, we say that x is the* **code of the $\boldsymbol{i}$-th TM** *if*

(i) $x = \text{Kod}(M)$ for a TM M, and
(ii) the set $\{y \in (\Sigma_{\text{bool}})^* \mid y \text{ is before } x \text{ with respect to the canonical order}\}$ contains exactly $i - 1$ words that are codes of Turing machines.

If $x = \text{Kod}(M)$ is the code of the i-th TM, then M is the **$\boldsymbol{i}$-th Turing machine $\boldsymbol{M_i}$**. *The positive integer i is the* **order of the TM $\boldsymbol{M_i}$**.

We observe that it is not hard to compute the code Kod(M_i) of the i-th Turing machine for any given positive integer i. Let *Gen* be a function from $\mathbb{N} - \{0\}$ to $(\Sigma_{\text{bool}})^*$ defined by $Gen(i) = \text{Kod}(M_i)$.

Lemma 4.34. *The function Gen is recursive, i.e., there exists an algorithm (a Turing machine) that computes* Kod(M_i) *for any given positive integer* i.

Proof. A program computing *Gen* can work as follows.

```
Input: an i ∈ IN − {0}
Step 1:
    x := 1 {x is a word over (Σbool)*}

        I := 0
Step 2:
    while I < i do
           begin perform Aver to decide whether x ∈ KodTM;
             if x ∈ KodTM then begin
               I := I + 1;
               y := x
             end;
           x :=the successor of x in the canonical order on (Σbool)*
        end
Step 3:
    output(y).
```

Exercise 4.35. Write a program that, for the code $\text{Kod}(M) \in (\Sigma_{\text{bool}})^*$ of a TM M, computes the order of the TM M.

4.6 Summary

The Turing machine is an abstract computing model whose computational power is equal to the computational power of real computers. The components of a TM are an infinite tape, a finite control, and a read/write head. The tape consists of cells (squares) and every cell contains a symbol of the working alphabet. So, a cell corresponds to a register of a real computer and the symbols of the working alphabet correspond to all possible computer words (allowed contents of a register). The tape is considered to be an input medium as well as the memory of the TM. The instructions (elementary actions) of a TM are called transitions. The arguments of a transition are the actual state of the finite control and the symbol read by the head on the tape. In the corresponding action, the TM can change its state, replace the symbol read by another one, and move its head one cell to the left or to the right. A computation is given by a sequence of such elementary actions. A TM accepts [rejects] a word x if it ends the computation on x in the special state

q_{accept} [q_{reject}], We say that a TM does not accept x if it rejects x or if the computation on x is infinite. For a given TM M, the language $L(M)$ is the set of all words accepted by M. A language is called recursively enumerable if $L = L(M)$ for a TM M. A language L is called recursive if $L = L(M)$ for a TM M that does not have any infinite computations (i.e., all computations of M finishes either in q_{accept} or in q_{reject}).

The model of a multitape Turing machine has instead of an infinite tape (used for an input as well as for memory) a finite tape as the input medium[24] and a finite number of infinite working tapes as a memory. The computing models of Turing machines and multitape Turing machines are equivalent in the sense that every TM can be simulated by an MTM and vice versa. These Turing machine models are equivalent to programs in any standard programming language.

The Church–Turing thesis says that a Turing machine without infinite computations (a Turing machine that will always halt) is the formalization of the intuitive notion "algorithm". Hence, all problems solvable by Turing machines are algorithmically (automatically) solvable, and all problems unsolvable by Turing machines are algorithmically unsolvable. The Church–Turing thesis is the only specific axiom of computer science and so it can never be proved. The only open possibility would be to revise it when somebody finds a stronger (more powerful) and nevertheless realistic model of algorithms.

Nondeterminism can be introduced to Turing machines in the same way as we have done for finite automata in the previous chapter. A nondeterministic Turing machine may have several different[25] computations on an input. The input x is accepted if there is at least one computation on x that ends in q_{accept}. Any nondeterministic Turing machine can be simulated by a (deterministic) Turing machine. Similarly as for finite automata, the simulation is based on a breadth-first search in the computation trees of the NTM.

Turing machines can be unambiguously coded[26] as words over $\{0, 1\}$. Since the words over $\{0, 1\}$ are linearly ordered with respect to the canonical order, one obtains a linear order of all Turing machines in this way. For any given positive integer, one can compute the code of the i-th TM. Vice versa, for any given TM, one can compute the order of this Turing machine.

The introduction of a formal model of algorithms was the first step that led to the founding of theoretical computer science. This progress was initialized by the seminal work [20] of Kurt Gödel. This paper presents the first ever proof of the existence of mathematical problems that cannot be solved algorithmically (by any method[27]). This result motivated Church [12], Kleene [38], Post [51], and Turing [68] to design formal models of the intuitive notion algorithm. All these models and many others discovered later are equivalent

[24] Similarly as a finite automaton
[25] Even infinitely many computations are possible
[26] Similarly as every program has its binary machine code
[27] Today, we would say by any algorithm.

to each other. The consequence of this experience is the Church–Turing thesis. The model of Turing machine [68] has become the basic model of algorithms (computers) in theoretical computer science although the original concept of this model is not related to computers. The aim of Turing was to formalize the methods (algorithms) for symbol manipulation. Instead of thinking about a computer he pictured a man (a human computer or a mathematician) who executes a calculation with a pen on a sheet. The one-dimensional (linear) tape of a TM is motivated by writing on a sheet row-by-row. The finitely many symbols used determine the working alphabet. To do it systematically, Turing partitioned the tape into cells such that each cell may contain exactly one symbol. The content of the tape (the size of the sheet) is considered to be unbounded. Turing assumed that the human brain is finite and therefore can only be in one of finitely many states. The finite set of states of a TM is the consequence of this consideration. Turing used similar arguments to adopt the assumption that one action of the human computer (mathematician) can influence a part of the tape whose size is bounded by a constant independent of the length of the whole content of the tape. Since any such activity can be performed by a sequence of elementary instructions, each of them acting on one symbol only, Turing decided to use transitions (as introduced in Section 4.2) as basic instructions of a TM.

The multitape Turing machine was introduced by Hartmanis and Stearns [25]. It became the basic computing model of the complexity theory. An exciting discussion on the topic of this chapter is presented by Harel [23].

Hundreds of talents show
the greatness of their epoche,
but only a genius realizes,
what they are lacking.

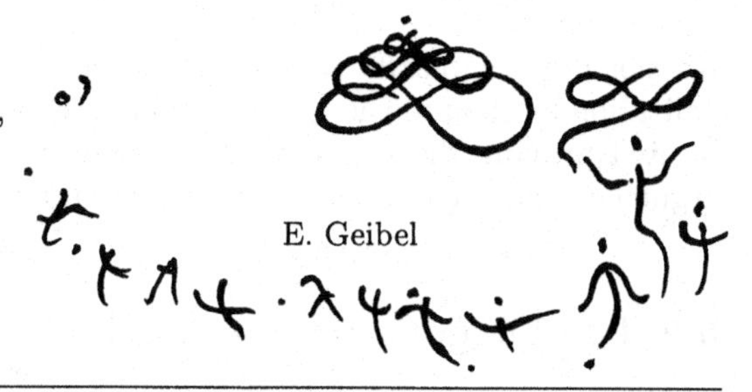

E. Geibel

5 Computability

5.1 Objectives

The computability theory is the first theory developed in computer science. It discovered methods for the classification of problems into algorithmically solvable and algorithmically unsolvable. This means that the computability theory provides techniques for proving the nonexistence of algorithms for solving concrete tasks. The main aim of this chapter is to learn these techniques.

We restrict our attention to decision problems in this chapter. Our first aim is to show that there are languages that cannot be accepted by any Turing machine. This can be easily seen when one is able to grasp that the number of languages is much larger than the number of Turing machines. Since both numbers are infinite, we have to learn how to prove that an infinite number is larger than another. To do this we present the diagonalization method from the set theory in Section 5.2. Moreover, the diagonalization technique enables us to prove that a specific language, called the diagonalization language, does not belong to the set of recursively enumerable languages, $\mathcal{L}_{\mathrm{RE}}$.

Our second aim is to introduce the reduction method. This method enables us to prove nonrecursivity of many concrete languages, provided that we already have at least one language that does not belong to $\mathcal{L}_{\mathrm{RE}}$. This is our main instrument for proving undecidability of languages. We apply this method to prove undecidability of some decision problems about Turing machines (programs) in Section 5.3. In this way, we learn that the well-motivated tasks of the verification of program correctness are not algorithmically solvable. Section 5.4 is devoted to the theorem of Rice that says that each nontrivial problem involving Turing machines (programs) is undecidable. In Section 5.5 we show that the method of reduction can also be used to prove undecidability of problems other than decision problems about Turing machines. As an example of such a problem we consider the Post correspondence problem that can be viewed as a domino game. Section 5.6 presents another method for proving undecidability of concrete problems. This method is based on the

Kolmogorov complexity and can be considered as an alternative to the diagonalization method in the sense that one can develop the computability theory by first proving the undecidability result using the Kolmogorov complexity argument and then applying the reduction method.

5.2 The Diagonalization Method

Our first aim is to show the existence of languages that are not recursively enumerable. To do this we use the following counting argument. We show that

> *the cardinality* $|\text{KodTM}|$ *of the set of Turing machines is smaller than the cardinality of the set of languages over* Σ_{bool}.

Remember that KodTM is the set of binary representations of all Turing machines as defined in Section 4.5.

The number of Turing machines is infinite and can be upper bounded by $|(\Sigma_{\text{bool}})^*|$, because $\text{KodTM} \subseteq (\Sigma_{\text{bool}})^*$. The cardinality of all languages over Σ_{bool} is $|\mathcal{P}((\Sigma_{\text{bool}})^*)|$, which is clearly an infinite number. To prove that

$$\left|(\Sigma_{\text{bool}})^*\right| < |\mathcal{P}((\Sigma_{\text{bool}})^*)|,$$

we need a method for comparing the sizes of two infinite numbers (the sizes of two infinite sets).

The following concept of Cantor for comparison of the cardinalities of two infinite sets touches the philosophical and axiomatical roots of mathematics and provides the fundamentals of the modern set theory.

Definition 5.1. Cantor's concept
Let A and B be sets. We say that

$$|\boldsymbol{A}| \leq |\boldsymbol{B}|,$$

if there exists a one-to-one[1] *function f from A to B. We say, that*

$$|\boldsymbol{A}| = |\boldsymbol{B}|,$$

if $|A| \leq |B|$ and $|B| \leq |A|$ (i.e., there exist a one-to-one, onto[2] *function between A and B). We say that*

$$|\boldsymbol{A}| < |\boldsymbol{B}|,$$

if $|A| \leq |B|$ and there does not exist any one-to-one mapping from B to A.

[1] A function f is a one-to-one mapping from A to B, when, for all $a, b \in A$, $a \neq b$ implies $f(a) \neq f(b)$.

[2] A function f is an onto mapping from A to B, when, for every $b \in B$, there exists an $a \in A$ such that $f(a) = b$.

First, we observe that for finite sets Definition 5.1 agrees with our understanding of the comparison of the cardinalities of two sets A and B (Figure 5.1). If, for all $x, y \in A$, $x \neq y$ implies $f(x) \neq f(y)$, then B has to have at least as many elements as A. If f is a one-to-one mapping and[3]

$$\{f(x) \mid x \in A\} = B$$

then f determines a pairing $(x, f(x))$ of elements of A and B, hence

$$|A| = |B|.$$

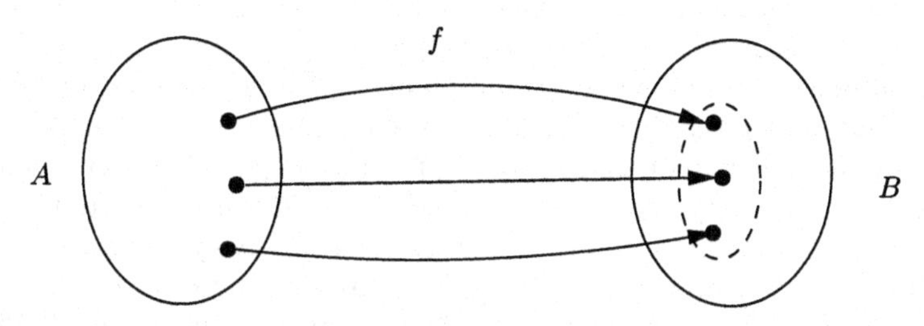

Fig. 5.1.

Following Definition 5.1, it is sufficient to prove that there is no one-to-one mapping from the set of languages over Σ_{bool} to the set of Turing machines. The consequence is that there is no mapping[4] from $\mathcal{P}((\Sigma_{\text{bool}})^*)$ to KodTM, that assigns a TM M to any language L in such a way that $L = L(M)$.

Exercise 5.2. Let A, B, and C be sets. Prove that

$$|B| \leq |A| \text{ and } |C| \leq |B| \text{ imply } |C| \leq |A|.$$

Exercise 5.3. Let A and B be sets. Prove that $A \subseteq B$ implies $|A| \leq |B|$.

Definition 5.1 also has consequences that may seem paradoxical in the "finite world". Let

$$\mathbb{N}_{\text{even}} = \{2i \mid i \in \mathbb{N}\}.$$

From Cantor's concept in Definition 5.1

$$|\mathbb{N}| = |\mathbb{N}_{\text{even}}|,$$

because the function $f : \mathbb{N} \rightarrow \mathbb{N}_{\text{even}}$ defined by

$$f(i) = 2i$$

[3] f is also onto.

[4] Because such a mapping must be one-to-one

for all $i \in \mathbb{N}$ is a one-to-one[5] mapping from $\mathbb{N}$ to $\mathbb{N}_{\text{even}}$. This is a paradox for the finite world because the finite world renders it impossible. The experience of the finite world says that an entity is always more than one of its parts, so a proper subset B of a finite set A cannot have the same size as A. But this is no reason for saying that Cantor's concept for comparing cardinalities of two sets is flawed. It only says that the world of infinite sets and numbers can be controlled by laws that do not correspond to our experience from the finite world. The result $|\mathbb{N}| = |\mathbb{N}_{\text{even}}|$ seems to be correct in the world of infinite objects because the one-to-one, onto mapping $f(i) = 2i$ determines the pairing $(i, 2i)$ of the elements from $\mathbb{N}$ and $\mathbb{N}_{\text{even}}$ and so both sets seem to be of the same cardinality. At this moment it may be reasonable to say that Cantor's concept lies on the axiomatic level of mathematics. Hence, nobody can prove that Cantor's concept is the only reasonable (correct) possibility for comparing sizes of infinite sets. Definition 5.1 is only an attempt to formalize the intuitive understanding of cardinality comparison in mathematics. One cannot exclude that another convenient formalization exists.[6] But independent of this possibility, the importance of Cantor's concept is its usefulness for proving the existence of languages that are recursively enumerable.

In what follows we consider $\mathbb{N}$ as the "smallest" infinite set. This poses the questions as to which infinite sets have the same cardinality as $\mathbb{N}$, and whether there exists an infinite set A with $|A| > |\mathbb{N}|$.

Definition 5.4. *A set A is called* **countable (denumerable)**,[7] *if A is finite or $|A| = |\mathbb{N}|$.*

The intuitive meaning of countability of a set A is that the elements of A can be ordered (denumerated) as the first, the second, the third, ..., etc. This is obvious because any one-to-one mapping $f : A \to \mathbb{N}$ determines a linear order[8] on A, which is a denumeration.[9] Therefore it is not surprising that $(\Sigma_{\text{bool}})^*$ and KodTM are countable.

Lemma 5.5. *Let Σ be an alphabet. Then Σ^* is countable.*

Proof. Let $\Sigma = \{a_1, \ldots, a_m\}$ be a finite, nonempty set. Let us fix a linear order $a_1 < a_2 < \cdots < a_m$ on Σ. This linear order on Σ determines the canonical

[5] Even one-to-one, onto mapping

[6] This is similar to our discussion about the Church–Turing thesis.

[7] An equivalent definition of countability (denumerability) of a set A is the following one.

$$A \text{ is denumerable } \Leftrightarrow \text{ there exists a one-to-one function } f : A \to \mathbb{N}.$$

This means that there does not exist any infinite set B with $|B| < |\mathbb{N}|$, i.e., $\mathbb{N}$ is one of the smallest infinite sets. We omit the proof of this fact here.

[8] An object $a \in A$ is before an object $b \in B$ if and only if $f(a) < f(b)$.

[9] A denumeration of a set A determines a linear order with the property that there are finitely many elements of A between any two elements of A.

order on Σ^* (Definition 2.16). The canonical order on Σ^* is a denumeration on Σ^* and hence it determines a one-to-one function from Σ^* to $\mathbb{N}$. □

Theorem 5.6. *The set* KodTM *of Turing machine codes is countable.*

Proof. Theorem 5.6 is a direct consequence of Lemma 5.5 and the fact KodTM $\subseteq (\Sigma_{\text{bool}})^*$. □

Exercise 5.7. Give explicitly the denumeration of $(\Sigma_{\text{bool}})^*$ that agrees with the canonical order on $(\Sigma_{\text{bool}})^*$.

Exercise 5.8. Prove that the set $\mathbb{Z}$ of natural numbers is countable.

Exercise 5.9. Let A be a countable set and let a be an element that does not belong to A. Prove that the set of $A \cup \{a\}$ is also countable.

Exercise 5.10. Let A and B be countable sets. Prove that $A \cup B$ is also countable.

The next result may be slightly surprising.[10] We prove that the cardinality of the set $\mathbb{Q}^+$ of positive rational numbers is equal to the cardinality of $\mathbb{N}$. The surprise is that one knows that the rational numbers have a high density on the real axis, namely, there are infinitely many rational numbers lying between any two different rational numbers. On the other hand, the natural numbers lies on the axis with the distance 1 and so there are always finitely many natural numbers between any two natural numbers. Since each positive rational number can be represented as $\frac{p}{q}$ for some $p, q \in \mathbb{N} - \{0\}$, one can guess that $|\mathbb{Q}^+|$ is approximately $|\mathbb{N} \times \mathbb{N}|$, which looks like infinity times infinity. In the finite world one could speak about the comparison between n^2 and n. However, $|\mathbb{Q}^+| = |\mathbb{N}|$ because one can denumerate[11] the elements of $\mathbb{Q}^+$. The following method for the denumeration of rational numbers is simple and transparent, and it also has applications in the computability theory. In what follows, let $\mathbb{N}^+ = \mathbb{N} - \{0\}$.

Lemma 5.11. $(\mathbb{N}^+) \times (\mathbb{N}^+)$ *is countable.*

Proof. Consider the infinite matrix $M_{\mathbb{N}^+ \times \mathbb{N}^+}$ depicted in Figure 5.2. The matrix $M_{\mathbb{N}^+ \times \mathbb{N}^+}$ has infinitely many rows and infinitely many columns that are labeled by the positive integers $1, 2, 3, \ldots$. The element $(i, j) \in (\mathbb{N}^+) \times (\mathbb{N}^+)$ lies on the intersection of the i-th row and the j-th column. Obviously, $M_{\mathbb{N}^+ \times \mathbb{N}^+}$ contains all elements of $(\mathbb{N}^+) \times (\mathbb{N}^+)$.

The attempt to order the elements of $(\mathbb{N}^+) \times (\mathbb{N}^+)$ by taking the elements of the first row, then the elements of the second row, etc., cannot be successful

[10] At least at first glance.

[11] Obviously not with respect to their values because if b would be a successor of a rational number a in such a denumeration, then one would have a contradiction with the existence of infinitely many rational numbers between a and b.

	1	2	3	4	5	6	...
1	(1,1)	(1,2)	(1,3)	(1,4)	(1,5)	(1,6)	...
2	(2,1)	(2,2)	(2,3)	(2,4)	(2,5)	(2,6)	...
3	(3,1)	(3,2)	(3,3)	(3,4)	(3,5)	(3,6)	...
4	(4,1)	(4,2)	(4,3)	(4,4)	(4,5)	(4,6)	...
5	(5,1)	(5,2)	(5,3)	(5,4)	(5,5)	(5,6)	...
6	(6,1)	(6,2)	(6,3)	(6,4)	(6,5)	(6,6)	...
⋮	⋮	⋮	⋮	⋮	⋮	⋮	⋱

Fig. 5.2.

because the first row is infinite and so the denumeration of its elements will never end, i.e., the denumeration of the elements of the second row will never start. A convenient possibility of denumerating the elements of the matrix $M_{\mathbb{N}^+ \times \mathbb{N}^+}$ is to use the zigzag line depicted in Figure 5.3. In this way one takes one finite diagonal after another one starting with the element $(1,1)$ in the upper-left corner. The resulting denumeration is

$$a_1 = (1,1),\ a_2 = (2,1),\ a_3 = (1,2),\ a_4 = (3,1),$$

$$a_5 = (2,2),\ a_6 = (1,3),\ a_7 = (4,1),\ \ldots .$$

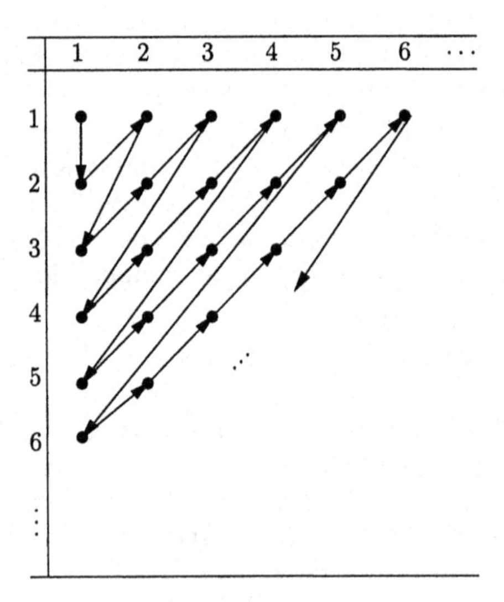

Fig. 5.3.

Formally, one defines the following linear order on $(\mathbb{N}^+) \times (\mathbb{N}^+)$:

$$(a,b) < (c,d) \Leftrightarrow a+b < c+d \text{ or } (a+b = c+d \text{ and } b < d).$$

The corresponding denumeration f is explicitly defined by

$$f(a,b) = \binom{a+b-1}{2} + b,$$

because the element (a,b) is the b-th element on the $(a+b-1)$-th diagonal and the number of elements in the first $a+b-2$ diagonals is

$$\sum_{i=1}^{a+b-2} i = \frac{(a+b-2)\cdot(1+a+b-2)}{2} = \binom{a+b-1}{2}.$$

Clearly, f is a one-to-one, onto mapping from $(\mathbb{N}^+) \times (\mathbb{N}^+)$ to $\mathbb{N}$.

□

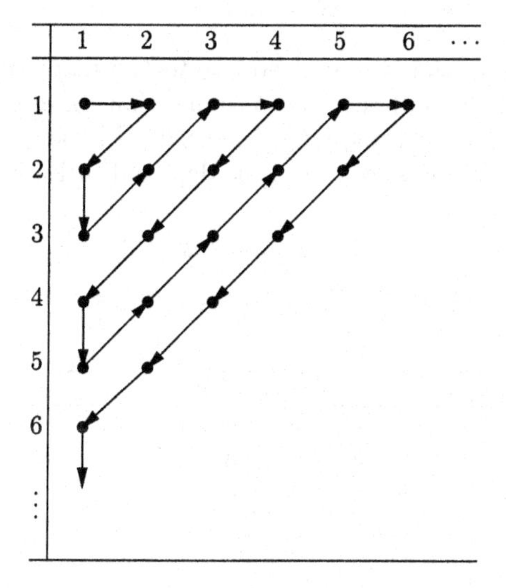

Fig. 5.4.

Exercise 5.12. To prove the countability of the set $(\mathbb{N}^+) \times (\mathbb{N}^+)$, one can also use the denumeration depicted in Figure 5.4. Determine the corresponding one-to-one, onto mapping from $(\mathbb{N}^+) \times (\mathbb{N}^+)$ to $\mathbb{N}$.

How many different one-to-one mappings from $(\mathbb{N}^+) \times (\mathbb{N}^+)$ to $\mathbb{N}$ exist?

Theorem 5.13. $\mathbb{Q}^+$ *is countable.*

Proof. Let h be the following mapping from $\mathbb{Q}^+$ to $(\mathbb{N}^+) \times (\mathbb{N}^+)$:

$$h\left(\frac{p}{q}\right) = (p,q)$$

for all p, q with the greatest common divisor 1.

Clearly, h is a one-to-one mapping. Since $(\mathbb{N}^+) \times (\mathbb{N}^+)$ is countable (Lemma 5.11), $\mathbb{Q}^+$ is countable too. □

Exercise 5.14. Prove that $\mathbb{N} \times \mathbb{N} \times \mathbb{N}$ is countable.

Exercise 5.15. Let A and B two countable sets. Prove that $A \times B$ is also a countable set.

Despite its density on the real axis, the positive rational numbers are denumerable. Following the assertion in Exercise 5.15, the set $(\mathbb{Q}^+)^i$ is countable for every positive integer i. Now, one could conjecture that all infinite sets are countable. But in what follows, we show that the set of real numbers, $\mathbb{R}$, is not countable. Hence $\mathbb{R}$ possesses a different kind of infinity from $\mathbb{N}$ and $\mathbb{Q}^+$.

Theorem 5.16. *The set* $[0, 1]$ *is not countable.*

Proof. We have to show that there is no one-to-one mapping from $[0, 1]$ to $\mathbb{N}^+$. We prove this in the indirect way. Assume, that $[0, 1]$ is countable, i.e., that there exists a one-to-one mapping f from $[0, 1]$ to $\mathbb{N}^+$. f determines a denumeration of real numbers from $[0, 1]$ as depicted in Figure 5.5. The i-th number from $[0, 1]$ is

$$a_i = 0.a_{i1}a_{i2}a_{i3}a_{i4}a_{i5}\ldots$$

(i.e., $f(a_i) = i$), where $a_{ij} \in \{0, 1, 2, \ldots, 9\}$ for $j = 1, 2, \ldots$.

$f(x)$	$x \in [0,1]$							
1	0.	$\boxed{a_{11}}$	a_{12}	a_{13}	a_{14}	$\ldots$		
2	0.	a_{21}	$\boxed{a_{22}}$	a_{23}	a_{24}	$\ldots$		
3	0.	a_{31}	a_{32}	$\boxed{a_{33}}$	a_{34}	$\ldots$		
4	0.	a_{41}	a_{42}	a_{43}	$\boxed{a_{44}}$	$\ldots$		
$\vdots$	$\vdots$	$\vdots$	$\vdots$	$\vdots$		$\ldots$		
i	0.	a_{i1}	a_{i2}	a_{i3}	a_{i4}	$\ldots$	$\boxed{a_{ii}}$	$\ldots$
$\vdots$	$\vdots$							

Fig. 5.5.

Now, we apply the so-called **diagonalization method** in order to show that at least one real number from $[0, 1]$ is missing in the table of Figure 5.5 and hence f is not a mapping from $[0, 1]$ to $\mathbb{N}^+$ (i.e., f does not provide any denumeration of real numbers from $[0, 1]$). The denumeration of real numbers in Figure 5.5 can be viewed as an infinite matrix

$$M = [a_{ij}]_{i=1,\ldots,\infty, j=1,\ldots,\infty}.$$

Our aim is to construct a real number

$$c = 0.c_1c_2c_3c_4c_5\dots$$

that differs from any row in M (i.e., from any real number in the denumeration given by f). The idea is to choose c in such a way that

$$c_i \neq a_{ii} \text{ and } c_i \notin \{0,9\}$$

for every $i \in \mathbb{N}^+$. To do that we simply consider the diagonal $a_{11}a_{22}a_{33}\dots$ of M and choose a digit

$$c_i \in \{1,2,3,4,5,6,7,8\} - \{a_{ii}\}$$

for every positive integer i. Hence, the representation of c is different from the representation of any a_i in Figure 5.5, namely, the representation of c differs from a_i at least in the i-th decimal digit a_{ii}. Since the representation of c does not contain the digits 0 and 9, this representation is unique[12] for c and so

$$c \neq a_i$$

for any positive integer i. Hence, c is not represented in Figure 5.5 and consequently f is not a denumeration of real numbers from $[0,1]$. □

We have shown that the set of Turing machines (algorithms) is countable. In order to prove the existence of problems that are not algorithmically solvable, it is sufficient to prove the uncountability of the set of all languages (decision problems) over $\{0,1\}$. We do this in two different ways. First, we show that

$$|[0,1]| \leq |\mathcal{P}((\Sigma_{\text{bool}})^*)|.$$

Then applying the diagonalization method, we directly construct a language that is not accepted by any Turing machine.

Theorem 5.17. *$\mathcal{P}((\Sigma_{\text{bool}})^*)$ is not countable.*

Proof. Since $[0,1]$ is not countable (Theorem 5.16), it suffices to show

$$|\mathcal{P}((\Sigma_{\text{bool}})^*)| \geq |[0,1]| \tag{5.1}$$

by finding a one-to-one mapping from $[0,1]$ to $\mathcal{P}((\Sigma_{\text{bool}})^*)$.

Every real number from $[0,1]$ can be represented in the binary form as follows. The representation

$$b = 0.b_1b_2b_3\dots$$

with $b_i \in \Sigma_{\text{bool}}$ for $i = 1,2,3,\dots$ codes the real number

[12] Note, that $1.00\overline{0}$ and $0.99\overline{9}$ are two different representations of the same number 1. Similarly, $0.142\overline{9}$ and $0.143\overline{0}$ represent the same number 0.143.

$$Number(b) = \sum_{i=1}^{\infty} a_i 2^{-i}.$$

We use this binary representation of real numbers to define the mapping $f : [0, 1] \to \mathcal{P}((\Sigma_{\text{bool}})^*)$ as follows. Let

$$a = 0.a_1 a_2 a_3 a_4 a_5 a_6 a_7 \ldots$$

be the binary representation of a real number in $[0, 1]$. Then, we define

$$f(a) = \{a_1, a_2 a_3, a_4 a_5 a_6, \ldots, a_{\binom{n}{2}+1} a_{\binom{n}{2}+2} \cdots a_{\binom{n+1}{2}}, \ldots\}.$$

Observe, that $f(a)$ is a language over Σ_{bool} that contains exactly one word of length n for any positive integer n. Therefore, any difference in a bit between two binary representations b and c implies

$$f(b) \neq f(c).$$

Hence, f is a one-to-one mapping, so the inequality (5.1) holds, i.e., $\mathcal{P}((\Sigma_{\text{bool}})^*)$ is not countable. □

Corollary 5.18. $|\text{KodTM}| < |\mathcal{P}((\Sigma_{\text{bool}})^*)|$ *and hence there are infinitely many languages over* Σ_{bool} *that are not recursively enumerable.*

Exercise 5.19. Prove $|[0, 1]| = |\mathcal{P}((\Sigma_{\text{bool}})^*)|$.

Now, we use the diagonalization method in order to construct a specific language that is not recursively enumerable. Let

$$w_1,\ w_2,\ w_3,\ w_4,\ w_5,\ \ldots$$

be the canonical order of all words over Σ_{bool}, and let

$$M_1,\ M_2,\ M_3,\ M_4,\ M_5\ \ldots$$

be the sequence of all Turing machines.[13] We define an infinite Boolean matrix (Figure 5.6)

$$A = [d_{ij}]_{i,j=1,\ldots,\infty}$$

by

$$d_{ij} = 1 \Leftrightarrow M_i \text{ accepts the } j\text{-th word } w_j.$$

In this way the i-th row

$$d_{i1} d_{i2} d_{i3} d_{i4} d_{i5} \ldots$$

of the matrix A determines the language

$$L(M_i) = \{w_j \mid d_{ij} = 1 \text{ for all } j \in \mathbb{N}^+\}.$$

	w_1	w_2	w_3	$\dots$	w_i	$\dots$
M_1	$\boxed{d_{11}}$	d_{12}	d_{13}	$\dots$	d_{1i}	$\dots$
M_2	d_{21}	$\boxed{d_{22}}$	d_{23}	$\dots$	d_{2i}	$\dots$
M_3	d_{31}	d_{32}	$\boxed{d_{33}}$	$\dots$	d_{3i}	$\dots$
$\vdots$	$\vdots$	$\vdots$	$\vdots$		$\vdots$	
M_i	d_{i1}	d_{i2}	d_{i3}	$\dots$	$\boxed{d_{ii}}$	$\dots$
$\vdots$	$\vdots$	$\vdots$	$\vdots$		$\vdots$	

Fig. 5.6.

Analogously to the construction of a real number that is not involved in the hypothetical denumeration of real numbers in Figure 5.5 (proof of Theorem 5.16), we construct the **diagonalization language** L_{diag}, which does not agree with any language $L(M_i)$. We define

$$\begin{aligned} \boldsymbol{L_{\text{diag}}} &= \{w \in (\Sigma_{\text{bool}})^* \mid w = w_i \text{ for an } i \in \mathbb{N}^+ \text{ and } \\ &\qquad M_i \text{ does not accept } w_i\} \\ &= \{w \in (\Sigma_{\text{bool}})^* \mid w = w_i \text{ for an } i \in \mathbb{N}^+ \text{ and } d_{ii} = 0\}. \end{aligned}$$

Theorem 5.20.

$$L_{\text{diag}} \notin \mathcal{L}_{\text{RE}} \ .$$

Proof. We prove $L_{\text{diag}} \notin \mathcal{L}_{\text{RE}}$ by contradiction. Assume $L_{\text{diag}} \in \mathcal{L}_{\text{RE}}$. Then, $L_{\text{diag}} = L(M)$ for a TM M. Since M is one of the machines in the denumeration of Turing machines, there exists a positive integer i such that $M = M_i$. But L_{diag} cannot be equal to $L(M_i)$ because

$$w_i \in L_{\text{diag}} \Leftrightarrow d_{ii} = 0 \Leftrightarrow w_i \notin L(M_i).$$

□

Exercise 5.21. Consider the language

$$\begin{aligned} L_{\text{2diag}} = \{w \in (\Sigma_{\text{bool}})^* \mid w = w_{2i} &\text{ for an } i \in \mathbb{N} - \{0\} \text{ and } \\ &M_i \text{ does not accept } w_{2i}\ (d_{i,2i} = 0)\}. \end{aligned}$$

Prove that $L_{\text{2diag}} \notin L_{\text{RE}}$.

Exercise 5.22. Let for any $k \in \mathbb{N}$

$$\begin{aligned} L_{k,\text{diag}} = \{w \in (\Sigma_{\text{bool}})^* \mid w = w_{i+k} &\text{ for an } i \in \mathbb{N} - \{0\} \text{ and } \\ &M_i \text{ does not accept } w_{i+k}\ (d_{i,i+k} = 0)\}. \end{aligned}$$

Prove that $L_{k,\text{diag}} \notin L_{\text{RE}}$ for any $k \in \mathbb{N}$.

---

[13] As usual, M_i denotes the i-th Turing machine.

5.3 The Reduction Method

The reduction method is the most common method for classifying decision problems with respect to their algorithmical solvability. The idea is very simple. Let A be a problem, for which one wants to show algorithmical unsolvability. If one finds a problem B such that

1. one knows already that B is not algorithmically solvable, and
2. the algorithmical solvability of A would imply the algorithmical solvability of B,

then one can conclude that A is not algorithmically solvable. This way of proving the algorithmical unsolvability of A is called a reduction from B to A.

Definition 5.23. *Let $L_1 \subseteq \Sigma_1^*$, and $L_2 \subseteq \Sigma_2^*$ be languages for some alphabets Σ_1 and Σ_2. We say that* **L_1 is (recursively) reducible to L_2, $L_1 \leq_{\mathrm{R}} L_2$,** *if*

$$L_2 \in \mathcal{L}_{\mathrm{R}} \Rightarrow L_1 \in \mathcal{L}_{\mathrm{R}}.$$

The notation

$$L_1 \leq_{\mathrm{R}} L_2$$

corresponds to the intuitive meaning that

"L_2 is at least as hard as L_1 with respect to algorithmical solvability",

because if L_2 is algorithmically solvable (i.e., $L_2 = L(A)$ for an algorithm A), then L_1 would also be algorithmically solvable (i.e., $L_1 = L(B)$ for an algorithm B).

We already know the diagonalization language L_{diag} as the example of a language that is not in $\mathcal{L}_{\mathrm{RE}}$ and hence not in $\mathcal{L}_{\mathrm{R}}$ either. To get further nonrecursive languages, we need concrete techniques for proving results of the kind $L_1 \leq_{\mathrm{R}} L_2$. In what follows we introduce two such techniques that correspond to the framework of the recursive reducibility.

The first technique corresponds to a very straightforward interpretation of the term reduction as converting of a problem P_1 to be solved into another problem P_2. The idea is that the solution for P_2 can be directly used as a solution for P_1. The corresponding technique is called mapping reduction because the whole reduction is simply the transformation of an input of P_1 to an input of P_2. Formally, we search for a TM (an algorithm) M that, for any input x of a decision problem (Σ_1, L_1), computes an input y of a decision problem (Σ_2, L_2) such that the solutions for input instances x (of (Σ_1, L_1)) and y (of (Σ_2, L_2)) are the same. This means that if one has an algorithm A for (Σ_2, L_2), then the concatenation of M and A (Figure 5.7) is an algorithm for (Σ_1, L_1).

Definition 5.24. *Let $L_1 \subseteq \Sigma_1^*$ and $L_2 \subseteq \Sigma_2^*$ be languages for some alphabets Σ_1 and Σ_2, respectively. We say that the language L_1 is* **mapping**[14] **reducible** *to the language L_2,* $\boldsymbol{L_1 \leq_{\mathrm{m}} L_2}$*, if there exists a* TM *M that computes a mapping*

$$f_M : \Sigma_1^* \to \Sigma_2^*$$

with the property (Figure 5.7)

$$x \in L_1 \Leftrightarrow f_M(x) \in L_2$$

for every $x \in \Sigma_1^*$. The function f_M is called the **reduction** *of L_1 to L_2 and we also say that* $\boldsymbol{M}$ **reduces the language** $\boldsymbol{L_1}$ **to the language** $\boldsymbol{L_2}$.

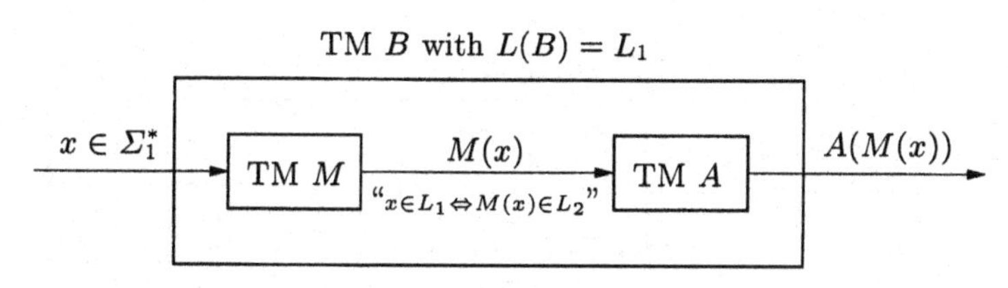

Fig. 5.7.

The following lemma shows that the relation $\leq_{\mathrm{m}}$ is a special case of the relation $\leq_{\mathrm{R}}$, which means that it is sufficient to prove $L_1 \leq_{\mathrm{m}} L_2$ so as to prove $L_1 \leq_{\mathrm{R}} L_2$.

Lemma 5.25. *Let $L_1 \subseteq \Sigma^*$ and $L_2 \subseteq \Sigma^*$ be languages for some alphabets Σ_1 and Σ_2. If $L_1 \leq_{\mathrm{m}} L_2$, then $L_1 \leq_{\mathrm{R}} L_2$.*

Proof. Let $L_1 \leq_{\mathrm{m}} L_2$. To prove $L_1 \leq_{\mathrm{R}} L_2$ it is sufficient to show the existence of an algorithm A (a TM A that always halts) that decides L_2 ($L_2 \in \mathcal{L}_{\mathrm{R}}$) under the assumption that an algorithm B deciding L_1 exists.

Let A be a TM that always halts and $L(A) = L_2$. Assuming $L_1 \leq_{\mathrm{m}} L_2$, there is a TM M that, for each $x \in \Sigma_1^*$, computes a word $M(x) \in \Sigma_2^*$ such that

$$x \in L_1 \Leftrightarrow M(x) \in L_2.$$

We construct a TM B that always halts and accepts L_1 (Figure 5.7). The TM B works on an input $x \in \Sigma_1^*$ as follows:

1. B simulates the work of M on x until the word $M(x)$ is written on the tape.
2. B simulates the work of A on $M(x)$.
   If A accepts $M(x)$, then B accepts its input x.
   If A rejects $M(x)$, then B rejects its input x.

[14] The mapping reducibility is also called "many-one-reducibility" in the literature.

Obviously, $L(B) = L_1$. Since A always halts,[15] B halts for every input too. Therefore $L_1 \in \mathcal{L}_R$. □

Exercise 5.26. Prove that $\leq_m$ is a transitive relation, i.e.,

$$(L_1 \leq_m L_2 \text{ and } L_2 \leq_m L_3) \Rightarrow L_1 \leq_m L_3).$$

We observe that the TM B with $L(B) = L_1$ uses the TM A as a subroutine. The subroutine A is executed exactly once (Figure 5.7) and B takes the output of A as its own output. But this is an unnecessary restriction. In general, one can build B in such a way that B executes A on several different inputs and uses the outputs of A to finally decide whether its input $x \in L_1$ or not (Figure 5.8).

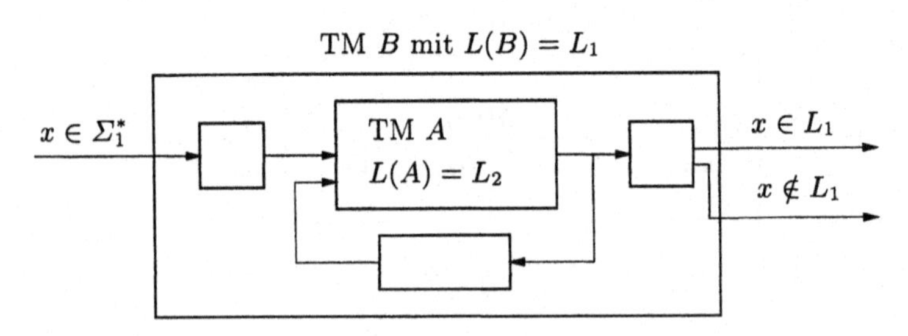

Fig. 5.8.

Now, we are ready to prove results of the kind "$L \notin \mathcal{L}_R$" for concrete languages L. We start with a general observation saying that L belongs to $\mathcal{L}_R$ if and only if the complement of L belongs to $\mathcal{L}_R$.

Lemma 5.27. *Let Σ be alphabet. For every language $L \subseteq \Sigma^*$*

$$L \leq_R L^C \text{ and } L^C \leq_R L.$$

Proof. It is sufficient to only prove $L^C \leq_R L$ for every language L, because $(L^C)^C = L$, so $L^C \leq_R L$ implies the relation $(L^C)^C \leq_R L^C$ (if one substitutes the language L^C for L in the relation $L^C \leq_R L$).

Let A be an algorithm that decides the decision problem (L, Σ). An algorithm B deciding L^C is described in Figure 5.9. B simply gives its input $x \in \Sigma^*$ to the subroutine A and negates the decision of A on x.

Now, we give an alternative proof of $L^C \leq_R L$ in the formalism of Turing machines. Let $A = (Q, \Sigma, \Gamma, \delta, q_0, q_{\text{accept}}, q_{\text{reject}})$ be a TM that always halts and accepts L_1. We construct a TM $B = (Q, \Sigma, \Gamma, \delta, q_0, q'_{\text{accept}}, q'_{\text{reject}})$ with

$$q'_{\text{accept}} = q_{\text{reject}} \text{ and } q'_{\text{reject}} = q_{\text{accept}}$$

[15] Remember that M computes $M(x)$ for every $x \in \Sigma_1^*$, so M also halts for any input.

by simply exchanging the roles of the accepting and rejecting states of A. Since A always halts, B also always halts and hence $L(B) = (L(A))^{\complement}$. Thus, the existence of an algorithm for L implies the existence of an algorithm for $L^{\complement}$. □

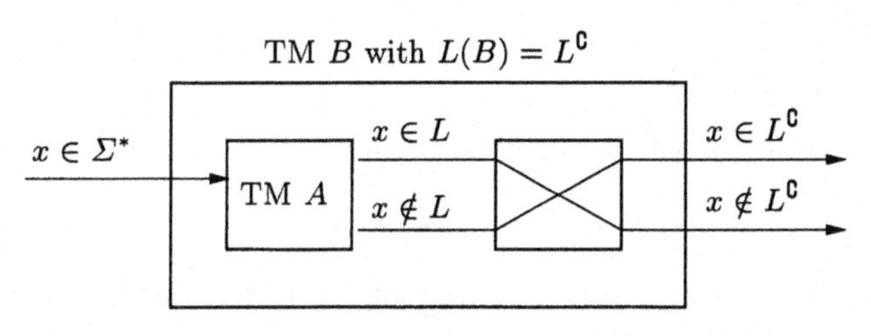

Fig. 5.9.

Corollary 5.28. $(L_{\text{diag}})^{\complement} \notin \mathcal{L}_{\text{R}}$.

Proof. We have proved in Theorem 5.17 that $L_{\text{diag}} \notin \mathcal{L}_{\text{RE}}$ and consequently $L_{\text{diag}} \notin \mathcal{L}_{\text{R}}$. Lemma 5.27 claims

$$L_{\text{diag}} \leq_{\text{R}} (L_{\text{diag}})^{\complement}.$$

Following the definition of the relation $\leq_{\text{R}}$,

$$(L_{\text{diag}})^{\complement} \in \mathcal{L}_{\text{R}} \text{ implies } L_{\text{diag}} \in \mathcal{L}_{\text{R}}.$$

Thus, we conclude that $(L_{\text{diag}})^{\complement} \notin \mathcal{L}_{\text{R}}$. □

Following the above assertions we are not able to conclude that $(L_{\text{diag}})^{\complement}$ does not belong to $\mathcal{L}_{\text{RE}}$. In fact, the contrary is true. Proving $(L_{\text{diag}})^{\complement} \in \mathcal{L}_{\text{RE}}$, we obtain $\mathcal{L}_{\text{R}} \subsetneq \mathcal{L}_{\text{RE}}$ (Figure 5.10).

Lemma 5.29. $(L_{\text{diag}})^{\complement} \in \mathcal{L}_{\text{RE}}$.

Proof. Following the definition of L_{diag} we obtain

$$(L_{\text{diag}})^{\complement} = \{x \in (\Sigma_{\text{bool}})^* \mid \text{if } x = w_i \text{ for an } i \in \mathbb{N} - \{0\}, \text{ then } M_i \text{ accepts } w_i\}.$$

A TM D that accepts $(L_{\text{diag}})^{\complement}$ can work as follows:

Input: an $x \in (\Sigma_{\text{bool}})^*$.
Phase 1. Compute i such that x is the i-th word w_i in the canonical order with respect to Σ_{bool}.
Phase 2. Generate the code Kod(M_i) of the i-th TM M_i.

Phase 3. Simulate the computation of M_i on the word $w_i = x$.
If M_i accepts w_i, then D accepts x too.
If M_i rejects[16] w_i, then D rejects $x = w_i$ too.
If the computation of M_i on w_i is infinite, (i.e., $w_i \notin L(M_i)$), then D simulates this infinite computation forever. Hence, D does not halt on x, so $x \notin L(D)$.

Clearly, $L(D) = (L_{\text{diag}})^{\complement}$. □

Corollary 5.30. *$(L_{\text{diag}})^{\complement} \in \mathcal{L}_{\text{RE}} - \mathcal{L}_{\text{R}}$, hence*

$$\mathcal{L}_{\text{R}} \subsetneq \mathcal{L}_{\text{RE}}.$$

In what follows we present further languages that are not recursive, but belong to $\mathcal{L}_{\text{RE}}$ (Figure 5.10).

Definition 5.31. *The* **universal language** *is the language*

$$L_{\text{U}} = \{\text{Kod}(M)\#w \mid w \in (\Sigma_{\text{bool}})^* \text{ and } M \text{ accepts } w\}.$$

Theorem 5.32. *There is a* TM U*, called the* **universal TM**, *such that*

$$L(U) = L_{\text{U}}.$$

Consequently, $L_{\text{U}} \in \mathcal{L}_{\text{RE}}$.

Proof. It is sufficient to construct a 2-MTM U with $L(U) = L_{\text{U}}$. U works as follows:

Input: An $z \in \{0, 1, \#\}^*$.
Phase 1. U checks whether z contains exactly one $\#$. If not, U rejects its input z. If yes, U continues with phase 2.
Phase 2. Let $z = y\#x$, $y, z \in (\Sigma_{\text{bool}})^*$. U verifies whether y is a code of a TM. If y does not represent any TM, then U rejects its input $z = y\#x$. If y codes a TM, then U continues with phase 3.
Phase 3. If $y = \text{Kod}(M)$ for a TM M, then U writes the initial configuration of M on x on its first working tape and continues with phase 4.
Phase 4. U simulates the computation of M on x step-by-step as follows:

> `while` the state of the configuration of M on the first working tape is different from q_{accept} and q_{reject} of M `do`
> simulate one computation step of M
> {U does it simply by reading the code Kod(M) on its input tape}
> `if` the state of M is q_{accept}
> `then` U accepts $z = \text{Kod}(M)\#x$
> `else` U rejects $z = \text{Kod}(M)\#x$

[16] Remember that to reject an input means to halt in the state q_{reject}.

Note, that the infinite computation of M on x causes an infinite computation[17] of U on Kod$(M)\#x$. Therefore, U does not accept Kod$(M)\#x$ in such a case. Hence, $L(U) = L_{\mathrm{U}}$. □

What does the fact $L_{\mathrm{U}} \in \mathcal{L}_{\mathrm{RE}}$ imply? It guarantees the existence of a Turing machine (a program) that can simulate an arbitrary TM (program) on any given input, but without halting assurance. This is a very natural property of Turing machines that is required for any formal[18] definition of algorithms. In the world of programming languages the universal Turing machine is nothing other than an interpreter. We cannot have a computer system that is unable to execute a syntactically correct program on a given input if the corresponding programming language[19] is a part of the system. In what follows, we prove that $L_{\mathrm{U}} \notin \mathcal{L}_{\mathrm{R}}$. The implication of this result is that the only general strategy (algorithm) for determining the result of the computation of a TM M on an input x is to simulate the computation of M on x. But, if the computation of M on x is infinite, one does not know at any instance of the simulation, whether the computation of M on x is infinite or M will halt after the next computation step and output the result. Therefore, one is not able to decide in finite time, whether $x \in L(M)$ or $x \in L(M)$ for any given M and x.

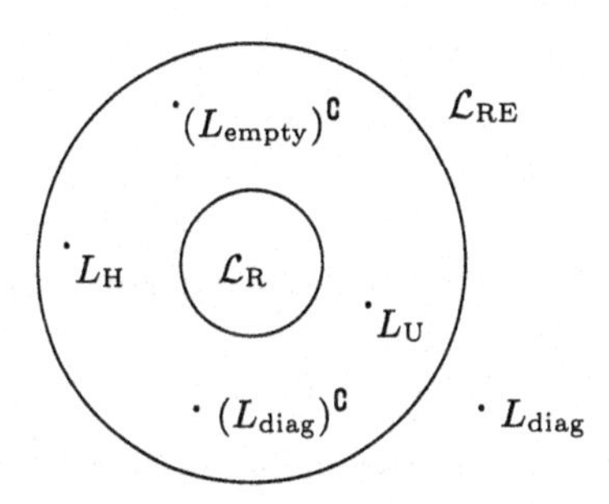

Fig. 5.10.

The following proofs are based on the reduction method. Most claims will be proved twice. First, we give a transparent proof on the level of algorithms as programs in a programming language by using the general reduction from Figure 5.8. Then we present a proof in the formalism of Turing machines by the mapping reduction (Figure 5.11).

[17] Simulation

[18] There is a system of axioms that must be satisfied by a computing model. The existence of the universal machine is one of the axioms.

[19] Remember that any programming language is a formal model of algorithms.

Theorem 5.33. $L_U \notin \mathcal{L}_R$.

Proof. It suffices to show that $(L_{\text{diag}})^{\complement} \leq_R L_U$, because Corollary 5.28 provides $(L_{\text{diag}})^{\complement} \notin \mathcal{L}_R$, so $L_U \notin \mathcal{L}_R$.

Let A be an algorithm (program) that decides L_U. We construct an algorithm B, that uses A as a subroutine for deciding $(L_{\text{diag}})^{\complement}$. The structure of the algorithm B is given in Figure 5.11. For any input $x \in (\Sigma_{\text{bool}})^*$, the subprogram C first computes an integer i, such that $x = w_i$. Then, C computes the code Kod(M_i) of the i-th TM. C provides its outputs w_i and Kod(M_i) as inputs[20] for the subroutine A. Finally, B inherits the output "accept" or "reject" of A as its own output. Obviously, $L(B) = (L_{\text{diag}})^{\complement}$ and B always halts, because both C and A always halt and provide their outputs. Thus, $(L_{\text{diag}})^{\complement} \leq_R L_U$ and the proof is completed.

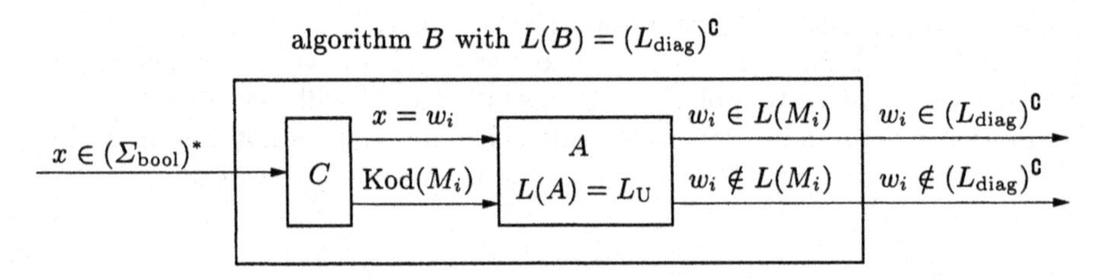

Fig. 5.11.

Now, we give an alternative proof in the formalism of Turing machines. We prove that

$$(L_{\text{diag}})^{\complement} \leq_m L_U.$$

We describe a TM M that computes a mapping f_M from $(\Sigma_{\text{bool}})^*$ to $\{0, 1, \#\}^*$ such that

$$x \in (L_{\text{diag}})^{\complement} \Leftrightarrow f_M(x) \in L_U.$$

M works as follows. For any input x, M first computes an i such that $x = w_i$. After that M computes the code Kod(M_i) of the i-th TM. M halts with the content Kod$(M_i)\#x$ on its tape. Since $x = w_i$, the definition of $(L_{\text{diag}})^{\complement}$ implies:

$$\begin{aligned} x = w_i \in (L_{\text{diag}})^{\complement} &\Leftrightarrow M_i \text{ accepts } w_i \\ &\Leftrightarrow w_i \in L(M_i) \\ &\Leftrightarrow f_M(x) = \text{Kod}(M_i)\#x \in L_U. \end{aligned}$$

[20] Formally, the input of A is Kod$(M_i)\#w_i$.

Exercise 5.34. Show that the following language

$$\{\text{Kod}(M)\#x\#0^i \mid x \in \{0,1\}^*,\ i \in \mathbb{N},\ M \text{ has at least } i+1 \text{ states, and during the computation of } M \text{ on } x \text{ the TM } M \text{ reaches the } i\text{-th state at least once}\}$$

is not recursive.

We see that the basic problems of the computability theory are strongly related to the halting of Turing machines (i.e., finiteness of computations). For the languages $(L_{\text{diag}})^{\complement}$ and L_{U} we have Turing machines (programs) that accept these languages, but there are no Turing machines that decide these languages (i.e., accept these languages and do not have any infinite computations). Therefore, we consider the following central problem.

Definition 5.35. *The* **halting problem** *is the decision problem* $(\{0,1,\#\}^*, L_{\text{H}})$, *where*

$$L_{\text{H}} = \{\text{Kod}(M)\#x \mid x \in \{0,1\}^* \textit{ and } M \textit{ halts on } x\}.$$

Exercise 5.36. Prove that $L_{\text{H}} \in \mathcal{L}_{\text{RE}}$.

The following result shows that there is no algorithm that can test whether a given program terminates or not.

Theorem 5.37. $L_{\text{H}} \notin \mathcal{L}_{\text{R}}$.

Proof. First, we give a proof on the level of programs and general reductions of Figure 5.8. We show

$$L_{\text{U}} \leq_{\text{R}} L_{\text{H}}.$$

Assume $L_{\text{H}} \in \mathcal{L}_{\text{R}}$, i.e., there is an algorithm A that decides L_{H}. We describe an algorithm B (Figure 5.12) that uses A as a subroutine in order to decide the universal language L_{U}. For any input w, B uses a subprogram C to verify whether w has the form $y\#x$, where $y = \text{Kod}(M)$ for a TM M and $x \in (\Sigma_{\text{bool}})^*$.

If y does not have this form, then B rejects w.

If $y = \text{Kod}(M)\#x$, then B gives y as the input to the subroutine A.

If A outputs "M does not halt on x" ("reject"), then B knows that $x \notin L(M)$ and can immediately reject its input $w = \text{Kod}(M)\#x$.

If A answers "M halts on x", then B simulates the work of M on x in the subroutine U. Since the computation of M on x is finite, U performs its simulation in a finite time.

If the answer of U is "M accepts x", then B accepts its input $y\#x = \text{Kod}(M)\#x$. If the output of U is "M rejects x", then B rejects its input $\text{Kod}(M)\#x$.

Clearly, $L(B) = L_{\text{U}}$ and B always halts. This completes the proof.

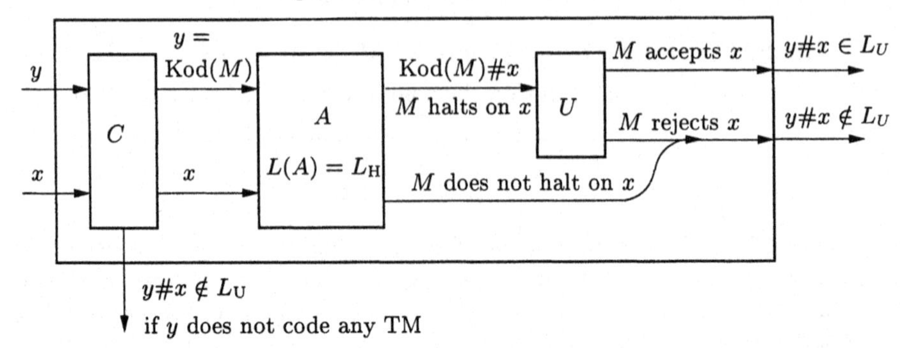

Fig. 5.12.

Next, we prove

$$L_U \leq_R L_H$$

in the formalism of Turing machines and thus provide an alternative proof of $L_U \leq_m L_H$ given above.

We describe a TM M, that reduces L_U to L_H. M works for any input w as follows. It verifies whether w is of the form

$$w = \text{Kod}(\overline{M})\#x$$

for a TM $\overline{M}$ and an $x \in (\Sigma_{\text{bool}})^*$.

(i) If w does not have this form, M generates the code Kod(M_1) of a TM M_1, that, for any input, runs in a cycle[21] $(\delta(q_0, ¢) = (q_0, ¢, \text{N}))$. Then, M halts with the tape content

$$M(w) = \text{Kod}(M_1)\#x.$$

(ii) If $w = \text{Kod}(\overline{M})\#x$, then M modifies the code of the TM $\overline{M}$ to the following TM M_2 with $L(M_2) = L(\overline{M})$. The TM M_2 works exactly as $\overline{M}$, except all transitions to the state q_{reject} of $\overline{M}$ are replaced by transitions to a new state p with $\delta(p, a) = (p, a, \text{N})$ for all $a \in \Sigma$. Hence, M_2 never rejects an input and runs an infinite computation for every input $y \notin L(\overline{M}) = L(M_2)$. M finishes its work with the tape content

$$M(w) = \text{Kod}(M_2)\#x.$$

Now, we prove for all $w \in \{0, 1, \#\}^*$, that

$$w \in L_U \Leftrightarrow M(w) \in L_H.$$

[21] I.e., M_1 does not halt for any input.

Let $w \in L_U$. Hence, $w = \text{Kod}(\overline{M})\#x$ for a TM $\overline{M}$ and a word $x \in \{0,1\}^*$, and $x \in L(\overline{M})$. Since $L(M_2) = L(\overline{M})$, M halts on x and accepts x. Hence, $M(w) = \text{Kod}(M_2)\#x$ belongs to L_H.

Let $w \notin L_U$. We distinguish two possibilities. If w does not have the form $\text{Kod}(\overline{M})\#x$ for any TM $\overline{M}$, then $M(w) = \text{Kod}(M_1)\#x$, where M_1 does not halt on any input. Hence, M_1 does not halt on x, so $M(w)$ does not belong to L_H. If w has the form $\text{Kod}(\overline{M})\#x$ for a TM $\overline{M}$ and $\text{Kod}(\overline{M})\#x \notin L_U$, then $x \notin L(\overline{M})$. In this case $M(w) = \text{Kod}(M_2)\#x$, where M_2 does not halt for any input from $(\Sigma_{\text{bool}})^* - L(\overline{M})$. Since $x \notin L(\overline{M})$, $\overline{M}$ does not halt on x and so $\text{Kod}(\overline{M})\#x$ does not belong to L_H. □

Exercise 5.38. Prove the following claims:

1. $L_U \leq_R (L_{\text{diag}})^{\complement}$,
2. $L_H \leq_R (L_{\text{diag}})^{\complement}$,
3. $L_{\text{diag}} \leq_R L_U$,
4. $(L_{\text{diag}})^{\complement} \leq_R L_H$,
5. $L_H \leq_R L_U$.

Now, we consider the following language

$$\boldsymbol{L_{\text{empty}}} = \{\text{Kod}(M) \mid L(M) = \emptyset\},$$

that contains the codes of Turing machines accepting the empty language (not accepting any input). Clearly,

$$(L_{\text{empty}})^{\complement} = \{x \in (\Sigma_{\text{bool}})^* \mid x \neq \text{Kod}(\overline{M}) \text{ for all TM } \overline{M} \text{ or } x = \text{Kod}(M) \text{ and } L(M) \neq \emptyset\}.$$

Lemma 5.39. $(L_{\text{empty}})^{\complement} \in \mathcal{L}_{\text{RE}}$.

Proof. We give two different proofs of the fact $(L_{\text{empty}})^{\complement} \in \mathcal{L}_{\text{RE}}$. The first proof shows that it is useful to have the model of nondeterministic Turing machines. The second proof shows how to apply the idea of the set theory used for proving $|\mathbb{N}| = |\mathbb{Q}^+|$ in Lemma 5.11.

Since, for any NTM M_1, there exists a TM M_2 such that $L(M_1) = L(M_2)$, it is sufficient to show that there exists an NTM M_1 with $L(M_1) = (L_{\text{empty}})^{\complement}$. The NTM M_1 works on every input x as follows.

Phase 1. M_1 deterministically verifies whether $x = \text{Kod}(M)$ for a TM M. If x does not code any TM, M_1 accepts x.

Phase 2. If $x = \text{Kod}(M)$ for a TM M, M nondeterministically generates a word $y \in (\Sigma_{\text{bool}})^*$ and deterministically simulates the computation of M on y.

Phase 3. If M accepts y (i.e., $L(M) \neq \emptyset$), then M_1 accepts its input $x = \text{Kod}(M)$.
If M rejects y, M_1 does not accept x in this computation.
If the computation of M on y is infinite, then M does not halt on x and thus does not accept the word $x = \text{Kod}(M)$ in this computation on x.

Because of phase 1, M_1 accepts all words that do not code Turing machines.

If $x = \text{Kod}(M)$ for a TM M and $L(M) \neq \emptyset$, then there exists a word y with $y \in L(M)$. Hence, there exists an accepting computation of M_1 on $x = \text{Kod}(M)$.

If $x \in L_{\text{empty}}$, then there is no accepting computation of M_1 on $x = \text{Kod}(M)$, so

$$L(M_1) = (L_{\text{empty}})^{\complement}.$$

Now, we give a second proof of the fact $(L_{\text{empty}})^{\complement} \in \mathcal{L}_{\text{RE}}$. Here, we directly construct a deterministic TM A that accepts $(L_{\text{empty}})^{\complement}$. A works on every input w as follows.

Phase 1'. If w is not a code of a TM, then A accepts w.
Phase 2'. If $w = \text{Kod}(M)$ for a TM M, then A works as follows:

A systematically generates (in the order given in Figure 5.3 or Figure 5.4) all pairs $(i, j) \in (\mathbb{N} - \{0\}) \times (\mathbb{N} - \{0\})$.

For each pair (i, j), A generates the i-th word w_i in the canonical order over the input alphabet of the TM M and simulates j computation steps of M on w_i.

If for a pair (k, l), the TM M accepts the word w_k in l steps, then A accepts its input $w = \text{Kod}(M)$. Otherwise, A works infinitely and hence does not accept w.

The kernel of the simulation idea is that when there exists a y with $y \in L(M)$, then $y = w_k$ for a positive integer k and the accepting computation of M on y has a finite length l. Hence, the exhaustive search over all pairs (i, j) in phase 2' of A assures that A will accepts w. □

Next we show that $(L_{\text{empty}})^{\complement} \notin \mathcal{L}_{\text{R}}$. This corresponds to proving the nonexistence of an algorithm that would verify whether a given program accepts an empty set. The consequence is that one is unable to algorithmically test the correctness of programs. Moreover, testing is impossible even for trivial tasks such as computing a constant function.

Lemma 5.40. $(L_{\text{empty}})^{\complement} \notin \mathcal{L}_{\text{R}}$.

Proof. We show

$$L_{\text{U}} \leq_{\text{m}} (L_{\text{empty}})^{\complement}.$$

We describe a TM A that reduces L_{U} to $(L_{\text{empty}})^{\complement}$ (Figure 5.13). For each input $x \in \{0, 1, \#\}$, A works as follows:

(i) If x does not have the form $\text{Kod}(M)\#w$ for a TM M and a word $w \in (\Sigma_{\text{bool}})^*$, then A writes the output $A(x) = \text{Kod}(B_x)$ on its tape, where B_x is a TM that works over Σ_{bool} and accepts the empty set[22] $\emptyset$ (i.e., $L(B_x) = \emptyset$).

[22] A can construct B_x easily by taking $\delta(q_0, ¢) = (q_{\text{reject}}, ¢, \text{N})$ for the transition function δ of B_x.

(ii) If $x = \text{Kod}(M)\#w$ for a TM M and a word $w \in (\Sigma_{\text{bool}})^*$, then A generates the code $\text{Kod}(B_x)$ of a TM B_x. For any input y of B_x, the TM B_x works independently[23] of y as follows:

a) B_x generates $x = \text{Kod}(M)\#w$ on its tape.
{The word x is finite, so it can be described by the states and the transition function of B_x}.

b) B_x simulates the work of M on w step-by-step.
If M accepts w, then B_x accepts its input y.
If M rejects w, then B_x rejects its input y.
If M does not halt on w, then B_x does not halt on y. Thus B_x does not accept y.

Now, we prove that

$$x \in L_{\text{U}} \Leftrightarrow A(x) = \text{Kod}(B_x) \in (L_{\text{empty}})^{\complement}$$

for all $x \in \{0, 1, \#\}^*$.

Let $x \in L_{\text{U}}$. Hence, $x = \text{Kod}(M)\#w$ for a TM M and $w \in L(M)$. In this case $L(B_x) = (\Sigma_{\text{bool}})^* \neq \emptyset$, so $\text{Kod}(B_x) \in (L_{\text{empty}})^{\complement}$.

Let $x \notin L_{\text{U}}$. Then, either x does not have the form $\text{Kod}(M')\#z$ for a TM M' and a $z \in \{0,1\}^*$ or $x = \text{Kod}(M)\#w$ for a TM M and $w \notin L(M)$. In both cases $L(B_x) = \emptyset$, so $\text{Kod}(B_x) \notin (L_{\text{empty}})^{\complement}$. □

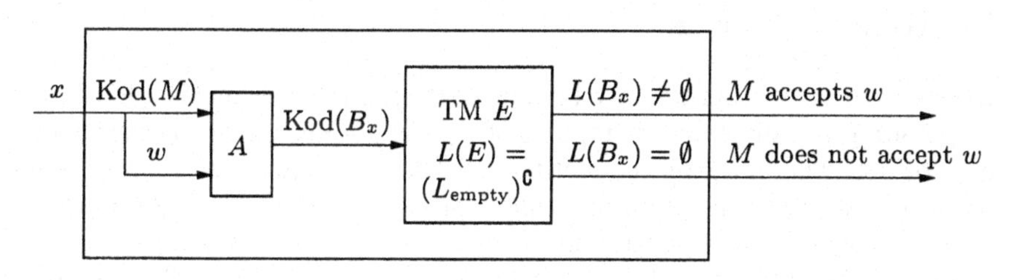

Fig. 5.13.

Corollary 5.41. $L_{\text{empty}} \notin \mathcal{L}_{\text{R}}$.

Proof. Lemma 5.27 claims that $L_{\text{empty}} \in \mathcal{L}_{\text{R}}$ implies $(L_{\text{empty}})^{\complement} \in \mathcal{L}_{\text{R}}$. □

Exercise 5.42.* Prove that the following languages are not in $\mathcal{L}_{\text{RE}}$:

1. L_{empty},
2. $(L_{\text{H}})^{\complement}$, and
3. $(L_{\text{U}})^{\complement}$.

[23] This means that B_x either accepts $\emptyset$ or $(\Sigma_{\text{bool}})^*$.

The next consequence of Lemma 5.40 is that the equivalence problem for Turing machines is undecidable. Thus, one cannot design a program that would be able to decide whether two given programs solve the same problem.

Corollary 5.43. *The language* $L_{\mathrm{EQ}} = \{\mathrm{Kod}(M)\#\mathrm{Kod}(\overline{M}) \mid L(M) = L(\overline{M})\}$ *is undecidable (i.e.,* $L_{\mathrm{EQ}} \notin \mathcal{L}_{\mathrm{R}}$*).*

Proof. The proof idea is simple because L_{empty} can be viewed as a special case of L_{EQ}. Formally, it is sufficient to show

$$L_{\mathrm{empty}} \leq_{\mathrm{m}} L_{\mathrm{EQ}}.$$

It is easy to construct a TM A that, for a given input $\mathrm{Kod}(M)$, constructs the output

$$\mathrm{Kod}(M)\#\mathrm{Kod}(C),$$

where C is a fixed, trivial TM with $L(C) = \emptyset$. Obviously,

$$\mathrm{Kod}(M)\#\mathrm{Kod}(C) \in L_{\mathrm{EQ}} \Leftrightarrow L(M) = L(C) = \emptyset \Leftrightarrow \mathrm{Kod}(M) \in L_{\mathrm{empty}}.$$

□

Exercise 5.44. Prove that the language $\{\mathrm{Kod}(M)\#\mathrm{Kod}(\overline{M}) \mid L(M) \subseteq L(M')\}$ does not belong to $\mathcal{L}_{\mathrm{R}}$.

5.4 Rice's Theorem

In the previous section we have learned that testing programs is a hard problem. For a program A and an input x of A, it is not decidable whether A halts on x or not. Therefore, one cannot test whether a program terminates for any input, i.e., whether the program is an algorithm. The trivial decision problem whether a given program does not accept any input (i.e., whether $L(M) = \emptyset$ for a TM M) is also undecidable. This leads us to suspect that there are not many test problems about programs that could be decidable. The aim of this section is to show that all (in a specific sense) nontrivial problems about programs (*TM*s) are undecidable. What the term "nontrivial" means in this context is specified in the following definition.

Definition 5.45. *A language* $L \subseteq \{\mathrm{Kod}(M) \mid M \text{ is a TM}\}$ *is a* **semantically nontrivial decision problem about Turing machines,** *if the following conditions hold:*

(i) There is a TM M_1*, such that* $\mathrm{Kod}(M_1) \in L$ *(i.e.,* $L \neq \emptyset$*).*
(ii) There exists a TM M_2*, such that* $\mathrm{Kod}(M_2) \notin L$ *(i.e., L does not contain the codes of all Turing machines).*
(iii) For any Turing machines A and B, $L(A) = L(B)$ *implies*

$$\mathrm{Kod}(A) \in L \Leftrightarrow \mathrm{Kod}(B) \in L.$$

Before proceeding on to the proof of the undecidability of semantically nontrivial decision problems, we still need to consider the following language

$$L_{\mathrm{H},\lambda} = \{\mathrm{Kod}(M) \mid M \text{ halt on } \lambda\}$$

as a special version of the halting problem.

Lemma 5.46.

$$L_{\mathrm{H},\lambda} \notin \mathcal{L}_{\mathrm{R}}.$$

Proof. We show

$$L_{\mathrm{H}} \leq_{\mathrm{m}} L_{\mathrm{H},\lambda}.$$

A TM A can reduce L_{H} to $L_{\mathrm{H},\lambda}$ as follows. For every input x that does not have the form $\mathrm{Kod}(M)\#w$, A generates a simple TM H_x that does not halt for any input.

If $x = \mathrm{Kod}(M)\#w$ for a TM M and a word w, then A generates the code $\mathrm{Kod}(H_x)$ of a TM H_x that works as follows:

1. Independent of its own input, the TM H_x generates the word $x = \mathrm{Kod}(M)\#w$ on the tape.
2. H_x simulates the work of M on w. If M halts on w, then H_x also halts and accepts its input. If the computation of M on w is infinite, then H_x does not halt.[24]

Clearly,

$$\begin{aligned} x \in L_{\mathrm{H}} &\Leftrightarrow x = \mathrm{Kod}(M)\#w \text{ and } M \text{ halt on } w \\ &\Leftrightarrow H_x \text{ always halts (for every own input)} \\ &\Leftrightarrow H_x \text{ halts on } \lambda \\ &\Leftrightarrow \mathrm{Kod}(H_x) \in L_{\mathrm{H},\lambda} \end{aligned}$$

for every $x \in \{0, 1, \#\}^*$. □

Theorem 5.47.* Rice's Theorem

Every semantically nontrivial decision problem over Turing machines is undecidable.

Proof. Let L be an arbitrary semantically nontrivial decision problem about Turing machines. We show

$$\text{either } L_{\mathrm{H},\lambda} \leq_{\mathrm{m}} L \text{ or } L_{\mathrm{H},\lambda} \leq_{\mathrm{m}} L^{\mathsf{C}}.$$

Let $M_\emptyset$ be a TM with $L(M_\emptyset) = \emptyset$. We distinguish two possibilities with respect to the membership of $\mathrm{Kod}(M_\emptyset)$ in L.

---

[24] In this way either H_x always halts (if M halts on w) or H_x does not halt for any input (if M does not halt on w).

I. Let $\text{Kod}(M_\emptyset) \in L$. In this case we show

$$L_{\text{H},\lambda} \leq_{\text{m}} L^{\complement}.$$

Because of Definition 5.45(ii), there exists a TM $\overline{M}$, such that $\text{Kod}(\overline{M}) \notin L$.

Now, we describe the work of a TM S (Figure 5.14) that reduces $L_{\text{H},\lambda}$ to $L^{\complement}$. For every input $x \in (\Sigma_{\text{bool}})^*$, S computes either

(i) $S(x) = \text{Kod}(M')$ with $L(M') = L(M_\emptyset) = \emptyset$ (i.e., $\text{Kod}(M') \notin L^{\complement}$) if $x \notin L_{\text{H},\lambda}$, or

(ii) $S(x) = \text{Kod}(M')$ with $L(M') = L(\overline{M})$ (i.e., $\text{Kod}(M') \in L^{\complement}$) if $x \in L_{\text{H},\lambda}$.

Thus, we see that the idea of the reduction is essentially based on the semantic nontriviality of L. S performs the computation as follows (Figure 5.14):

Input: An $x \in (\Sigma_{\text{bool}})^*$.

Phase 1. S checks whether $x = \text{Kod}(M)$ for a TM M. If x is not a code of a TM, then S writes $S(x) = \text{Kod}(M_\emptyset)$ as its output on the tape.

Phase 2. If $x = \text{Kod}(M)$ for a TM M, then S generates the code $\text{Kod}(M')$ of a TM M' that works as follows.

(a) The input alphabet of M' is $\Sigma_{\overline{M}}$ (the input alphabet of the TM $\overline{M}$ with $\text{Kod}(\overline{M}) \notin L$, i.e., with $\text{Kod}(\overline{M}) \in L^{\complement}$).

(b) For every input $y \in (\Sigma_{\overline{M}})^*$, M' generates the word $x = \text{Kod}(M)$ on the tape to the right of y (i.e., y is not rewritten) and simulates the work of M on λ.

If M does not halt on λ (i.e., $\text{Kod}(M) \notin L_{\text{H},\lambda}$), then M' does not halt on y either, so $y \notin L(M')$.

{Since the simulation of the computation of M on λ by M' runs independently of the input y of M', $L(M') = \emptyset = L(M_\emptyset)$, so $\text{Kod}(M') \in L$ (i.e., $\text{Kod}(M') \notin L^{\complement}$).}

If M halts on λ (i.e., if $\text{Kod}(M) \in L_{\text{H},\lambda}$), then M' generates the code $\text{Kod}(\overline{M})$ of the TM $\overline{M}$ on its tape. After that M' simulates the work of $\overline{M}$ on its own input $y \in (\Sigma_{\overline{M}})^*$. M' accepts y if and only if $\overline{M}$ accepts y.

{Hence, $L(M') = L(\overline{M})$, so $\text{Kod}(M') \notin L$ (i.e., $\text{Kod}(M') \in L^{\complement}$).}

We see that

$$x \in L_{\text{H},\lambda} \Leftrightarrow S(x) \in L^{\complement}$$

for all $x \in (\Sigma_{\text{bool}})^*$ and hence $L_{\text{H},\lambda} \leq_{\text{m}} L^{\complement}$.

II. Let $\text{Kod}(M_\emptyset) \notin L$.

Because of Definition 5.45(i), there exists a TM $\tilde{M}$ such that $\text{Kod}(\tilde{M}) \in L$. Now, one can prove $L_{\text{H},\lambda} \leq_{\text{m}} L$ in the same way that we proved $L_{\text{H},\lambda} \leq_{\text{m}} L^{\complement}$ in part I. The Turing machine $\tilde{M}$ plays the role of $\overline{M}$ in this proof.

□

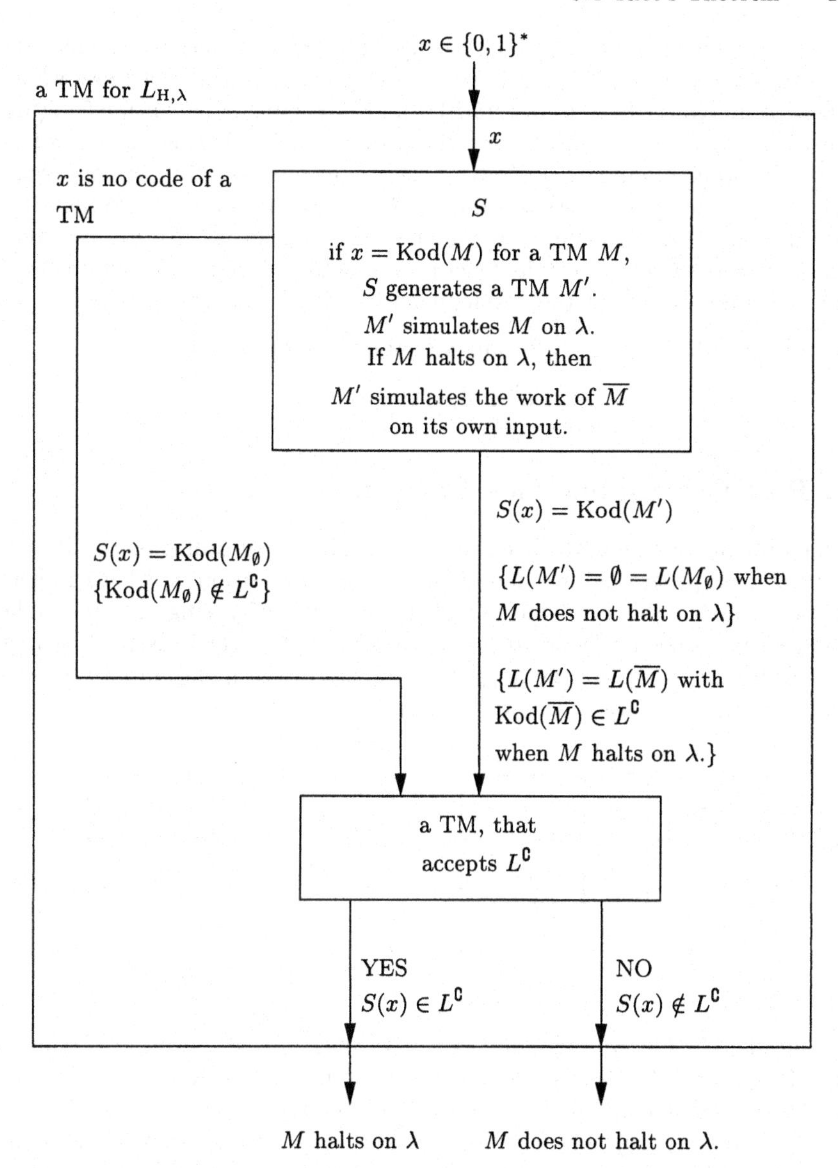

Fig. 5.14.

Exercise 5.48. Perform a detailed proof of $L_{\mathrm{H},\lambda} \leq_{\mathrm{m}} L$ in case II where $\mathrm{Kod}(M_\emptyset) \notin L$.

Rice's theorem has the following consequence. Let L be an arbitrary recursive language, and let

$$\mathrm{Kod}_L = \{\mathrm{Kod}(M) \mid M \text{ is a TM and } L(M) = L\}$$

be the language of the codes of all Turing machines that accept the language L. Since L is recursive, $\text{Kod}_L \neq \emptyset$. Clearly, there exist Turing machines whose code is not in Kod_L (Definition 5.45(ii)) and Kod_L satisfies Definition 5.45(iii). Thus, Kod_L is a semantically nontrivial problem about Turing machines and Rice's Theorem implies that $\text{Kod}_L \notin \mathcal{L}_\text{R}$. The interpretation of this result is that one is unable to test whether a given program is a correct algorithmic solution of the problem specified. Hence, program verification is a very hard task and therefore a well-structured, modular design of programs is of enormous importance for the development of reliable software products.

Exercise 5.49. Prove that, for any alphabet Σ and every $L \subseteq \Sigma^*$,

$$L, L^\complement \in \mathcal{L}_\text{RE} \Leftrightarrow L \in \mathcal{L}_\text{R}.$$

5.5 Post Correspondence Problem

In the previous sections we have showed that almost all problems about Turing machines (programs) are undecidable. A good question is, whether there exist also undecidable problems outside of the world of Turing machines. The answer to this question is positive and the aim of this section is to show how the reduction method can be applied to transform the undecidability results about Turing machines to the world of games.

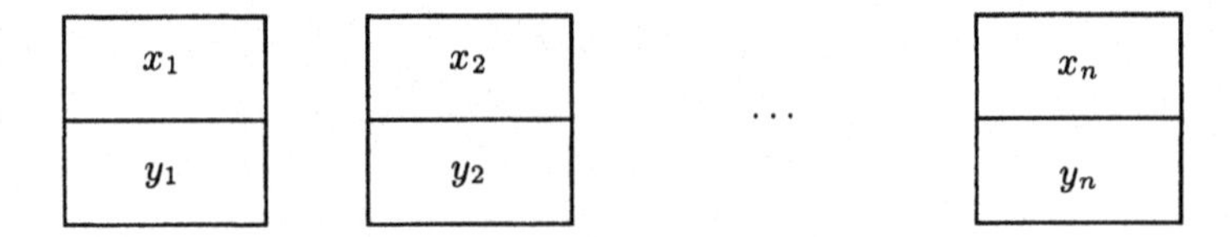

Fig. 5.15.

Consider the following domino game. One has a finite collection of domino types (Figure 5.15), where each domino type represents a pair (x, y) of words x and y over a fixed alphabet Σ. For each domino type (x, y) one has arbitrary many[25] dominos (x, y). The question is whether it is possible to place the same dominos side by side in such a way that the upper text (the word determined by the concatenation of the first elements of the dominos) is the same as the lower text (determined by the concatenation of the second elements of the dominos). We illustrate this using the following example.

Let

$$s_1 = (1, \#0),\ s_2 = (0, 01),\ s_3 = (\#0, 0),\ s_4 = (01\#, \#)$$

be the allowed domino types over $\{0, 1, \#\}$. A graphic representation of dominos s_1, s_2, s_3, and s_4 is given in Figure 5.16.

[25] Infinitely many

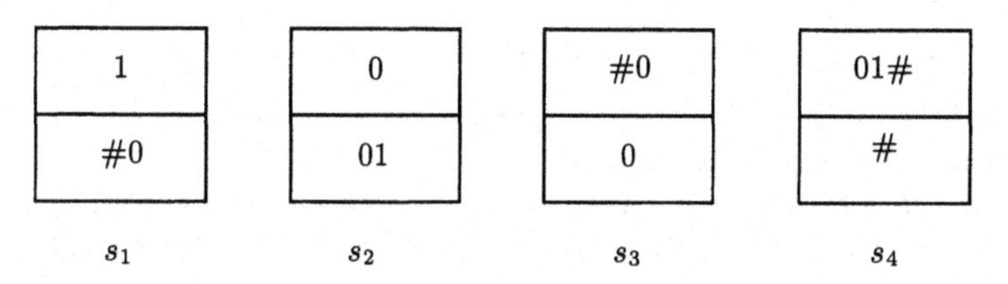

Fig. 5.16.

This domino game has a solution, namely the sequence s_2, s_1, s_3, s_2, s_4. This sequence determines the word (text) 01#0001# as depicted in Figure 5.17 and Figure 5.18 in two different ways.

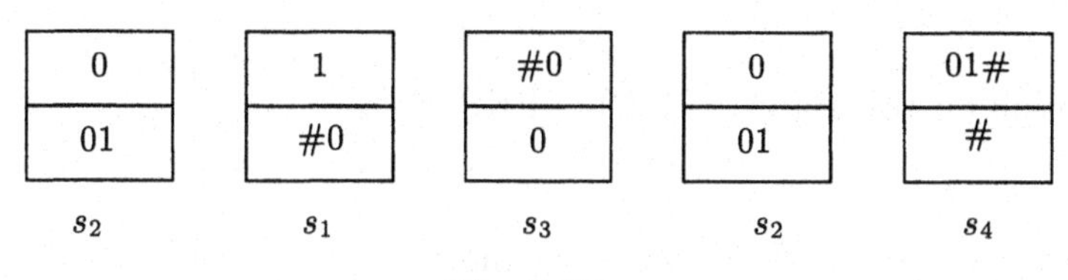

Fig. 5.17.

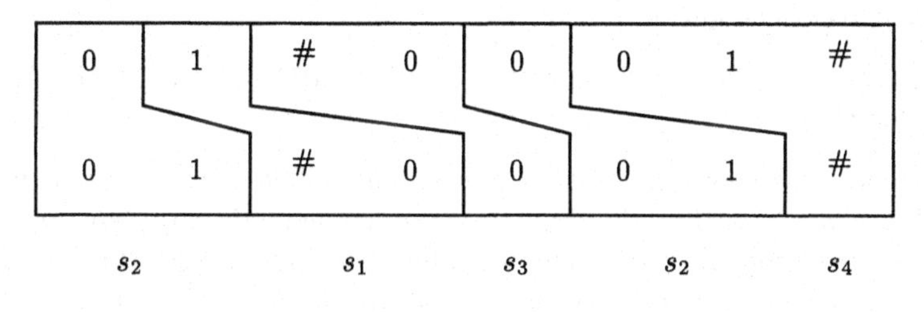

Fig. 5.18.

There is no solution for the domino game

$$s_1 = (00, 001),\ s_2 = (0, 001),\ s_3 = (1, 11),$$

because all second elements (words) of dominos are longer then their corresponding first counterparts.

This domino game is called the Post correspondence problem.

Definition 5.50. *Let Σ be an alphabet. An* **instance of the Post correspondence problem** *over Σ is a pair (A, B), where*

$$A = w_1, \ldots, w_k \text{ and } B = x_1, \ldots, x_k$$

for a positive integer k, $w_i, x_i \in \Sigma^*$ for $i = 1, \ldots, k$. For each $i \in \{1, \ldots, k\}$, the pair (w_i, x_i) is called a **domino**.

We say that the instance (A, B) of the Post correspondence problem has a **solution**, *if there exist a positive integer k and positive integers $i_1, i_2, \ldots, i_m$, such that*

$$w_{i_1} w_{i_2} \ldots w_{i_m} = x_{i_1} x_{i_2} \ldots x_{i_m}.$$

The **Post correspondence problem (PKP)** *is to decide whether a given instance of* PKP *has a solution or does not have any solution.*

Lemma 5.51. *If an instance (A, B) of* PKP *has a solution, then (A, B) has infinitely many solutions.*

Proof. If

$$i_1, i_2, \ldots, i_k$$

is a solution for the instance (A, B) of PKP, then

$$(i_1, i_2, \ldots, i_k)^j$$

is a solution of (A, B) for every positive integer j. □

Exercise 5.52. Define PKP as a decision problem $(L_{\text{PKP}}, \Sigma_{\text{bool}})$ for a language $L_{\text{PKP}} \subseteq (\Sigma_{\text{bool}})^*$. Prove that $L_{\text{PKP}} \in \mathcal{L}_{\text{RE}}$.

Exercise 5.53. Prove that the instance $((10, 011, 101), (101, 11, 011))$ of PKP does not have any solution.

Exercise 5.54. Does the instance $((1, 1110111, 101), (111, 1110, 01))$ of PKP have a solution?

Our goal is to show that PKP is undecidable. The significance of this undecidability proof is that the domino game has enough expressive power to simulate Turing machine computations. Due to technical limitations, we introduce a modified version of PKP, where the first domino was predetermined.

Definition 5.55. *Let Σ be an alphabet. An* **instance of the modified Post correspondence problem** *over Σ is a pair (C, D), where*

$$C = u_1, \ldots, u_k, \text{ and } D = v_1, \ldots, v_k$$

for a positive integer k and $u_i, v_i \in \Sigma^*$ for $i = 1, \ldots, k$.

We say that the instance (C, D) of the modified Post correspondence problem has a solution, if there exists a positive integer m and m positive integers $j_1, j_2, \ldots, j_m$, such that

$$u_1 u_{j_1} u_{j_2} \ldots u_{j_m} = v_1 v_{j_1} v_{j_2} \ldots v_{j_m}.$$

The **modified Post correspondence problem (MPKP)** *is to decide whether a given instance of* MPKP *over Σ has a solution.*

Consider the following instance (A, B) of PKP. Let

$$A = 0, 11, 1,\ B = 001, 1, 11,$$

$$i.e.,\ s_1 = (0, 001),\ s_2 = (11, 1) \text{ and } s_3 = (1, 11).$$

Clearly, $s_2 s_3$ is a solution of this PKP instance. If one considers (A, B) as an instance of MPKP, then one observes that (A, B) does not have any solution. This means that there may be a difference whether (A, B) is considered as an instance of PKP or as an instance of MPKP.

Exercise 5.56. Prove that the instance $((0, 11, 1), (001, 1, 11))$ of MPKP does not have any solution.

Lemma 5.57. *If* PKP *is decidable, then also* MPKP *is decidable.*

Proof. We prove this assertion by the reduction method. Let (A, B) be an instance of MPKP. We construct an instance (C, D) of PKP, such that

$$\text{MPKP}(A, B) \text{ has a solution} \Leftrightarrow \text{PKP}(C, D) \text{ has a solution.}$$

Let

$$A = w_1,\ \ldots,\ w_k,\ B = x_1,\ \ldots,\ x_k,$$

where Σ is an alphabet, k is a positive integer, $w_i, x_i \in \Sigma^*$ for $i = 1, \ldots, k$. Let $\$, ¢ \notin \Sigma$. We will construct an instance (C, D) of PKP over the alphabet $\Sigma_1 = \Sigma \cup \{¢, \$\}$.

First we define two homomorphisms h_{L} and h_{R} from Σ^* to Σ_1^* as follows. For every $a \in \Sigma$,

$$h_{\mathrm{L}}(a) = ¢a \text{ and } h_{\mathrm{R}}(a) = a¢.$$

We see that h_{L} inserts the symbol ¢ on the left side of any symbol and h_{R} inserts the symbol ¢ on the right side of any symbol. For instance, for the word 0110,

$$h_{\mathrm{L}}(0110) = ¢0¢1¢1¢0 \text{ and } h_{\mathrm{R}}(0110) = 0¢1¢1¢0¢.$$

We set

$$C = y_1,\ y_2,\ \ldots,\ y_{k+2} \text{ and } D = z_1,\ z_2,\ \ldots,\ z_{k+2},$$

where

$$\begin{array}{rlrl}
y_1 &= ¢h_{\mathrm{R}}(w_1) & z_1 &= h_{\mathrm{L}}(x_1)\\
y_2 &= h_{\mathrm{R}}(w_1) & z_2 &= h_{\mathrm{L}}(x_1)\\
y_3 &= h_{\mathrm{R}}(w_2) & z_3 &= h_{\mathrm{L}}(x_2)\\
&\vdots & &\vdots\\
y_{i+1} &= h_{\mathrm{R}}(w_i) & z_{i+1} &= h_{\mathrm{L}}(x_i)\\
&\vdots & &\vdots\\
y_{k+1} &= h_{\mathrm{R}}(w_k) & z_{k+1} &= h_{\mathrm{L}}(x_k)\\
y_{k+2} &= \$ & z_{k+2} &= ¢\$.
\end{array}$$

For example, for the instance

$$((0, 11, 1), (001, 1, 11))$$

of MPKP, the constructed instance of PKP is

$$((¢0¢, 0¢, 1¢1¢, 1¢, \$), (¢0¢0¢1, ¢0¢0¢1, ¢1, ¢1¢1, ¢\$)).$$

It is obvious that there is an algorithm (a TM) that can construct (C, D) from any given (A, B). It remains to show that either both MPKP(A, B) and PKP(C, D) have solutions, or neither MPKP(A, B) nor PKP(C, D) has a solution.

1. First, we prove that every solution for MPKP(A, B) determines a solution for PKP(C, D). Let $i_1, i_2, \ldots, i_m$ be a solution for MPKP(A, B). Hence

$$u = w_1 w_{i_1} w_{i_2} \ldots w_{i_m} = x_1 x_{i_1} x_{i_2} \ldots x_{i_m}.$$

The sequence of indices

$$2,\ i_1 + 1,\ i_2 + 1,\ \ldots,\ i_m + 1,$$

for PKP(C, D) corresponds to applying h_{R} on $w_1 w_{i_1} w_{i_2} \ldots w_{i_m}$ and h_{L} on $x_1 x_{i_1} x_{i_2} \ldots x_{i_m}$. Thus

$$¢h_{\mathrm{R}}(u) = ¢y_2 y_{i_1+1} \ldots y_{i_m+1} = z_2 z_{i_1+1} \ldots z_{i_m+1}¢ = h_{\mathrm{L}}(u)¢.$$

Hence, the difference between

$$h_{\mathrm{R}}(u) = y_2 y_{i_1+1} \ldots y_{i_m+1} \text{ and } h_{\mathrm{L}}(u) = z_2 z_{i_1+1} \ldots z_{i_m+1}$$

lies only in the first symbol[26] ¢ and the last symbol[27] ¢. Following the construction of (C, D), $y_1 = ¢y_2$ and $z_1 = z_2$. Thus, we can replace the first index 2 by the index 1. In this way we obtain the sequence

$$1,\ i_1 + 1,\ i_2 + 1,\ \ldots,\ i_m + 1,$$

for which

$$y_1 y_{i_1+1} y_{i_2+1} \ldots y_{i_m+1} = z_1 z_{i_1+1} z_{i_2+1} \ldots z_{i_m+1}¢.$$

Now, the only difference between the left side and the right side is the additional symbol ¢ at the end of the upper text. To obtain a solution for PKP(C, D), we add the $(k + 2)$-th domino $(¢, ¢\$)$. Thus,

$$y_1 y_{i_1+1} y_{i_2+1} \ldots y_{i_m+1} y_{k+2} = z_1 z_{i_1+1} z_{i_2+1} \ldots z_{i_m+1} z_{k+2}$$

and consequently

$$1,\ i_1 + 1,\ i_2 + 1,\ \ldots,\ i_m + 1,\ k + 2$$

is a solution of PKP(C, D).

[26] In $h_{\mathrm{L}}(u)$, but not in $h_{\mathrm{R}}(u)$
[27] In $h_{\mathrm{R}}(u)$, but not in $h_{\mathrm{L}}(u)$

2. We have to show that the existence of a solution for PKP(C, D) implies the existence of a solution for the original MPKP(A, B) instance. First, we observe that all words z_i begin with the symbol ¢ and that y_1 is the only word among the y_is that begins with the symbol ¢. Hence, every solution for PKP(C, D) has to start with the first domino. On the other hand, the only domino whose both words end with the same symbol is the $(k+2)$-th domino type. Therefore, every solution for PKP(C, D) must end with the index $k + 2$.
   Let
   $$1,\ j_1,\ j_2,\ \dots,\ j_m,\ k+2$$
   be a solution for PKP(C, D). We claim that
   $$1, j_1 - 1, j_2 - 1, \dots, j_m - 1$$
   is a solution of MPKP(C, D). The arguments are that
   (i) deleting the symbols ¢ and \$ from $y_1 y_{j_1} y_{j_2} \cdots y_{j_m} y_{k+2}$ results in the word $w_1 w_{j_1-1} w_{j_2-1} \cdots w_{j_m-1}$, and
   (ii) deleting the symbols ¢ and \$ from $z_1 z_{j_1} z_{j_2} \dots z_{j_m} z_{k+2}$ results in the word $x_1 x_{j_1-1} x_{j_2-1} \cdots x_{j_m-1} x_{k+2}$.
   Since $1, j_1, \dots j_m, k+2$ is a solution of PKP(C, D),
   $$y_1, y_{j_1} y_{j_2} \cdots y_{j_m} y_{k+2} = z_1 z_{j_1} z_{j_2} \dots z_{j_m} z_{k+2}.$$
   Together with (i) and (ii), this implies
   $$w_1 w_{j_1-1} w_{j_2-1} \cdots w_{j_m-1} = x_1 x_{j_1-1} x_{j_2-1} \cdots x_{j_m-1}.$$
   Hence, $1, j_1 - 1, j_2 - 1, \dots, j_m - 1$ is a solution for MPKP(A, B).

□

Exercise 5.58. Prove that the decidability of MPKP implies the decidability of PKP.

Now, we prove the undecidability of MPKP by showing that this domino game can be used to simulate computations of Turing machines.

Lemma 5.59.* *The decidability of* MPKP *implies the decidability of* L_U.

Proof. Let $x \in \{0, 1, \#\}^*$. We construct an instance (A, B) of MPKP such that
$$x \in L_U \Leftrightarrow \text{MPKP}(A, B) \text{ has a solution.}$$
If x does not have the form Kod$(M)\#w$ for a TM M and a word $w \in \{0, 1\}^*$ (i.e., $x \notin L_U$), then we set $A = 0$ and $B = 1$. Obviously, the domino game with only the domino type $(0, 1)$ does not have any solution.

Let $x = \text{Kod}(M)\#w$ for a TM $M = (Q, \Sigma, \Gamma, \delta, q_0, q_{\text{accept}}, q_{\text{reject}})$ and a word $w \in \Sigma^*$. Without loss of generality one may assume that M moves its head in every computation step. We describe the construction of MPKP(C, D) in four steps. Every step determines a group of domino types with a specific meaning. The rough idea of the construction is using the B-words (the lower part) to simulate the computation of M on w and using the A-words to follow this simulation with the delay of one configuration. Due to this delay one can use dominos where the upper part corresponds to an argument (the state and the symbol read) and the lower part corresponds to the result after applying the transition function δ of M on this argument. Using special dominos the delayed upper part can overtake the lower part, if and only if the computation ends in q_{accept}.

Let # be a new symbol that is not contained in Γ.

1. The first group contains only one domino type

$$(\#, \#q_0¢w\#).$$

   This domino starts the simulation of the the computation of M on w. The lower part corresponds to the initial configuration of M on w and the upper part has the "delay" $q_0¢w\#$.
2. The second group contains the following $|\Gamma| + 1$ domino types:

$$(X, X) \text{ for all } X \in \Gamma$$
$$(\#, \#).$$

   This group is used to copy these parts of configurations (symbol by symbol) from the lower text to the upper text that do not change in the next computation step.
3. The third group is built to simulate the computation step of M. For all $q \in Q - \{q_{\text{accept}}\}$, $p \in Q$, $X, Y, Z \in \Gamma$, we take the following domino types:

$$\begin{aligned}(qX, Yp) &\text{ if } \delta(q, X) = (p, Y, \text{R})\\ (ZqX, pZY) &\text{ if } \delta(q, X) = (p, Y, \text{L})\\ (q\#, Yp\#) &\text{ if } \delta(q, \sqcup) = (p, Y, \text{R})\\ (Zq\#, pZY\#) &\text{ if } \delta(q, \sqcup) = (p, Y, \text{L}).\end{aligned}$$

   These dominos make it possible to copy the configuration parts containing a state from the lower text to the upper text and to generate the next configuration in the lower part.
4. The fourth group enables the upper text to reach (overcome) the lower part if the computation of M on w has finished in the state q_{accept}. Using the following dominos, the state q_{accept} can "swallow" the symbols from $\Gamma \cup \{¢\}$. For all $X, Y \in \Gamma \cup \{¢\}$, we take

$$(Xq_{\text{accept}}Y, q_{\text{accept}})$$
$$(Xq_{\text{accept}}, q_{\text{accept}})$$
$$(q_{\text{accept}}Y, q_{\text{accept}}).$$

If q_{accept} in the lower text has "absorbed" all tape symbols, the delay of the upper part is reduced to $q_{\text{accept}}\#$. The following domino

$$(q_{\text{accept}}\#\#, \#)$$

enables us to finally balance the lengths of the upper text and the lower text.

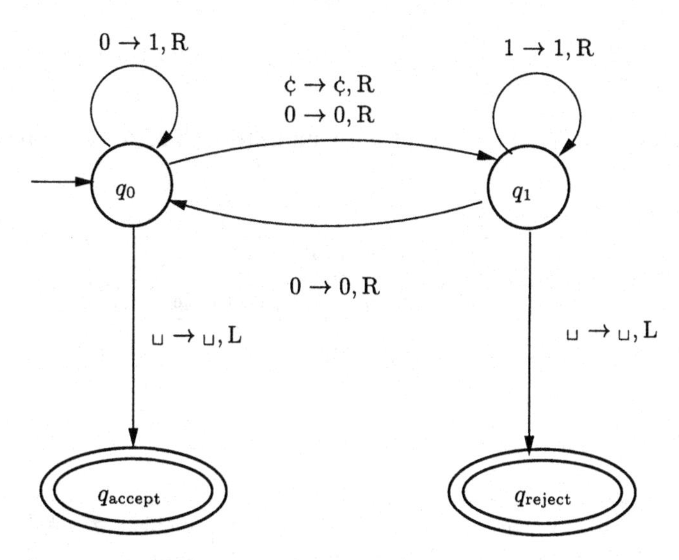

Fig. 5.19.

Let us illustrate the construction of the MPKP(A, B) instance for the TM

$$M = (\{q_0, q_1, q_{\text{accept}}, q_{\text{reject}}\}, \{0, 1\}, \{¢, 0, 1, \sqcup\}, \delta_M, q_{\text{accept}}, q_{\text{reject}})$$

in Figure 5.19. The constructed domino types for the input word $w = 01$ are the following ones:

(i) The first group:

$$(\#, \#q_0¢01\#).$$

(ii) The second group:

$$(0,0), (1,1), (¢, ¢), (\$, \$), (\#, \#).$$

(iii) The third group:

$$(q_0 1, 1q_0), (q_0 0, 0q_1), (q_0 ¢, ¢q_1),$$
$$(1q_0\#, q_{\text{accept}}1\#), (0q_0\#, q_{\text{accept}}0\#)$$
$$(q_1 1, 1q_1), (q_1 0, 0q_0), (1q_1\#, q_{\text{reject}}1\#), (0q_1\#, q_{\text{reject}}0\#).$$

(iv) The fourth group:

$$(0q_{\text{accept}}0, q_{\text{accept}}), (1q_{\text{accept}}1, q_{\text{accept}}), (1q_{\text{accept}}0, q_{\text{accept}}),$$
$$(0q_{\text{accept}}1, q_{\text{accept}}), (0q_{\text{accept}}, q_{\text{accept}}), (1q_{\text{accept}}, q_{\text{accept}}),$$
$$(q_{\text{accept}}0, q_{\text{accept}}), (q_{\text{accept}}1, q_{\text{accept}}),$$
$$(¢q_{\text{accept}}, q_{\text{accept}}), (q_{\text{accept}}¢, q_{\text{accept}}),$$
$$(¢q_{\text{accept}}a, q_{\text{accept}}), (aq_{\text{accept}}¢, q_{\text{accept}}) \text{ for all } a \in \{0,1\}$$
$$(q_{\text{accept}}\#\#, \#).$$

The beginning of the simulation of the computation of M on 01 is presented in Figure 5.20.

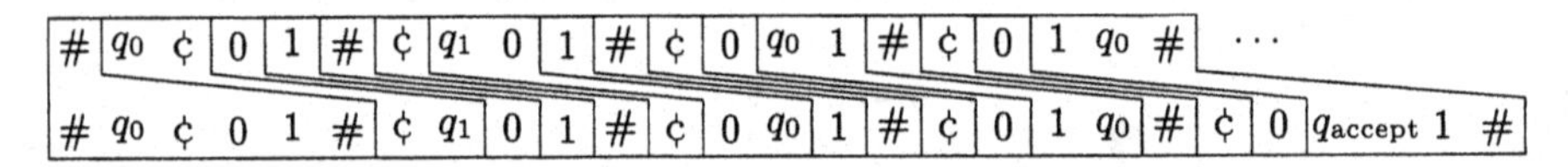

Fig. 5.20.

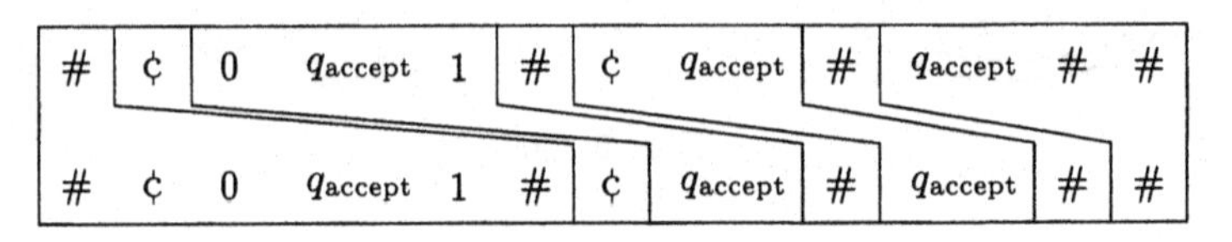

Fig. 5.21.

Figure 5.21 shows the phase of shortening, where the upper text overcomes the lower text.

It is obvious that the MPKP(A, B) instance can be algorithmically constructed from the given code of a TM M.

It remains to prove that

$$M \text{ accepts } w \Leftrightarrow \text{MPKP}(A, B) \text{ has a solution.}$$

To save unnecessary technical details, we argue for the validity of this equivalence informally.

If $w \in L(M)$, then there exists an accepting computation of M on w. We use this computation to build a solution to the MPKP(A, B) instance. The solution starts with the first domino type $(\#, \#q_0¢w\#)$ of the first group. Then, we use the dominos of the second group in order to copy symbols of the lower text to the upper text, if these symbols remain unchanged in the next computation step. The dominos of the third group are used to derive the successor configuration in the lower text. After any symbol #, the text below is always one configuration longer than the text above.[28] If the text below contains the whole computation that finishes in the state q_{accept}, then the dominos of the fourth group enable the text above to overcome the text below. Thus, we obtain a solution for MPKP(A, B).

Let $w \notin L(M)$. Every potential solution of MPKP(A, B) has to start with the first domino $(\#, \#q_0¢w\#)$. Since the dominos of the groups 2 and 3 allow only such changes of the first configuration that correspond to a computation step of M, the symbol q_{accept} will never occur in the text below. Then the text below always remains longer than the upper text, so MPKP(A, B) does not have any solution. □

Exercise 5.60. Write the MPKP(A, B) for the TM in Figure 4.6.

Exercise 5.61. Let (A, B) be an instance of MPKP for a TM M and a word w according to the construction presented in the proof of Lemma 5.59. Prove the following claims by induction:

If $w \in L(M)$, then there exists a sequence of indices, such that

(i) the lower text contains the complete computation of M on w, where the configurations are separated by the symbol #, and
(ii) the upper text is a proper prefix of the lower text and the upper text contains the whole computation except the last configuration.

Theorem 5.62. PKP *is undecidable.*

Proof. Lemma 5.59 implies that MPKP is not decidable and Lemma 5.57 claims that PKP is at least as hard as MPKP. □

Exercise 5.63.* Consider a restricted version of PKP, where all dominos are over the one-element alphabet $\{0\}$. Is this problem decidable?

5.6 The Kolmogorov-Complexity Method

In Section 5.2 we applied the diagonalization method in order to find the first algorithmically unsolvable problem L_{diag}. By proving the nonrecursivity of L_{diag}, we obtained the starting point (basis) for building the computability theory. We used this basis for proving undecidability of further decision problems by the reduction method in Sections 5.3, 5.4, and 5.5. The aim of this section is to give an alternative way of building the computability theory. Without assuming the existence of undecidable problems (i.e., without using any of the undecidability results proved in the previous sections) we use the

[28] Remember, that the text above is always a prefix of the text below.

Kolmogorov-complexity theory to show that there is no algorithm that can compute the Kolmogorov complexity $K(x)$ of a given word $x \in (\Sigma_{\text{bool}})^*$. Taking this unsolvability result as an alternative starting point[29] we again show using the reduction method that the halting problem is undecidable. Having this result we can continue in the same way as presented in Sections 5.3 and 5.4 to prove the undecidability of other concrete decision problems.

Theorem 5.64. *The problem of computing the Kolmogorov complexity $K(x)$ for any given $x \in (\Sigma_{\text{bool}})^*$ is not algorithmically solvable.*[30]

Proof. We prove Theorem 5.64 by contradiction. Let A be an algorithm that, for a given $x \in (\Sigma_{\text{bool}})^*$, computes $K(x)$. Let x_n be the first word with respect to the canonical order over Σ_{bool} with

$$K(x_n) \geq n.$$

Now, for any positive integer n, we describe an algorithm B_n, that

(i) uses A as a subroutine, and
(ii) computes x_n (for an empty input λ).

B_n works as follows.

```
B_n:    begin x := λ
          Compute K(x) by the algorithm A;
          while K(x) < n do
            begin x := the successor of x in the canonical order;
              Compute K(x) by the algorithm A
            end;
          output(x)
        end
```

Obviously, for every positive integer n, B_n computes the word x_n. We observe that all algorithms B_n are identical, except the number n. Let c be the length of the machine code of B_n excluding n. Then, the binary length of B_n is at most

$$\lceil \log_2(n+1) \rceil + c$$

for all $n \in \mathbb{N}$ and the constant[31] c, which is independent of n. Since B_n generates the word x_n,

$$K(x_n) \leq \lceil \log_2(n+1 \rceil + c$$

[29] Instead of the diagonalization language

[30] If one considers K as a function from $(\Sigma_{\text{bool}})^*$ to $(\Sigma_{\text{bool}})^*$ instead of a function from $(\Sigma_{\text{bool}})^*$ to $\mathbb{N}$ (i.e., $K(x)$ is the binary representation of the Kolmogorov complexity of x), then Theorem 5.64 says that the function K is not recursive.

[31] Note also that the binary length of the algorithm A is a constant with respect to n.

for all positive integers n. But following the definition of x_n, we have

$$K(x_n) \geq n$$

for all $n \in \mathbb{N} - \{0\}$. The inequality

$$\lceil \log_2(n+1) \rceil + c \geq K(x_n) \geq n$$

can be satisfied for at most finitely many positive integers. Therefore, we have a contradiction to the assumption that there is an algorithm A computing $K(x)$ for any x. □

Exercise 5.65. Prove that the problem of computing the first word x_n with $K(x_n) \geq n$ for any given positive integer n is not algorithmically solvable.

To show that one can use the reduction method to advance from Theorem 5.64 to the undecidability of fundamental languages (such as L_{U}, L_{H}, L_{empty}, etc.), we reduce the problem of computing the function K to the halting problem. The following lemma provides an alternative proof of the fact $L_{\mathrm{H}} \notin \mathcal{L}_{\mathrm{R}}$ from Theorem 5.37.

Lemma 5.66. *If $L_{\mathrm{H}} \in \mathcal{L}_{\mathrm{R}}$, then there exists an algorithm that computes the Kolmogorov complexity $K(x)$ for any given $x \in (\Sigma_{\mathrm{bool}})^*$.*

Proof. Let $L_{\mathrm{H}} \in \mathcal{L}_{\mathrm{R}}$ and let H be an algorithm that decides L_{H}. The following algorithm A (Figure 5.22) computes $K(x)$ for every $x \in (\Sigma_{\mathrm{bool}})^*$.

A generates the words $w_1, w_2, w_3, \ldots$ in the canonical order. For every positive integer i, A checks whether w_i is a binary code of a Pascal program M. If w_i is not the code of any program, then A continues with w_{i+1}. If $w_i = \mathrm{Kod}(M)$ for a Pascal program M, then A applies H in order to learn whether M halts on λ or not.

If M halts on λ, then M simulates the work of M on λ.

If M halts with the output $u = M(\lambda)$, then A checks whether $u = x$ or $u \neq x$. If $u = x$, then A outputs

$$K(x) = |w_i|.$$

If $u \neq x$, then A continues with w_{i+1}.

If $|w_i|$ is the output of A for an input x, then w_i is the shortest word with the property

$$w_i = \mathrm{Kod}(M)$$

for a Pascal program M and M generates x. Therefore, A computes the Kolmogorov complexity $K(x)$ for any given $x \in (\Sigma_{\mathrm{bool}})^*$. □

Exercise 5.67. Prove the following assertions:

(i) If $L_{\mathrm{U}} \in \mathcal{L}_{\mathrm{R}}$, then there exists an algorithm that computes $K(x)$ for any $x \in (\Sigma_{\mathrm{bool}})^*$.

(ii) If $L_{\mathrm{empty}} \in \mathcal{L}_{\mathrm{R}}$, then there exists an algorithm for computing $K(x)$ for any $x \in (\Sigma_{\mathrm{bool}})^*$.

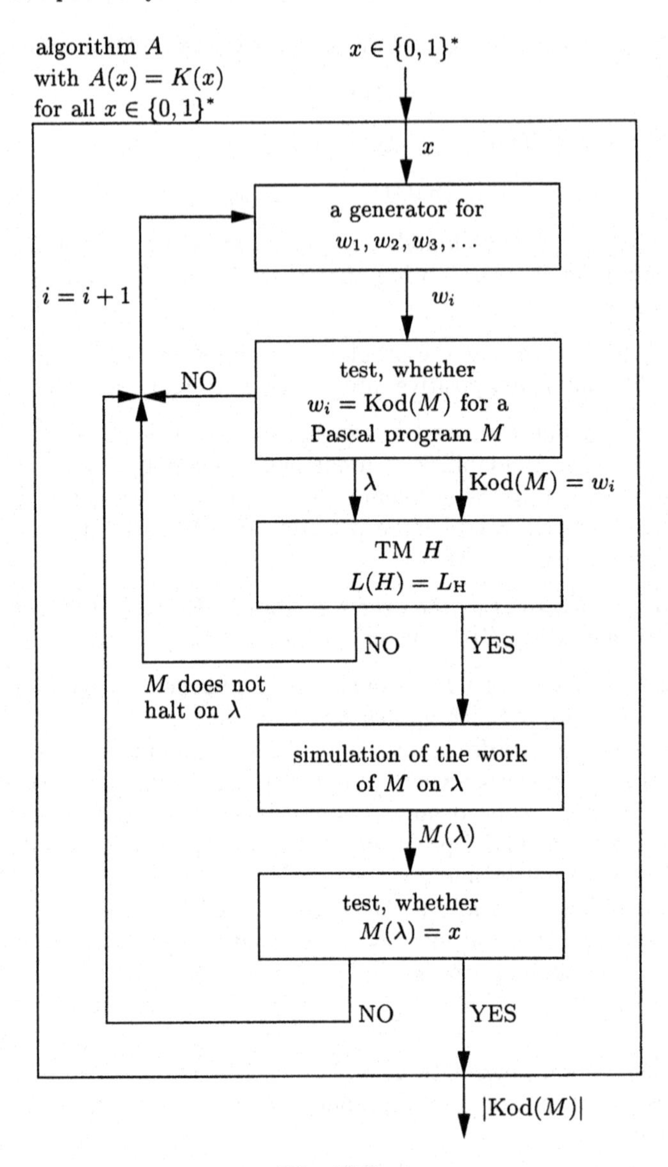

Fig. 5.22.

5.7 Summary

The fundamentals of proof methods in the computability theory lie in the set theory. The main concepts (instruments) are

- The comparison of infinite numbers (the cardinalities of infinite sets)

- The diagonalization method
- The reduction method

A set A has a cardinality that is at least as large as the cardinality of a set B, if there exists a one-to-one mapping from B to A. For any infinite set, $|A| = |A \times A|$ and hence $|\mathbb{N}| = |\mathbb{Q}^+|$. The smallest infinite set is $\mathbb{N}$, and every set C with $|C| \leq |\mathbb{N}|$ is called countable. Every set B with $|B| > |\mathbb{N}|$ is called uncountable.

The set of all Turing machines (programs) is countable and the set of all languages (problems) over the alphabet $\{0, 1\}$ is not countable. Therefore, most computing problems are not solvable.

If $|A| < |B|$ for two infinite sets A and B, the diagonalization techniques can help to construct an element from $B - A$.

A language L accepted by a Turing machine is called recursively enumerable. A language L with $L = L(M)$ for a TM M that halts in one of the states q_{accept} and q_{reject} for every input, is called recursive (or decidable). The diagonalization method makes it possible to construct the so-called diagonalization language L_{diag}, which is not recursively enumerable.

Applying the reduction method one can show that some problems are at least as hard[32] as L_{diag} and hence they are not recursive. The most important examples of undecidable decision problems are the universal language and the halting problem. The universal language contains all words that code a TM M and a word w, and $w \in L(M)$. The problem is to decide whether a given TM M accepts a given input w. The halting problem is to decide, whether a given TM M halts on a given input w.

Rice's Theorem says that each nontrivial decision problem about Turing machines (programs) is not decidable. Hence, there are are no algorithms for testing correctness and termination of programs. Thus, one cannot algorithmically decide whether a given Turing machine (program) is an algorithm[33] for a given problem, even if the problem is trivial like computing a constant function. Applying the reduction method one can also extend the proofs of undecidability beyond the world of problems about Turing machines (programs). A nice example of such a problem is the Post correspondence problem that can be viewed as a special domino game.

The first motivation for studying decidability and undecidability of mathematical problems came from the famous mathematicians David Hilbert. At the beginning of the 20th century he formulated a research project for mathematics, whose aim was to develop a formalism (a mathematical theory) in which one can solve all mathematical problems. Kurt Gödel [20] proved in 1931 that the objectives of Hilbert are unrealistic, because each nontrivial[34] mathematical theory is undecidable (i.e., that one can formulate problems in

[32] With respect to algorithmical solvability

[33] Remember that an algorithm terminates (halts) for every input with the correct result.

[34] Nontrivial means that the theory contains at least formal arithmetics.

a given theory that cannot be solved in the framework of this theory). This seminal work [20] led to the formalization of the notion of an "algorithm" and subsequently to the building of the fundamentals of the computability theory.

The undecidability of the universal language was established by Turing [68]. In 1953, Rice [57] published the result that is now known as Rice's Theorem. The undecidability of the Post correspondence problem was proved by Post [52].

For a more extensive and detailed study of elementary fundamentals of the computability theory, the corresponding chapters of the textbooks [27, 28, 65] are strongly recommended.

If one applies the mapping reduction in order to define the equivalence of problems with respect to recursivity, then one partitions the set of nonrecursive languages into infinitely many classes $\mathcal{L}_i$, $i = 1, 2, \ldots$. The meaning of the fact "a language L belongs to $\mathcal{L}_i$" is that L remains undecidable even when all languages from the classes $\mathcal{L}_1, \mathcal{L}_2, \ldots, \mathcal{L}_{i-1}$ were to be decidable. An interesting point is that there are problems of practical interest that belong to higher classes of this hierarchy. To deepen the knowledge in the computability theory we warmly recommend the classical textbooks of Trakhtenbrot [67] and Rogers [59].

There is no greater damage
than the time lost.

Michelangelo

6 Complexity Theory

6.1 Objectives

The computability theory provides methods for classifying problems with respect to their algorithmic solvability. A problem is algorithmically solvable when there exists an algorithm that solves this problem. A problem is not algorithmically solvable if there does not exist any algorithm that solves it.

The complexity theory can be viewed as a continuation of the computability theory in the sense that one tries to partition the class of algorithmically solvable problems into several subclasses with respect to the achievable efficiency of solving them. Since the 1960s when the use of computers was no longer restricted to a few research institutes, many practitioners have learned that the existence of an algorithm (program) for a problem is not sufficient for solving this problem by a computer. Many practical, algorithmically solvable problems were discovered, for which all designed algorithms had run so long that the computers had crashed before any result was computed. The resulting question was whether this is a consequence of our incapability to find an efficient algorithmic solution to the given problem or a consequence of an inherent property of the given problem that does not allow for any efficient algorithmic solution. These considerations led to the idea of measuring the hardness of computing tasks with respect to the amount of computer resources necessary and sufficient to compute them. This in turn led to the classification of algorithmically solvable problems according to their computational degree of hardness.

The complexity theory is the theory of quantitative laws and limits of algorithmic information processing. This theory also has a physical dimension. For instance, one could consider an algorithmically solvable problem to be "practically unsolvable" (intractable) if the execution of any algorithm solving the problem for realistic input instances needs more energy than the energy of the whole known universe.

The main goals of the complexity theory are as follows:

(i) To estimate the computational complexity (time complexity being the number of computer instructions or space complexity being the size of the computer memory) necessary and sufficient to solve concrete algorithmic problems.
(ii) To specify the notion of the class of "practically (efficient) solvable" problems and to develop methods for classifying algorithmic problems into "practically solvable" (tractable) problems and "practically unsolvable" (intractable) problems.
(iii) To compare the efficiency (computational power) of deterministic, nondeterministic and randomized algorithms.

During our first encounter with the complexity theory in this chapter we will confine ourselves to the following objectives.

In Section 6.2 we learn how to measure the computational complexity of Turing machines and programs. We also look at some properties of these complexity measures.

In Section 6.3 the problem of determining the hardness of algorithmic problems is discussed. The fundamental complexity classes of decision problems (languages) are defined as classes of languages that are decidable within a given complexity. We also discuss how to specify the class of practically solvable (tractable) decision problems, and provide the reasons to take the first "approximation" of the class of tractable problems as the class of problems solvable in time polynomial in the length of inputs.

Section 6.4 shows how to measure the complexity of nondeterministic Turing machines and introduces the fundamental nondeterministic complexity classes.

Section 6.5 is devoted to the comparison of the efficiency of nondeterministic computations and deterministic ones. This comparison touches the philosophical fundamentals of mathematics. We show that the time complexity of nondeterministic computations corresponds to the complexity of deterministic verification of the correctness of a given mathematical proof of a claim, whereas the deterministic time complexity corresponds to the complexity of inventing a mathematical proof of the given claim. Consequently, the question of whether nondeterministic algorithms can be more efficient than deterministic ones is equivalent to the question whether verifying the correctness of mathematical proofs is easier than their algorithmic creation.

Section 6.6 presents the concept of NP-completeness that is currently the main instrument for showing that specific problems are hard in the sense that they do not belong to the class of tractable problems. This is the first example showing that such a computationally nonrealistic concept as the concept of nondeterministic computation may be useful as a powerful instrument for investigating deterministic computation.

6.2 Complexity Measures

For the definition of the basic complexity measures, we use the model of multitape Turing machines. The main reasons for this choice are that this model is, on one hand, simple enough and, on the other hand, corresponds to the basic framework given by the von Neumann computer model. We will see later that the model of multitape Turing machines is robust enough such that the fundamental results about complexities defined by MTMs hold for the complexity of the performance of computer programs implemented in an arbitrary programming language. Especially, all classification results with respect to Turing machine tractability are of general validity.

Here, we define two fundamental complexity measures – the time complexity and the space complexity. The time complexity of a computation is the number of elementary instructions (Turing machine steps) executed in this computation. Consequently there is a linear relation between the time complexity of a computation and the energy needed to executed the computation. The space complexity is the size of memory used, where the size is measured in the number of computer words. For Turing machine models the size of computer words is determined by the working alphabet because the symbols of a working alphabet represent the allowed contents of computer words (registers).

Definition 6.1. *Let M be an* MTM *or a* TM *that always halts. Let Σ be the input alphabet of M. Let $x \in \Sigma^*$ and let $D = C_1, C_2, \ldots, C_k$ be a computation of M on x. The* **time complexity $\mathbf{Time}_M(x)$ of the computation of M on x** *is defined as*

$$\mathbf{Time}_M(x) = k - 1,$$

i.e., the number of steps of D.

The **time complexity of M** *is the function* $\mathrm{Time}_M : \mathbb{N} \to \mathbb{N}$*, defined as*

$$\mathbf{Time}_M(n) = \max\{\mathrm{Time}_M(x) \mid x \in \Sigma^n\}.$$

Note, that $\mathrm{Time}_M(n)$ is defined such that every input of size n (i.e., every input from Σ^n) is decided by M in time of at most $\mathrm{Time}_M(n)$ and that there exists an input x of length n with $\mathrm{Time}_A(x) = \mathrm{Time}_A(n)$. Phrased differently, $\mathrm{Time}_M(n)$ is the time complexity of the longest computation on an input of length n. Therefore this kind of complexity measurement is called the worst-case complexity. The worst-case complexity may not be a good measure, for instance, when describing a TM with very different complexities on inputs of the same length. In such a case, one might contemplate using the average case analysis instead, where the complexity is averaged over all inputs of length n. However, the worst-case complexity is generally preferred for two major reasons. First, determining the average time complexity of an algorithm is usually a much harder problem than determining $\mathrm{Time}_M(n)$. In many cases one might even be unable to perform the average case analysis.

Secondly, people have an affinity for guarantees. The worst-case complexity measurement assures that M can solve any problem instance x of length n in time $\text{Time}_M(n)$. Hence, we will restrict our attention to the worst-case complexity in this book.

Definition 6.2. *Let $k \in \mathbb{N} - \{0\}$. Let M be a k-tape-TM that always halts. Let*

$$C = (q, x, i, \alpha_1, i_1, \alpha_2, i_2, \ldots, \alpha_k, i_k)$$
$$\text{with } 0 \leq i \leq |x| + 1 \text{ and } 0 \leq i_j \leq |\alpha_j| + 1 \text{ for } j = 1, \ldots, k$$

be a configuration of M. The **space complexity of C** *is* [1]

$$\mathbf{Space}_M(C) = \max\{|\alpha_i| \mid i = 1, \ldots, k\}.$$

Let $C_1, C_2, \ldots, C_l$ be a computation of M on x. The **space complexity of M on x** *is*

$$\mathbf{Space}_M(x) = \max\{\text{Space}_M(C_i) \mid i = 1, \ldots, l\}.$$

The **space complexity of M** *is a function $\text{Space}_M : \mathbb{N} \to \mathbb{N}$ defined by*

$$\mathbf{Space}_M(n) = \max\{\text{Space}_M(x) \mid x \in \Sigma^n\}.$$

It may seem surprising that we define the space complexity of a configuration to be the maximum over the lengths of nonblank contents of all working tapes instead of taking the sum of the lengths of all working tapes. We point out that it does not matter which of these possibilities is chosen. Lemma 4.13 states that for any $k \in \mathbb{N}$, k tapes can be simulated by one tape of length bounded by the maximum of the lengths of these k tapes. The following lemma is a direct consequence of this observation.

Lemma 6.3. *Let k be a positive integer. For any k-tape-TM A that always halts, there exists an equivalent 1-tape-TM B, such that*

$$\text{Space}_B(n) \leq \text{Space}_A(n).$$

This property of space complexity is credited to the fact that the cardinality of the working alphabet of M (the length of the computer words of a real computer) does not have any influence on the definition of $\text{Space}_M(n)$. Hence, our definition of space complexity is not suitable for measuring differences of a multiplicative constant factor.

Lemma 6.4. *Let k be a positive integer. For any k-tape-TM A there exists a k-tape-TM B such that $L(A) = L(B)$ and*

$$\text{Space}_B(n) \leq \frac{\text{Space}_A(n)}{2} + 2.$$

---

[1] Note that the space complexity does not depend on the size of the working alphabet.

Proof. We will only describe the idea behind the proof. Let Γ_A be the working alphabet of A. We construct the working alphabet Γ_B of B in such a way that Γ_B contains all symbols from $\Gamma_A \times \Gamma_A$. If

$$\alpha_1, \alpha_2, \ldots, \alpha_m$$

is the content of the i-th working tape of A, $i \in \{1, 2, \ldots, k\}$, and the head is adjusted on α_j for a $j \in \{1, 2, \ldots, m\}$, then the content of the i-th tape of B is the word

$$¢\binom{\alpha_1}{\alpha_2}\binom{\alpha_3}{\alpha_4}\cdots\binom{\alpha_{m-1}}{\alpha_m}, \text{ if } m \text{ even, and}$$

$$¢\binom{\alpha_1}{\alpha_2}\binom{\alpha_3}{\alpha_4}\cdots\binom{\alpha_{m-1}}{\sqcup}, \text{ if } m \text{ odd.}$$

a The head of the i-th working tape of B is adjusted on the pair[2] that contains α_i and B stores in its state on which of the two symbols of this pair the head of the i-th tape of A is adjusted. Thus, B can unambiguously represent any configuration of A in space $1 + \lceil \text{Space}_A(n)/2 \rceil$.

It is not difficult to find a strategy that simulates the work of A step-by-step on B. Here, the head movements of A are modeled by either a corresponding head movement on B or a state change. A state change refers to the change of the indicator that tracks which symbols of the read pairs are read by A. In this way, one can construct the MTM B with

$$L(A) = L \text{ and } \text{Space}_B(n) \leq \frac{\text{Space}_A(n)}{2} + 2.$$

□

Applying Lemma 6.4 iteratively, one can construct, for any constant k and any MTM M, an equivalent MTM whose space complexity is bounded by

$$\frac{\text{Space}_A(n)}{k} + 2.$$

Observe that this can happen only because the size of the working alphabet of Turing machines is irrelevant in the measurement of space complexity.

One can do a reduction of time complexity in a similar way as the reduction of space complexity in Lemma 6.4. If one stores a larger number of symbols of an MTM A in one symbol of an MTM B then the transition function of B defined over complex symbols determines more complex operations than the transition function of B. This provides the possibility to simulate several steps of A by a few steps of B, achieving a constant speed-up of computations. We formulate this result in the following exercise.

[2] This is $\binom{\alpha_i}{\alpha_{i+1}}$ for i odd, and $\binom{\alpha_{i-1}}{\alpha_i}$ for i even.

Exercise 6.5. Prove the following claim:

For any MTM M there exists an equivalent MTM A such that

$$\mathrm{Time}_A(n) \leq \frac{\mathrm{Time}_M(n)}{2} + 2n.$$

The assertions of Lemma 6.4 and Exercise 6.5 show that the introduced complexity measurement is rough. But this is no drawback to our objectives because the differences between the complexity measures of different computing models are often larger than differences expressible by a constant multiplicative factor. We want robust classification results that are valid for all reasonable computing models. For this reason we are interested in the asymptotic growth of Time_M and Space_M and not in the exact estimation of the values of these complexity functions. For the asymptotic analysis of complexity functions we use the standard Ω, O, Θ and o-notation.

Definition 6.6. *For any function $f : \mathbb{N} \to \mathbb{R}^+$ we define*

$$\boldsymbol{O(f(n))} = \{r : \mathbb{N} \to \mathbb{R}^+ \mid \exists n_0 \in \mathbb{N}, \exists c \in \mathbb{N},\ \textit{such that}\ \forall n \geq n_0 : r(n) \leq c \cdot f(n)\}.$$

For any function $r \in O(f(n))$ we say that r **does not asymptotically grow faster than f**.

For any function $g : \mathbb{N} \to \mathbb{R}^+$ we define

$$\boldsymbol{\Omega(g(n))} = \{s : \mathbb{N} \to \mathbb{R}^+ \mid \exists n_0 \in \mathbb{N}, \exists d \in \mathbb{N},\ \textit{such that}\ \forall n \geq n_0 : s(n) \geq \frac{1}{d} \cdot g(n)\}.$$

For any function $s \in \Omega(g(n))$ we say that s **grows asymptotically at least as fast as g**.

For any function $h : \mathbb{N} \to \mathbb{R}^+$ we define

$$\begin{aligned}\boldsymbol{\Theta(h(n))} &= \{q : \mathbb{N} \to \mathbb{R}^+ \mid \exists c, d, n_0 \in \mathbb{N},\ \textit{such that}\ \forall n \geq n_0 : \\ &\qquad \frac{1}{d} \cdot h(n) \leq q(n) \leq c \cdot h(n)\} \\ &= O(h(n)) \cap \Omega(h(n)).\end{aligned}$$

If $q \in \Theta(h(n))$ we say that q and h **are asymptotically equivalent**.

Let f and g be functions from $\mathbb{N}$ to $\mathbb{R}^+$. If

$$\lim_{n\to\infty} \frac{f(n)}{g(n)} = 0,$$

then we say that g **grows asymptotically faster than f** *and we write* $\boldsymbol{f(n) = o(g(n))}$.

Exercise 6.7. Which of the following claims are true? Give detailed reasons for your choice.

(i) $2^n \in \Theta(2^{n+a})$ for any positive constant $a \in \mathbb{N}$.
(ii) $2^{b \cdot n} \in \Theta(2^n)$ for any positive constant $b \in \mathbb{N}$.
(iii) $\log_b n \in \Theta(\log_c n)$ for all positive real numbers $b, c > 1$.
(iv) $(n+1)! \in O(n!)$.
(v) $\log(n!) \in \Theta(n \cdot \log n)$.

Exercise 6.8. Prove the following assertions for all functions f and $g : \mathbb{N} \to \mathbb{R}^+$.

(i) $f \in O(g)$ and $g \in O(h) \Rightarrow f \in O(h)$
(ii) $f \in O(g) \Leftrightarrow g \in \Omega(f)$
(iii) $f \in \Theta(g) \Leftrightarrow g \in \Theta(f) \Leftrightarrow \Theta(f) = \Theta(g)$

Consider the Turing machine M in Figure 4.5 that accepts the language L_{middle}. M repetitively runs from the left tape boundary to the right tape boundary and back, during which it moves the left tape boundary one square to the right and the right boundary one square to the left. Using this strategy, M estimates the position of the square in the middle of the tape. Clearly the time complexity of M is in $O(n^2)$.

Exercise 6.9. Give an exact analysis of $\text{Time}_M(n)$ of the TM M from Figure 4.5.

Exercise 6.10. Design a 1-tape-TM B with $L(B) = L_{\text{middle}}$ and $\text{Time}_B(n) \in O(n)$.

Exercise 6.11. Give an asymptotic analysis of the time complexity of the TM A from Figure 4.6, that accepts the language $L_{\mathcal{P}}$.

The 1-tape-TM A from Figure 4.10 accepts the language $L_{\text{equal}} = \{w\#w \mid w \in (\Sigma_{\text{bool}})^*\}$ by copying the prefix of the input word up to # to the working tape and then comparing the contents of the input tape and of the working tape. We see that $\text{Time}_A(n) \leq 3 \cdot n \in O(n)$ and $\text{Space}_M(n) \in O(n)$.

Exercise 6.12. Design a 2-tape-TM M with $L(M) = L_{\text{equal}}$, $\text{Time}_M(n) \in O(\frac{n^2}{\log_2 n})$ and $\text{Space}_M(n) \in O(\log_2 n)$.

Above we defined and considered time complexity and space complexity of multitape Turing machines. Since the main goal of complexity theory is to classify algorithmic problems according to their computational difficulty we are interested in defining time and space complexities of problems. A natural idea could be to define the time complexity of a problem U as the time complexity of an asymptotically "optimal" MTM (algorithm) M for U. The optimality of M for U can be defined by requiring

$$\text{Time}_A(n) \in \Omega(\text{Time}_M(n))$$

for each MTM A that also solves U (i.e., there does not exist any MTM for U that solves U asymptotically faster than M). Although the idea of saying that the time complexity of U is the complexity of the best algorithm solving U seems to be reasonable, unfortunately it does not work. The next theorem explains why.

Theorem 6.13. *There exists a decision problem $(L, \Sigma_{\text{bool}})$, such that for every* MTM *A that decides $(L, \Sigma_{\text{bool}})$, there exists an* MTM *$B$ that decides $(L, \Sigma_{\text{bool}})$ and*

$$\text{Time}_B(n) \leq \log_2(\text{Time}_A(n))$$

for infinitely many positive integers $n \in \mathbb{N}$.

Theorem 6.13 tells us that there are problems for which one can essentially improve any algorithm for them. This implies the existence of an infinite sequence of algorithm improvements. Thus, there are no asymptotically best (optimal) algorithms for such problems and hence it is impossible to define the complexity of problems in the above-proposed way. What should we do now? We can classify problems without defining their complexity, simply by defining lower and upper bounds on the problem complexity as proposed in the following definition.

Definition 6.14. *Let L be a language and let f and g be functions from $\mathbb{N}$ to $\mathbb{R}^+$. We say that $O(g(n))$ is an* **upper bound on the time complexity of L** *if there exists an* MTM *(an algorithm) A such that*

$$L = L(A) \text{ and } \text{Time}_A(n) \in O(g(n)).$$

We say that $\Omega(f(n))$ is a **lower bound on the time complexity of L** *if*

$$\text{Time}_B(n) \in \Omega(f(n))$$

holds for every MTM *(algorithm) B with $L(B) = L$.*

An MTM *(an algorithm) C is* **optimal** *for L if and only if*

$$L(C) = L \text{ and } \Omega(\text{Time}_C(n)) \text{ is a lower bound on the time complexity of } L.$$

To establish an upper bound on the complexity of a problem U it is sufficient to find an algorithm that solves U and to analyze its complexity. Establishing a nontrivial lower bound on the complexity of U is a very hard task because it requires proving that each of the infinitely many known and unknown algorithms solving U must have its time complexity in $\Omega(f(n))$ for some function f. This is a nonexistence proof because one has to prove the nonexistence of any algorithm solving U with the time complexity asymptotically smaller than $f(n)$. The best illustration of the hardness of proving lower bounds on problem complexity is the fact that we know thousands of algorithmic problems for which

(i) the time complexity of the best known algorithm is exponential in the input size, and

(ii) no super-linear lower bound like $\Omega(n \log n)$ is known for any of them.

Thus, we conjecture, for many of these problems, that there does not exist any algorithm solving them within polynomial time of the input size, but we are unable to prove that one really needs more than $O(n)$ time to solve them. Missing a sufficiently powerful mathematical machinery to prove lower bounds on complexity makes classifying problems with respect to their hardness really difficult. In Section 6.6 we will show how to overcome this difficulty by making a reasonable assumption that enables us to prove the nonexistence of polynomial-time algorithms for given algorithmic problems (languages). We give a detailed discussion about the validity of this assumption in Sections 6.5 and 6.6.

Above we have explained how to measure complexity on abstract machine models. Before finishing this section about complexity measurement we would like to discuss the complexity measurement of programs in an arbitrary programming language. We distinguish two basic complexity measurements, namely the **uniform-cost measurement** and the **logarithmic-cost measurement**.

The approach based on uniform-cost measure is the simpler, but rougher one. The measurement of time complexity includes determining the overall number of elementary[3] instructions executed in the considered computation, and the measurement of space complexity includes determining the number of variables used in the computation. The advantage of this measurement is that it is simple in the sense that it simplifies the complexity analysis of a program (an algorithm). The drawback of uniform-cost measurement is that it is not always adequate because it assumes cost 1 for any arithmetic operation over two integers regardless of their size. When the operands are integers whose binary representations consist of several hundreds of bits, none of them can be stored in one computer word (16 or 32 bits). Then the operands must be stored in several computer words (i.e., one needs several space units to save them) and the execution of the arithmetic operation over these two large integers corresponds to the execution of a special program performing an operation over large integers by several operations over integers of the computer word size. Thus, the uniform-cost measurement may be applied to the cases where one can assume that all variables contain values whose representation sizes are bounded by a fixed constant (hypothetical computer word length) during the entire computation.

To see how a serious anomaly can appear when using the uniform-cost measurement, we present the following example. Let k and $a \geq 2$ be two positive integers of sizes that do not exceed the size of a hypothetical computer word. Consider the task of computing the number a^{2^k}. This value can be computed by the following program.

[3] Elementary instructions are arithmetic instructions over integers, comparison of two integers, reading, writing, loading integers and symbols, etc.

$$\texttt{for } i = 1 \texttt{ to } k \texttt{ do } a := a \cdot a.$$

This program computes the value a^{2^k} by executing the following k multiplications:

$$a^2 := a \cdot a, \ a^4 := a^2 \cdot a^2, \ a^8 := a^4 \cdot a^4, \ \ldots, \ a^{2^k} := a^{2^{k-1}} \cdot a^{2^{k-1}}.$$

Thus, the uniform-cost time complexity is in $O(k)$. The uniform-cost space complexity is 3 because three variables are sufficient to execute the program. This contrasts with the fact that one needs at least 2^k bits to represent the result a^{2^k} and to write 2^k bits any machine needs $\Omega(2^k)$ operations on its machine words. Since this is true for every positive integer k, we have an exponential gap between the uniform-cost time complexity and any realistic time complexity, and an unbounded gap between the uniform-cost space complexity and any realistic space complexity.

The solution to such a situation, where the values of variables grow unboundedly, is to use the **logarithmic-cost measurement**. With respect to this measurement the cost of every elementary operation is the sum of the sizes of the binary representations of the operands[4] and the time complexity of a computation is the sum of the costs of all operations executed in the computation. The space complexity is the sum of the lengths of the representations of the values of all variables used. The logarithmic-cost measurement is always realistic. Its drawback is that a proper logarithmic-cost measurement may be too complicated or excessively time consuming in many situations.

6.3 Complexity Classes and the Class P

To define complexity classes we use the computing model of the multitape Turing machine. The complexity classes considered here are language classes, i.e., sets of decision problems.

Definition 6.15. *For all functions f, g from $\mathbb{N}$ to $\mathbb{R}^+$ we define:*

$$\begin{aligned}
\textbf{TIME}(\boldsymbol{f}) &= \{L(B) \mid B \textit{ is an } \text{MTM} \textit{ with } \text{Time}_B(n) \in O(f(n))\},\\
\textbf{SPACE}(\boldsymbol{g}) &= \{L(A) \mid A \textit{ is an } \text{MTM} \textit{ with } \text{Space}_A(n) \in O(g(n))\},\\
\textbf{DLOG} &= \text{SPACE}(\log_2 n),\\
\textbf{P} &= \bigcup_{c \in \mathbb{N}} \text{TIME}(n^c),\\
\textbf{PSPACE} &= \bigcup_{c \in \mathbb{N}} \text{SPACE}(n^c),\\
\textbf{EXPTIME} &= \bigcup_{d \in \mathbb{N}} \text{TIME}(2^{n^d}).
\end{aligned}$$

[4] To be theoretically exact, one has to add the binary length of the addresses of the variables (operands) in memory to the complexity.

In what follows we study some fundamental relations between these complexity classes and some basic properties of time and space complexity.

Lemma 6.16. *For any function* $t : \mathbb{N} \to \mathbb{R}^+$

$$\text{TIME}(t(n)) \subseteq \text{SPACE}(t(n)).$$

Proof. Each MTM M, which works in time $\text{Time}_M(n)$, can visit at most $\text{Time}_M(n)$ squares of any working tape. So, $\text{Space}_M(n) \leq \text{Time}_M(n)$ for every MTM M. □

Corollary 6.17.

$$\text{P} \subseteq \text{PSPACE}$$

To get further relationships between complexity classes we need the notion of constructible functions. The idea behind this notion is that we would like to construct multitape Turing machines that are able to take care of themselves, in the sense that they count the amount of computational resources that they use and never use more than a prescribed upper bound. To be able to do this self-control, the upper bounds on complexity must be given by some well-behaved functions.

Definition 6.18. *A function* $s : \mathbb{N} \to \mathbb{N}$ *is called* **space constructible**, *if there exists a 1-tape-*TM M, *such that*

(i) $\text{Space}_M(n) \leq s(n)$ *for all* $n \in \mathbb{N}$, *and*
(ii) for each input 0^n, $n \in \mathbb{N}$, M *generates the word* $0^{s(n)}$ *on its working tape and halts in* q_{accept}.

A function $t : \mathbb{N} \to \mathbb{N}$ *is called* **time constructible**, *if there exists an* MTM A, *such that*

(i) $\text{Time}_A \in O(t(n))$, *and*
(ii) for any input 0^n, $n \in \mathbb{N}$, A *generates the word* $0^{t(n)}$ *on the first working tape and halts in* q_{accept}.

Common monotone functions with $f(n) \geq \log_2(n+1)$ $[f(n) \geq n]$ are space [time] constructible. For instance, a 1-tape-TM A can construct the function $\lceil \log_2(n+1) \rceil$ as follows. A reads 0^n from the left to the right on the input tape, during which A writes the actual position of the reading head on the input tape in binary coding on the working tape. This can be done easily by adding 1 to the content of the working tape for each step to the right on the input tape. If the head on the input tape has reached $\sqcup$, then the binary word on the working tape has exactly the length $\lceil \log_2(n+1) \rceil$. After which it is sufficient to exchange all 1s of the working tape by 0s to obtain the word $0^{\lceil \log_2(n+1) \rceil}$ on the working tape.

We note that it does not matter whether a 1-tape-TM M or an MTM is used for the definition of space constructibility because any MTM can be simulated by a 1-tape-TM within the same space complexity.

In what follows we describe the work of an MTM M that constructs the function $\lceil \sqrt{n} \rceil$. The idea of M is to consecutively test for $i = 1, 2, \ldots$ whether

$$i^2 \leq n < (i+1)^2.$$

A test for a fixed i can be done by M in the following way. M writes 0^i on the first two tapes. To check whether $i \cdot i \geq n$, M tries to go $i \cdot i$ steps from ¢ to the right on the input tape. M can execute this as follows. At the beginning the head on the input tape and the head on the first working tape move simultaneously to the right and the head on the second working tape does not move. When the head on the first working tape has reached ␣, it returns to the left endmarker ¢ and the head on the second working tape moves one square to the right. Then, the heads on the input tape and the first working tape can continue to move simultaneously to the right. In this way the number of the steps to the right on the input tape is the product of the lengths of the tape contents of the two working tapes and so M finds the smallest i with $i \cdot i > n$.

Exercise 6.19. Give formal descriptions (e.g., diagrams) of the multitape Turing machines used above for the construction of the functions $\lceil \log_2(n+1) \rceil$ and $\lceil \sqrt{n} \rceil$.

Note that the above-described multiplication of two tape lengths can be used to show that the function $f(n) = n^q$ is time constructible for any $q \in \mathbb{N}$.

Exercise 6.20. Prove that the following functions are space constructible.

(i) $\lceil \sqrt{n} \rceil^q$ for any positive integer q
(ii) $\lceil n^{\frac{1}{3}} \rceil$
(iii) $\lceil n^{\frac{q}{2}} \rceil$ for any positive integer $q \geq 2$
(iv) 2^n

Exercise 6.21. Show that the following functions are time constructible.

(i) n^j for any $j \in \mathbb{N} - \{0\}$
(ii) $c \cdot n$ for any $c \in N - \{0\}$
(iii) 2^n
(iv) c^n for any $c \in \mathbb{N} - \{0, 1\}$

Exercise 6.22. Let $s(n)$ and $t(n)$ be space [time] constructible. Prove that the function $t(n) \cdot s(n)$ is also space [time] constructible.

Let s be a space-constructible function. The following lemma shows that to prove the existence of an MTM M accepting a language $L(M) = L$ within space complexity $s(n)$ (i.e., an MTM M with $L = L(M)$ and $\text{Space}_M(x) \leq s(|x|)$ for every input $x \in \Sigma^*$), it suffices to construct an MTM A that accepts L and does not use more than $s(|x|)$ space on words from L only (i.e., the computations on words from L^{C} may have an unbounded space complexity).

Lemma 6.23. *Let* $s : \mathbb{N} \to \mathbb{N}$ *be a space-constructible function. Let* M *be an* MTM *with* $\text{Space}_M(x) \leq s(|x|)$ *for all* $x \in L(M)$*. Then there exists an* MTM A *with* $L(A) = L(M)$ *and*

$$\text{Space}_A(n) \leq s(n),$$

i.e., $\text{Space}_A(y) \leq s(|y|)$ *for all* y *over the input alphabet of* M.

Proof. Let M be a k-tape-TM for a $k \in \mathbb{N} - \{0\}$ with $\text{Space}_M(x) \leq s(|x|)$ for all $x \in L(M)$. Let B be a 1-tape-TM that constructs s. We describe the work of a $(k+1)$-tape-TM A with $L(A) = L(M)$ and $Space_A(n) \leq s(n)$. Let x be an input.

1. A views x as $0^{|x|}$ and simulates the computation of B on $0^{|x|}$ on the $(k+1)$-th working tape. The simulation finishes when the word $0^{s(|x|)}$ has been written on the $(k+1)$-th tape.
2. A writes a special symbol $\# \notin \Gamma_M$ on the position $s(|x|)$ of all working tapes.
3. A performs a step-by-step simulation of the work of M on x by using the first k working tapes. If simulating M the machine A was required to move to the right from the symbol # on a working tape, then A halts in q_{reject}. If A was able to simulate the whole computation of M on x, that A accepts x iff M accepts x.

Clearly, $\text{Space}_A(z) \leq s(|z|)$ for all inputs z. We show now, that

$$L(A) = L(M).$$

If $x \in L(M)$, then $\text{Space}_M(x) \leq s(|x|)$. Thus A simulates the whole computation of M on x and finishes in q_{accept}. Hence $x \in L(A)$.

If $y \notin L(M)$, we distinguish two cases.

If $\text{Space}_M(y) \leq s(|y|)$, then A simulates the whole computation of M on y and rejects y.

If $\text{Space}_M(y) > s(|y|)$, then A stops the simulation in the first moment when M tries to use more than $s(|y|)$ squares on a working tape, upon which A stops in q_{reject} and hence $y \notin L(A)$. □

The following Lemma 6.23 provides an analogous assertion to Lemma 6.23 for time complexity.

Lemma 6.24. *Let* $t : \mathbb{N} \to \mathbb{N}$ *be a time-constructible function. Let* M *be an* MTM *with* $\text{Time}_M(x) \leq t(|x|)$ *for all* $x \in L(M)$*. Then, there exists an* MTM A *with* $L(A) = L(M)$ *and*

$$\text{Time}_A(n) \in O(t(n)).$$

Exercise 6.25. Prove Lemma 6.24.

Lemma 6.23 and Lemma 6.24 show that for a space-constructible function s and a time-constructible function t, the complexity classes SPACE(s) and TIME(t) do not change when we exchange our definition

$$\text{Space}_M(n) = \max\{\text{Space}_M(x) \mid x \in \Sigma^n\} \text{ and}$$
$$\text{Time}_M(n) = \max\{\text{Time}_M(x) \mid x \in \Sigma^n\}$$

of classes $Space_M$ and $Time_M$ for the following definition

$$\text{Space}_M(n) = \max\{\text{Space}_M(x) \mid x \in L(M) \text{ and } |x| = n\} \text{ and}$$
$$\text{Time}_M(n) = \max\{\text{Time}_M(x) \mid x \in L(M) \text{ and } |x| = n\}.$$

The next results shows an important relationship between space complexity and time complexity.

Theorem 6.26. *For every space-constructible s with $s(n) \geq \log_2 n$,*

$$\text{SPACE}(s(n)) \subseteq \bigcup_{c \in \mathbb{N}} \text{TIME}(c^{s(n)}).$$

Proof. Let $L \in \text{SPACE}(s(n))$. Following Lemma 6.4 there exists a 1-tape-TM $M = (Q, \Sigma, \Gamma, \delta, q_0, q_{\text{accept}}, q_{\text{reject}})$, such that $L = L(M)$ and $\text{Space}_M(n) \leq s(n)$.

For every configuration $C = (q, w, i, x, j)$ of M we define the **internal configuration of C** as

$$\text{In}(C) = (q, i, x, j).$$

An internal configuration In(C) contains only those parts of C that can change during a computation. Hence, In(C) does not contain the content w of the input tape that remains unchanged during the whole computation on w.

Let InConf(n) denote the set of all possible internal configurations that can occur during computations of M on input words of length n for $n \in N$. Our idea is to show that to check whether M accepts an input w or not, it is sufficient to simulate at most $|\text{InConf}(|w|)|$ steps of the computation of M on w, because each computation longer than $|\text{InConf}(|w|)|$ is an infinite computation. Using this fact one can construct an MTM A simulating M that works in time

$$O(|\text{InConf}(|w|)).$$

First, let us estimate $|\text{InConf}(n)|$. For every internal configuration

$$(q, i, x, j) \in \text{InConf}(n),$$

we have $0 \leq i \leq n+1$, $|x| \leq \text{Space}_M(n) \leq s(n)$ and $0 \leq j \leq \text{Space}_M(n) \leq s(n)$. Hence

$$\begin{aligned}|\text{InConf}_M(n)| &\leq |Q| \cdot (n+2) \cdot |\Gamma|^{\text{Space}_M(n)} \cdot \text{Space}_M(n) \\ &\leq (\max\{2, |Q|, |\Gamma|\})^{4 \cdot s(n)} \\ &\leq c^{s(n)}\end{aligned}$$

for $c = (\max\{2, |Q|, |\Gamma|\})^4$.

Let w be an input word of length n. Obviously, every computation $D = C_1, C_2, C_3, \ldots$ of M on w that is longer than $|\text{InConf}_M(n)|$ must contain two configurations C_i and C_j such that $\text{In}(C_i)$ and $\text{In}(C_j)$ are identical. Applying the definition of an internal configuration we obtain that C_i and C_j are also identical. Since M is deterministic

$$D = C_1, \ldots, C_{i-1}, C_i, C_{i+1}, \ldots, C_{j-1}, C_i, C_{i+1}, \ldots, C_{j-1}, C_i \ldots$$

is an infinite computation with the cycle $C_i, C_{i+1}, \ldots, C_j$. Hence, any finite computation of M on an input w has length at most $|\text{InConf}_M(|w|)|$. This means that the length of each accepting computation on an input of length n is at most $|\text{InConf}_M(n)|$.

Now, we describe the work of a 3-tape-TM A with $L(A) = L(M)$ and $\text{Time}_A(n) \in O(k^{s(n)})$ for a constant k. For any input w, A computes as follows.

1. A simulates the construction of s and writes $0^{s(|w|)}$ on the first working tape.
   {Since a 1-tape-TM constructing s uses at most $s(|w|)$ squares of the working tape, there exists a constant d such that $\text{Time}_B(n) \leq d^{s(n)}$. Hence, A generates $0^{s(|n|)}$ in time $d^{s(n)}$.}
2. A writes $0^{c^{s(|w|)}}$ on the second working tape in $c^{s(|w|)}$ steps.
   {A uses the third tape to speed this up.}
3. A simulates the work of M on w step-by-step on the first working tape. After each simulation step A erases one 0 on the second working tape. If there is no more 0's on the second tape and the simulation has not finished, then A halts in q_{reject}.
   If the computation of M on w consists of at most $c^{s(|w|)}$ steps, then the simulation succeeds and A accepts iff M accepts.

As already observed, computation phases 1 and 2 of A run in time $O(d^{s(|w|)})$ and $O(c^{s(|w|)})$ respectively. Phase 3 of A runs in $O(c^{s(|w|)})$ time. Hence

$$\text{Time}_A(n) \in O((\max\{c, d\})^{s(|w|)}).$$

If $x \in L(M)$, then the length of the computation of M on x is at most $c^{s(|w|)}$, hence after a successful simulation A accepts x too. If $x \notin L(M)$, A rejects x (it does not matter whether the computation of M is infinite or a rejecting one). □

Corollary 6.27.

$$\text{DLOG} \subseteq \text{P} \textit{ and } \text{PSPACE} \subseteq \text{EXPTIME}$$

Applying Corollaries 6.17 and 6.27 we obtain the following fundamental hierarchy of deterministic complexity classes:

$$\text{DLOG} \subseteq \text{P} \subseteq \text{PSPACE} \subseteq \text{EXPTIME}.$$

The following hierarchy theorems belong to the fundamental results of complexity theory. Their proofs are based on a more elaborated version of the diagonalization method and so we omit their presentation here.

Theorem 6.28.* *Let s_1 and s_2 be two functions from $\mathbb{N}$ to $\mathbb{N}$ that satisfy the following properties:*

(i) $s_2(n) \geq \log_2 n$,
(ii) s_2 *is space constructible, and*
(iii) $s_1(n) = o(s_2(n))$.

Then

$$\text{SPACE}(s_1) \subsetneq \text{SPACE}(s_2).$$

Theorem 6.29.* *Let t_1 and t_2 be functions from $\mathbb{N}$ to $\mathbb{N}$ that satisfy the following properties:*

(i) t_2 *is time constructible and*
(ii) $t_1(n) \cdot \log_2(t_1(n)) = o(t_2(n))$.

Then

$$\text{TIME}(t_1) \subsetneq \text{TIME}(t_2).$$

The complexity hierarchies presented in Theorem 6.28 and 6.29 show the existence of arbitrarily "hard" problems. For instance, there are computing problems that are not in TIME(2^n) (i.e., all algorithms solving such problems have a time complexity larger than 2^n). Table 6.1 documents the fact that the execution of algorithms for hard problems may exceed the boundary of what is physically possible. For time complexity functions $10n$, $2n^2$, n^3, 2^n and $n!$ Table 6.1 shows how many operations have to be executed for input lengths 10, 50, 100 and 300. Where the numbers are too large, the length of these numbers are stated instead.

Table 6.1.

n / $f(n)$	10	50	100	300
$10n$	100	500	1000	3000
$2n^2$	200	5000	20 000	180 000
n^3	1000	125 000	1 000 000	27 000 000
2^n	1024	16 digits	31 digits	91 digits
$n!$	$\approx 3.6 \cdot 10^6$	65 digits	158 digits	615 digits

Let us assume that we have a PC that executes 10^6 operations per second. Then an algorithm A with $\text{Time}_A(n) = n^3$ can compute the results for the largest input length $n = 300$ within 3 seconds. If $\text{Time}_A(n) = 2^n$, then A needs for inputs of length 50 more than 3 years, and for $n = 100$ more than $3 \cdot 10^{15}$ years of computation. Comparing the values of 2^n and $n!$ for a realistic input size between 100 and 300 with the suggested number of seconds since the "Big Bang" (21 digits), then it is obvious that the execution of algorithms of exponential complexity on realistic inputs is beyond the border of physical reality.

Moreover, we call attention to the following properties of functions n^3 and 2^n. If t is the time you can wait for the results, then developing a better computer that executes twice as many instructions per time unit has the following consequences.

(i) For an algorithm working in time n^3, the size of tractable input instances can be increased by a factor of $\sqrt[3]{2}$ from $t^{1/3}$ to $\sqrt[3]{2} \cdot t^{1/3}$ (i.e., one can compute on $\sqrt[3]{2}$ times larger sizes of input instances than before).
(ii) However, for an algorithm running in time 2^n, the size of tractable input instances can only be increased by a meager one bit.

Thus, an algorithm of an exponential time complexity cannot be considered practical, and algorithms of a polynomial-time complexity $O(n^c)$ for small cs are considered to be practical. Hence, any problem that does not belong to TIME(2^n) can be considered intractable and any problem in TIME(n^3) may be considered tractable (practically solvable).

As already mentioned, the main objectives of the complexity theory are

to find a formal specification of the class of practically solvable (tractable) problems

and

to develop methods that enable one to classify algorithmic problems with respect to their membership in the class of tractable problems.

The first efforts in searching for a reasonable specification of the intuitive notion of tractable problems resulted in the following definition.[5] In what follows, any algorithm A with $\text{Time}_A(n) \in O(n^c)$ for a constant c is called a **polynomial-time algorithm**.

A problem is practically solvable (tractable) if and only if there exists a polynomial-time algorithm that solves it. The complexity class P *is the class of tractable decision problems.*

The two main reasons for connecting polynomial-time computations with the intuitive notion of practical solvability are the following:

[5] Currently, this specification of algorithmic tractability is not accepted in this restrictive form. We will give a more detailed discussion on this matter in the next chapter.

(i) The first reason is practical and based on experience. We have already observed that algorithms with exponential complexity are not practical and that polynomial-time algorithms with a small degree of polynomials are practical. But what about a running time of n^{1000}? Of course, an algorithm with n^{1000} time complexity is unlikely to be of any use, because $n^{1000} > 2^n$ for all reasonable input sizes n. Nevertheless, experience has proved the reasonability of considering polynomial-time computations to be tractable. In almost all cases, once a polynomial-time algorithm has been found for an algorithmic problem that formerly appeared to be hard, some key insight into the problem has been gained. With this new knowledge new polynomial-time algorithms with a lower degree than the former polynomial were designed. There are only a few known exceptions of non-trivial problems for which the best polynomial-time algorithm is not of practical utility.

(ii) The second reason is of a theoretical nature. Any definition of an important problem class has to be robust such that the defined class is invariant with respect to all reasonable models of computation. We cannot allow that a problem is tractable for a programming language (e.g. JAVA) but not for MTM. Such a situation would arise if we define TIME(n^6) as the set of tractable problems. The crucial fact is that the class P is robust with respect to all reasonable computing models. The proof of this assertion is based on the notion of polynomial-time reducibility between models of computation. A computing model $\mathcal{A}$ is **polynomial-time reducible** to a computing model $\mathcal{B}$, if there exists a polynomial p such that for every algorithm $A \in \mathcal{A}$ there is an algorithm $B \in \mathcal{B}$ that solves the same problem as A and

$$\text{Time}_B(n) \in O(p(\text{Time}_A(n))).$$

If one considers $\mathcal{A}$ as the class of TMs and $\mathcal{B}$ as the class of MTMs, then Lemma 4.11 provides a polynomial-time reduction from $\mathcal{A}$ to $\mathcal{B}$ for $p(n) = n$. The simulation in Lemma 4.11 gives a polynomial-time reduction form $\mathcal{B}$ to $\mathcal{A}$.

In fact, for all pairs of reasonable computing models, the known polynomial-time reducibilities work with $p(n) \in O(n^3)$. Thus, if one designs a polynomial-time algorithm for a program U in JAVA, then there is a polynomial-time algorithm for U in any reasonable formalism. On the other hand, if one proves that there is no polynomial-time Turing machine accepting a language L, then one can be sure that there is no polynomial-time computer program deciding L. We see that this kind of robustness is very important and it has to be required for any reasonable specification of the class of tractable problems.

Exercise 6.30. Analyze the simulation of Lemma 4.13 in order to prove the following assertion.

For every MTM A with $\text{Time}_A(n) \geq n$ there exists an equivalent TM B such that

$$\text{Time}_B(n) \in O((\text{Time}_A(n))^2).$$

6.4 Nondeterministic Complexity Measures

Nondeterministic Turing machines (algorithms) can have many[6] different computations on an input, and these complexities may substantially differ from each other. So how does one define the complexity of a nondeterministic machine (algorithm) M on an input w? When working with nondeterminism, we take the optimistic view that a nondeterministic machine always takes the "best" possible choice from the given possibilities. The best choice implies not only that this choice provides the correct solution, but also that it results in a computation of minimal complexity. Interpreting nondeterminism in this optimistic way, one can define the complexity of the work of a nondeterministic algorithm M on an input w as the complexity of the most efficient computation of M on w yielding the correct result. In the case of decision problems (language recognition) we consider only the complexity on inputs that are in the language.

Definition 6.31. *Let M be an* NTM *or a nondeterministic* MTM. *Let $x \in L(M) \subseteq \Sigma^*$.*

The **time complexity of M on x, $\text{Time}_M(x)$**, *is the length of the shortest accepting computation of M on x.*

The **time complexity of M** *is the function* $\text{Time}_M : \mathbb{N} \to \mathbb{N}$, *defined as*

$$\mathbf{Time}_M(n) = \max\{\text{Time}_M(x) \mid x \in L(M) \cap \Sigma^n\}.$$

Let $C = C_1, C_2, \ldots, C_m$ be an accepting computation of M on x. Let $\text{Space}_M(C_i)$ *be the space complexity of the configuration C_i.*

We define

$$\mathbf{Space}_M(C) = \max\{\text{Space}_M(C_i) \mid i = 1, 2, \ldots, m\}.$$

The **space complexity of M on x** *is*

$$\mathbf{Space}_M(x) = \min\{\text{Space}_M(C) \mid C \textit{ is an accepting computation of } M \textit{ on } x\}.$$

The **space complexity of M** *is the function* $\text{Space}_M : \mathbb{N} \to \mathbb{N}$ *defined as*

$$\mathbf{Space}_M(n) = \max\{\text{Space}_M(x) \mid x \in L(M) \cap \Sigma^n\}.$$

[6] Even infinitely (but countable) many

Definition 6.32. *For all functions* $f, g : \mathbb{N} \to \mathbb{R}^+$,

$$\mathbf{NTIME}(f) = \{L(M) \mid M \textit{ is a nondeterministic } \text{MTM} \textit{ with } Time_M(n) \in O(f(n))\},$$

$$\mathbf{NSPACE}(g) = \{L(M) \mid M \textit{ is a nondeterministic } \text{MTM} \textit{ with } \text{Space}_M(n) \in O(g(n))\},$$

$$\mathbf{NLOG} = \text{NSPACE}(\log_2 n),$$

$$\mathbf{NP} = \bigcup_{c \in \mathbb{N}} \text{NTIME}(n^c), \textit{ and}$$

$$\mathbf{NPSPACE} = \bigcup_{c \in \mathbb{N}} \text{NSPACE}(n^c).$$

First, we show the relations between nondeterministic time and nondeterministic space. Observe that these relations are the same as in the deterministic case (Lemma 6.16 and Theorem 6.26).

Lemma 6.33. *For all space-constructible functions* t *and* s,

(i) $\text{NTIME}(t) \subseteq \text{NSPACE}(t)$, *and*
(ii) $\text{NSPACE}(s) \subseteq \text{NTIME}(c^{s(n)})$ *for a constant* c.

Proof. First, we prove (i) and then (ii).

(i) Assume $L \in \text{NTIME}(t)$, i.e., there exists a nondeterministic MTM M with
$$L(M) = L \text{ and } \text{Time}_M(n) \leq d \cdot t(n)$$
for a constant d and all sufficiently large n. Hence, for any sufficiently long $x \in L(M)$, there is an accepting computation C_x of M on x of length of at most $d \cdot t(|x|)$. Since M can visit at most $d \cdot t(n)$ squares of a tape in $d \cdot t(n)$ steps, we have
$$\text{Space}_M(C_x) \leq d \cdot t(|x|).$$
Thus
$$\text{Space}_M(n) \leq d \cdot t(n)$$
for all sufficiently large n and thus $L \in \text{NSPACE}(t)$.

(ii) Let $L \in \text{NSPACE}(s)$ and let s be space constructible. Then there exists a nondeterministic MTM A with
$$L = L(M) \text{ and } \text{Space}_A(x) \leq d \cdot s(|x|)$$
for a constant d and all sufficiently long words $x \in L(A) = L$. This means that for all sufficiently long $x \in L(A)$ there exists an accepting computation C_x with
$$\text{Space}_A(C_x) \leq d \cdot s(|x|).$$

Let C_x be the shortest accepting computation of A on x with this property. Using the same argument as in the proof of Theorem 6.26, we obtain that there exists a constant k such that the length of C_x is at most

$$|\text{InConf}_A(|x|)| \leq k^{d \cdot s(|x|)}$$

for all sufficiently large $x \in L(A)$. If the length of C_x is greater than $|\text{InConf}_A(|x|)|$, there would exist $i, j \in \mathbb{N} - \{0\}$, $i \neq j$, such that

$$C_x = C_1, C_2, \ldots, C_{i-1}, C_i, C_{i+1}, \ldots, C_j, C_{j+1}, \ldots, C_m,$$

and C_i and C_j are identical. Then

$$C'_x = C_1, C_2, \ldots, C_{i-1}, C_i, C_{j+1}, \ldots, C_m$$

is also an accepting computation of A on x. Since C'_x is shorter than C_x we have a contradiction to the assumption that C_x is the shortest accepting computation of A on x with $\text{Space}_A(C_x) \leq d \cdot s(|x|)$.
Hence, for every sufficiently large $x \in L(A)$ there exists an accepting computation of A of length of at most

$$k^{d \cdot s(|x|)} = c^{s(|x|)},$$

where $c = k^d$ for constants k and d. Thus

$$\text{Time}_A(n) \in O(c^{s(n)})$$

and we have $L \in \text{NTIME}(c^{s(n)})$.

□

Exercise 6.34. Let M be a nondeterministic MTM with $\text{Time}_M(n) \leq t(n)$ for a time-constructible function t. Prove that there exists a nondeterministic MTM A with

(i) $L(A) = L(M)$ and
(ii) there exists a constant d such that for every $w \in \Sigma^*$ all computations of A on w have a length of at most $d \cdot t(|w|)$.

The following theorem shows fundamental relationships between deterministic complexity measures and nondeterministic ones.

Theorem 6.35. *For any function* $t : \mathbb{N} \to \mathbb{R}^+$ *and any time- and space-constructible function* $s : \mathbb{N} \to \mathbb{N}$ *with* $s(n) \geq \log_2 n$

(i) $\text{TIME}(t) \subseteq \text{NTIME}(t)$,
(ii) $\text{SPACE}(t) \subseteq \text{NSPACE}(t)$, *and*
(iii) $\text{NTIME}(s(n)) \subseteq \text{SPACE}(s(n)) \subseteq \bigcup_{c \in \mathbb{N}} \text{TIME}(c^{s(n)})$.

Proof. The assertions (i) and (ii) are obvious, because every MTM is also a nondeterministic MTM.

The relation

$$\text{SPACE}(s(n)) \subseteq \bigcup_{c \in \mathbb{N}} \text{TIME}(c^{s(n)})$$

has been proved in Theorem 6.26. Thus, to prove (iii), it is sufficient to show

$$\text{NTIME}(s(n)) \subseteq \text{SPACE}(s(n)).$$

Let $L \in \text{NTIME}(s(n))$, i.e., there exists a nondeterministic k-tape-TM $M = (Q, \Sigma, \Gamma, \delta_M, q_0, q_{\text{accept}}, q_{\text{reject}})$ with

$$L = L(M) \text{ and } \text{Time}_M(n) \in O(s(n)).$$

Obviously,

$$r = r_M = \max\{|\delta_M(U)| \mid U = (q, a, b_1, \ldots, b_k) \in Q \times (\Sigma \cup \{¢, \$\}) \times \Gamma^k\}$$

is the upper bound on the number of possible distinct actions of M from any configuration. Let $T_{M,x}$ be the computation tree of M on an input $x \in \Sigma^*$. If one orders the nondeterministic choices of δ_M for any argument U, then one can assign these orders $1, 2, \ldots, r$ as labels to the corresponding edges of $T_{M,x}$ (Figure 6.1). In this way we have unambiguously assigned a word

$$z = z_1 z_2 \ldots z_l \in \{1, 2, \ldots, r\}^*$$

to any computation of M on x with l nondeterministic decisions. Then having M and x, a sequence z either unambiguously determines a prefix of some computations of M on x (for instance, $z = r322$ in $T_{M,x}$ in Figure 6.1 determines the computation prefix where one takes the r-th possibility for the first choice, the second possibility for the third choice, and the second possibility for the fourth choice) or is nonsense (for instance, no computation in Figure 6.1 corresponds to $z = 24\ldots$ because there is no fourth possibility for the second choice).

Without loss of generality (Exercise 6.34) we assume that there exists a constant d such that all computations of M on an input w have a length of at most $d \cdot s(|w|)$. Hence, no computation of M on w uses more than $d \cdot s(|w|)$ space.

Now, we describe a $(k+2)$-tape-TM A that simulates all computations of M of length of at most $|\text{InConf}_M(n)|$.

For any input $w \in \Sigma^*$, A works as follows:

1. A writes $0^{s(|w|)}$ onto the $(k+2)$-th tape.
2. A writes $0^{d \cdot s(|w|)}$ onto the $(k+1)$-th tape and erases the content of the $(k+2)$-th tape.

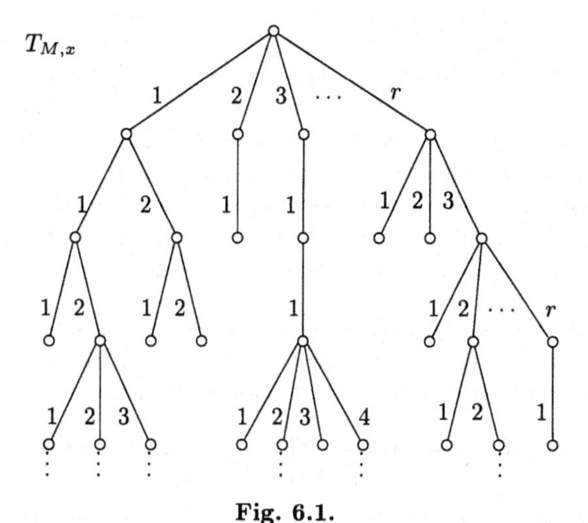

Fig. 6.1.

3. A consecutively generates all words $z \in \{1, 2, \ldots, r_M\}^*$ with a length of at most $d \cdot s(|w|)$ in the canonical order on the $(k+2)$-th tape. For each generated z it simulates the corresponding computation of M on w (if any) by its first k working tapes. If M has reached its accepting state in any of the simulated computations, then A accepts w. If all simulated computations of M are not accepting (the state q_{accept} was not reached in any of these computations), then M rejects w.

Clearly, $L(A) = L(M)$, because the length of any computation of M on inputs of length n is bounded by $d \cdot s(n)$, thus for any input w, A checks all computations of M on w. Since $\text{Space}_M(n) \leq \text{Time}_M(n)$, we have $\text{Space}_M(n) \leq d \cdot s(n)$, therefore A never uses more than $\text{Space}_M(n) \leq d \cdot s(n)$ squares on the first k working tapes. The $(k+1)$-th tape uses exactly $d \cdot s(n)$ squares for $0^{d \cdot s(|w|)}$. The same holds for the $(k+2)$-th tape that always contains one word from $\{1, 2, \ldots, r\}^*$ with a length of at most $d \cdot s(n)$ during the simulation. Thus

$$\text{Space}_A(n) \leq d \cdot s(n).$$

This completes the proof of Theorem 6.35. □

Corollary 6.36.

$$\text{NP} \subseteq \text{PSPACE}$$

Exercise 6.37. Analyze the time complexity of the $(k+2)$-tape-TM A described in the proof of Theorem 6.35.

Unfortunately, we are unaware of any more efficient deterministic simulation of nondeterministic algorithms other than to systematically simulate all computations of the nondeterministic algorithm on the given input. This was the case in the proof of Theorem 6.35 where we performed the breadth-first search in the computation tree as well as in Theorem 4.27 where the depth-first search in the computation tree was used. Both simulations take an exponential time in the time complexity of the nondeterministic machines, because the number of computations in a computation tree may be exponential in its depth. Currently, most researchers believe that there is no possibility of simulating nondeterminism within a polynomial increase of time complexity, but nobody has been able to prove the nonexistence of such a simulation.[7]

Exercise 6.38. Estimate the time complexity of the deterministic MTM from the proof of Theorem 4.27, i.e., estimate the time complexity of the breadth-first search deterministic simulation of nondeterminism.

The following theorem presents the most efficient known simulation of nondeterministic space by deterministic time. Observe that this simulation is as efficient as the simulation of deterministic space by deterministic time in Theorem 6.26.

Theorem 6.39.* *For any space-constructible function s with $s(n) \geq \log_2 n$,*

$$\text{NSPACE}(s(n)) \subseteq \bigcup_{c \in \mathbb{N}} \text{TIME}(c^{s(n)}).$$

Proof. Let M be a nondeterministic MTM with

$$L(M) = L \text{ and } \text{Space}_M(n) \in O(s(n)).$$

Without loss of generality we may assume:

(i) There exists a constant d such that for any input w, all computations of M on w have a space complexity of at most

$$d \cdot s(n).$$

(ii) For each input $w \in L(M)$, M has only one accepting configuration

$$(q_{\text{accept}}, w, 0, \lambda, \ldots, \lambda, 0),$$

i.e., before reaching q_{accept}, M erases the contents of all its working tapes and adjusts all heads on the left endmarker ¢.

Following (i) we know (Lemma 6.33) that there exists a constant c such that, for any input w, the number of all distinct configurations with w on the input tape is at most

$$|\text{InConf}(|w|)| \leq c^{s(|w|)}.$$

One can order all these configurations in a sequence

[7] This problem will be discussed in detail in the next two sections.

$$C_0, C_1, \ldots, C_{|\text{InConf}(|w|)|}$$

with respect to the canonical order. We shall construct a (deterministic) MTM A with $L(A) = L$.

For any input w, A computes as follows:

1. A constructs the adjacency matrix $M(w)$ of the directed graph $G(w)$, whose vertices are the $|\text{InConf}(|w|)|$ many configurations with w on the input tape and space complexity bounded by $d \cdot s(n)$. There is a directed edge from C_i to C_j if and only if $C_i \vdash_{\overline{M}} C_j$ (i.e., C_j can be reached from C_i in one step of M).
2. Let C_k be the only accepting configuration of M with w on the input tape. Let C_0 be the initial configuration of M on w. Clearly, M accepts w if and only if there is a directed path from C_0 to C_k in $G(w)$. Using a standard approach (Floyd algorithm, for instance) A verifies whether the vertex C_k is reachable from the vertex C_0. If C_k is reachable from C_0, then A accepts w. A rejects w otherwise.

Clearly, $L(A) = L(M)$. Now, we analyze the time complexity of A. To construct $M(w)$, A has to compute

$$|\text{InConf}(|w|)| \cdot |\text{InConf}(|w|)| \leq c^{d \cdot s(|w|)} \cdot c^{d \cdot s(|w|)} \leq c^{2d \cdot s(|w|)}$$

elements m_{ij} of the matrix $M(w)$. To determine m_{ij}, one has to generate C_i and C_j. Any configuration can be generated in time

$$d \cdot s(|w|) \cdot |\text{InConf}(|w|)| \leq c^{2d \cdot s(|w|)}.$$

Whether C_j can be reached from C_i in one step of M can be verified in time $2d \cdot s(|w|)$. Thus, the time complexity of part (i) of the computation of M is at most

$$c^{2d \cdot s(|w|)} \cdot (2c^{2d \cdot s(|w|)} + 2d \cdot s(|w|)) \leq c^{12d \cdot s(|w|)}.$$

The verification of the path existence from C_0 to C_k in $G(w)$ can be done within polynomial time with respect to $|\text{InConf}(|w|)|$. An MTM can perform this task in $O(|\text{InConf}(|w|)|^4)$ steps. Since

$$(c^{d \cdot s(|w|)})^4 = c^{4d \cdot s(|w|)},$$

it is obvious, that

$$\text{Time}_A(n) \in O(c^{12 \cdot d \cdot s(n)}).$$

□

Corollary 6.40.

NLOG $\subseteq$ P *and* NPSPACE $\subseteq$ EXPTIME.

A slightly more involved search[8] in the graph $G(w)$ of all possible configurations on w provides the following result.

Theorem 6.41.* Savitch's Theorem
Let s with $s(n) \geq \log_2 n$ be a space-constructible function. Then

$$\mathrm{NSPACE}(s(n)) \subseteq \mathrm{SPACE}(s(n)^2).$$

Corollary 6.42.

$$\mathrm{PSPACE} = \mathrm{NPSPACE}.$$

For all of the above-presented deterministic simulations of nondeterministic computations, it is still unknown whether a more efficient simulation exists. The following **fundamental complexity hierarchy** of sequential computation

$$\mathrm{DLOG} \subseteq \mathrm{NLOG} \subseteq \mathrm{P} \subseteq \mathrm{NP} \subseteq \mathrm{PSPACE} \subseteq \mathrm{EXPTIME}$$

is the consequence of the above simulation results. It is open whether each of these inclusions is proper or not. However, some proper inclusions must be present because DLOG $\subsetneq$ PSPACE and P $\subsetneq$ EXPTIME are direct consequences of the hierarchy theorems. The verification of proper inclusions in the fundamental complexity hierarchy has been the central open problem of theoretical computer science over the past 30 years.

6.5 The Class NP and Proof Verification

The central open problem of the fundamental complexity hierarchy is the relation between P and NP. The question of

$$\text{whether } \mathrm{P} = \mathrm{NP} \text{ or } \mathrm{P} \subsetneq \mathrm{NP}$$

is the most famous research problem of computer science and recently it was put among the most important open problems of mathematics. There are several reasons for this large interest. One relates polynomial time and so the class P to the practical solvability. On the other hand, we know more than 4000 algorithmic problems in NP such that there is no (deterministic) polynomial-time algorithm for any of them. One would like to know whether all these problems are in P or in NP$-$P (i.e., outside of P). Another reason for our interest in the study of the classes P and NP is related to formal proofs as the fundamental concept of mathematics. In some framework, the time complexity of deterministic computations corresponds to the complexity of algorithmically creating (finding) a mathematical proof for a given theorem, while the time complexity of nondeterministic computations corresponds to

[8] The search is executed without constructing $G(w)$, because this would require too much space.

the complexity of an algorithmic verification of the correctness of a given proof.

Therefore, the comparison of P and NP is equivalent to the question of

whether verifying a given formal proof is "easier" than creating it.

The main aim of this section is to show the relationship between the class NP and the algorithmical polynomial-time proof verification.

First, we outline the relationship between computations and proofs. Let C be an accepting computation of a TM on an input x. Without any doubt, one can view C as the proof of the claim "$x \in L(M)$". Analogously, a rejecting computation of a (deterministic) TM M on a word w is the proof of the claim "$x \notin L(M)$". This point of view is not too far from the classical view on mathematical proofs. Consider L to be a language that contains all correct theorems[9] of a mathematical theory. The proof[10] of "$x \in L(M)$" is nothing other than the proof of the truth (validity) of the assertion x, and the proof of "$x \notin L(M)$" is the proof of untruth of the assertion x. For instance, consider $L = \text{SAT}$, where

$$\text{SAT} = \{x \in (\Sigma_{\text{logic}})^* \mid x \text{ codes a satisfiable formula in CNF}\}.$$

Then, the claim "$\Phi \in \text{SAT}$" is equivalent to the claim "Φ is a satisfiable formula".

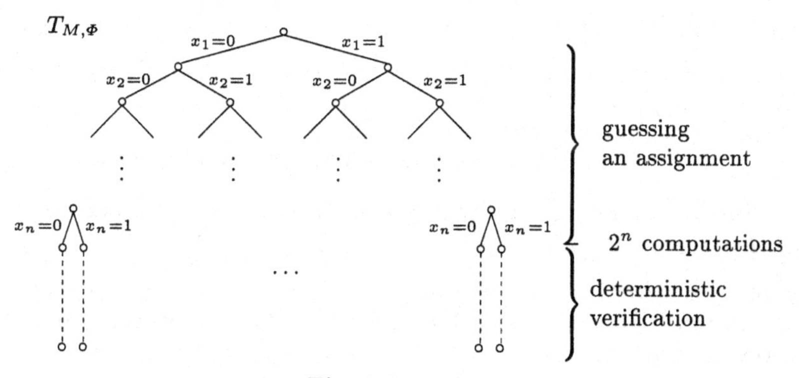

Fig. 6.2.

Now, let us search for a relation between nondeterministic computations and proof verification. A typical nondeterministic computation starts with a guess and continues with the verification of the correctness of this guess. The guess of a computation can be a proof of "$x \in L(M)$". We illustrate this consideration by the satisfiability problem. Consider the following NTM M

---

[9] More precisely, the representations of all theorems

[10] An accepting computation of M on x is a proof of "$x \in L(M)$".

with $L(M) = \text{SAT}$. For a formula Φ over n Boolean variables $x_1, \ldots, x_n$, M guesses an assignment $\alpha_1, \ldots, \alpha_n$ for the variables $x_1, \ldots, x_n$ in the first n steps of the computation. Then M computes the value $\Phi(\alpha_1, \ldots, \alpha_n)$ in order to verify whether the guessed assignment $\alpha_1, \ldots, \alpha_n$ satisfies the formula Φ (Figure 6.5). If $\alpha_1, \ldots, \alpha_n$ satisfies Φ, then it is obvious that the proof of the claim "Φ is satisfiable" can be efficiently created if the witness $\alpha_1, \ldots, \alpha_n$ is given. The proof is nothing other than the evaluation of Φ for the values $\alpha_1, \ldots, \alpha_n$. Therefore, $\alpha_1, \ldots, \alpha_n$ is called a certificate or a witness of the claim "Φ is satisfiable". The time complexity of M is the time complexity of guessing a witness plus the time complexity of the verification. Since the complexity of guessing is less than the input length, the complexity of M is asymptotically equal to the complexity of the verification.

Our aim is to show that any polynomial-time nondeterministic Turing machine accepting a language L can be transformed to an equivalent NTM that first nondeterministically guesses a candidate w for a proof (a witness) of "$x \in L$" and then deterministically verifies whether w is really a proof (witness) of the claim "$x \in L$". In this way one reduces the time complexity of nondeterministic algorithms to the time complexity of deterministic proof verification. To do this, we need the following formal concept.

Definition 6.43. *Let $L \subseteq \Sigma^*$ be a language and let $p : \mathbb{N} \to \mathbb{N}$ be a mapping. We say that a* MTM *(an algorithm) A working over inputs from $\Sigma^* \times (\Sigma_{\text{bool}})^*$ is a* **p-verifier for L**, *denoted by* $\mathbf{V(A) = L}$, *if the following three conditions hold:*

(i) $\text{Time}_A(w, x) \leq p(|w|)$ *for every input $(w, x) \in \Sigma^* \times (\Sigma_{\text{bool}})^*$.*
(ii) For every $w \in L$ there exists an $x \in (\Sigma_{\text{bool}})^*$, such that

$$|x| \leq p(|w|) \text{ and } (w, x) \in L(A) \text{ (i.e., } A \text{ accepts } (w, x)).$$

The word x is called the **proof** *or the* **witness** *of the claim "$w \in L$".*
(iii) For every $y \notin L$, $(y, z) \notin L(A)$ for all $z \in (\Sigma_{\text{bool}})^*$.

If $p(n) \in O(n^k)$ for some positive integer k, then we say that A is a **polynomial-time verifier**. *We define the* **class of polynomially verifiable languages** *as*

$$\mathbf{VP} = \{V(A) \mid A \text{ is a polynomial-time verifier}\}.$$

Note, that for a p-verifier A, $L(A)$, and $V(A)$ are different languages. Following Definition 6.43 we have

$$V(A) = \{w \in \Sigma^* \mid \text{ there exists an } x \text{ with } |x| \leq p(|w|), \text{ such that } (w, x) \in L(A)\}.$$

Thus, a verifier A for a language L is a deterministic algorithm that verifies for each input (w, x) whether x is a witness (proof) of the claim "$w \in L$". The word w belongs to $V(A)$ if there exists a proof x for "$w \in L$" with $|x| \leq p(|w|)$.

Example 6.44. A polynomial-time verifier A for SAT can work as follows. For any input (w, x), A first checks whether w is a code of a formula Φ_w in CNF. If not, A rejects the input. Otherwise, A estimates the number n of Boolean variables in Φ_w and checks whether the length of $x \in \{0,1\}^*$ is at least n. If $|x| < n$ then A rejects the input. Otherwise, $|x| \geq n$ and A considers the first n bits of x as an assignment to the Boolean variables of Φ_w. Obviously, A accepts (w, x) if and only if this assignment satisfies Φ_w. We observe that A is a polynomial-time verifier for SAT because for any w, there exists a witness x such that

$$|x| \leq |w| \text{ and } (w, x) \in L(A),$$

and one can efficiently[11] evaluate a Boolean formula in CNF for a given assignment of its variables.

Example 6.45. A k-clique of a graph G with n vertices, $k \leq n$, is a complete subgraph of k vertices on G. Let

$$\textbf{CLIQUE} = \{x\#y \mid x, y \in \{0,1\}^*, x \text{ codes a graph } G_x, \\ \text{that contains a } \mathit{Number}(y)\text{-clique}\}.$$

A polynomial-time verifier B for CLIQUE can work as follows. For any input (w, z), B checks whether $w = x\#y$, where x is a representation of a graph G_x and $y \in (\Sigma_{\text{bool}})^*$. If this is not true, B rejects (w, z). Otherwise, B estimates the number n of vertices of G_x. Let $v_1, \ldots, v_n$ be the vertices of G_x. Now, B verifies whether $\mathit{Number}(y) \leq n$ and $|z| \geq \lceil \log_2 n \rceil \cdot \mathit{Number}(y)$. If not, B rejects its input (w, z). Otherwise, B interprets the prefix of z of the length $\lceil \log_2 n \rceil \cdot \mathit{Number}(y)$ as a code of $\mathit{Number}(y)$ numbers from $\{1, 2, \ldots, n\}$. B verifies whether all these $\mathit{Number}(y)$ are pair-wise different. If not, B rejects the input. Let $i_1, i_2, \ldots, i_{\mathit{Number}(y)}$ be $\mathit{Number}(y)$ different positive integers. Then, B verifies whether the vertices $v_{i_1}, v_{i_2}, \ldots, v_{i_{\mathit{Number}(y)}}$ build a complete subgraph of G_x. Obviously, B accepts (w, z) iff this verification was successful.

Exercise 6.46. Design a polynomial-time verifier for the language

$$\text{COMPOSITE} = \{x \in (\Sigma_{\text{bool}})^* \mid \mathit{Number}(x) \text{ is composite}\}.$$

Exercise 6.47. Design a polynomial-time verifier and a polynomial-time NTM for the language HC (the problem of a Hamiltonian cycle from Example 2.35).

The following theorem shows that every polynomial-time NTM can be transformed into an equivalent[12] polynomial-time NTM that does all its nondeterministic decisions at the very beginning of its computations and then verifies the correctness of its decisions.

[11] A MTM can do it in quadratic time, but using a convenient data structure an algorithm can even do it in linear time.

[12] Nondeterministic Turing machines are called equivalent if they accept the same language

Theorem 6.48.

$$\mathrm{VP} = \mathrm{NP}.$$

Proof. We prove VP = NP by proving NP $\subseteq$ VP and VP $\subseteq$ NP.

(i) First, we show NP $\subseteq$ VP.
Let $L \in \mathrm{NP}$, $L \subseteq \Sigma^*$ for an alphabet Σ. Then there exists a polynomial-time NTM M with $L = L(M)$ and $\mathrm{Time}_M(n) \in O(n^k)$ for a positive integer k. Without loss of generality one may assume that, for any argument of its transition function, M has a choice from at most two possibilities. We describe a verifier A, that works for any input $(x, c) \in \Sigma^* \times (\Sigma_{\mathrm{bool}})^*$ as follows:
   a) A interprets c as a navigator for the simulation of nondeterministic choices of M. A simulates the work of M (step-by-step) on w. If M has a choice of two possibilities, then A takes the first one if the next bit of c is 0, and the second possibility if the next bit of x is 1. In this way A simulates exactly one of the computations of M on x.
   {This is the same simulation strategy as used in the proof of Theorem 6.35 where one simulates nondeterminism by the depth-first search in the computation trees.}
   b) If M still has a choice and A has used already all bits of c, then A halts and rejects the input (x, c).
   c) If A succeeds in simulating a complete computation of M on x, then A halts and accepts (x, c) if and only if M accepts x in the computation determined[13] by c.

Now, we have to show that A is a polynomial-time verifier with

$$L(A) = L(M).$$

If $x \in L(M)$, then a shortest accepting computation $C_{M,x}$ of M on x runs in time $O(|x|^k)$. Hence, there exists a witness (navigator) c that determines $C_{M,x}$ and $|c| \leq |C_{M,x}| \in O(|x|^k)$. Since A simulates the computation $C_{M,x}$ step-by-step, the computation of A on (x, c) runs in time $O(|x|^k)$.
If $x \notin L(M)$, then there does not exist any accepting computation of M on x, so A rejects the inputs (x, d) for all $d \in (\Sigma_{\mathrm{bool}})^*$.
We conclude that A is an $O(n^k)$-verifier[14] for $L(M)$.

(ii) We show VP $\subseteq$ NP.
Let $L \subseteq \Sigma^*$ (for an alphabet Σ) be a language from VP. Hence, there exists a polynomial-time verifier A such that $V(A) = L$. One can design a polynomial-time NTM M that simulates A as follows.
Input: A $x \in \Sigma^*$
Phase 1 M nondeterministically generates a word $c \in (\Sigma_{\mathrm{bool}})^*$.
Phase 2 M simulates step-by-step the work of A on (x, c).

[13] In the simulated computation
[14] That is, that $V(A) = L(M)$

Phase 3 M accepts x if A accepts (x, c), and M rejects x if A rejects (x, c).

Clearly, $L(M) = V(A)$ and

$$\mathrm{Time}_M(x) \leq 2 \cdot \mathrm{Time}_A(x, c)$$

for every $x \in L(M)$ and a shortest witness c of $x \in L(M)$. Hence, M is a polynomial-time NTM and hence $L \in \mathrm{NP}$.

□

Following Theorem 6.48 the class NP is the class of all languages L such that for every $x \in L$ there exits a witness (proof) c_x of the claim $x \in L$ with the following properties:

(i) c_x is of a length polynomial in the length of x.
(ii) One can verify whether c_x is a witness of $x \in L$ in polynomial time.

6.6 NP-Completeness

In contrast to the computability theory that provides a well-developed methodology for classifying problems into algorithmically solvable and algorithmically unsolvable, one has been unable to develop any successful mathematical method for classifying algorithmic problems with respect to practical solvability (to the membership to P) in the complexity theory. Sufficiently powerful techniques for proving lower bounds on the complexity of concrete problems are missing. The following fact shows how far we are from proving that a concrete problem from NP cannot be solved on polynomial time. The highest known lower bound on the time complexity of multitape Turing machines for solving a concrete problem from NP is the trivial lower bound[15] $\Omega(n)$ (i.e., we are unable to prove a lower bound $\Omega(n \cdot \log n)$ for a problem from NP), though the best known algorithms for thousands of problems in NP run in exponential time. Hence, our experience lets us believe that $\Omega(2^n)$ is a lower bound on the time complexity of many problems, but we are unable to prove a higher lower bound than $\Omega(n)$ for them.

To at least partially overcome the gap between the mathematical reality (the unsatisfiable state of the proof techniques development) and the belief in the existence of exponential lower bounds for some problems, some computer scientists proposed to develop a method for the classification of problems with respect to the class P (practical solvability) by allowing one to assume the validity of an unproved, but believable assertion. The consideration has resulted in the creation of the concept of NP-completeness that enables one to prove that some problems are not solvable in polynomial time under the

[15] The time one needs to read the whole input at least.

assumption that P is a proper subset of NP. The aim of this section is to introduce this concept.

First, let us discuss the credibility of the assumption P $\subsetneq$ NP. A theoretical argument for assuming P $\subsetneq$ NP has been explained in Section 6.5. It is not believed that proof verification is as hard as proof creation. Additionally, despite intensive efforts, no simulation technique of nondeterminism by determinism that would essentially differ from a systematic investigation of all nondeterministic computations on a given input has yet been discovered. Since the computation trees of nondeterministic algorithms may contain exponentially many different computations in the tree depth, and the computation tree depth is the time complexity of the nondeterministic algorithm, it seems that the exponential blow up of the time complexity by a deterministic simulation is unavoidable.

A practical reason for the assumption P $\subsetneq$ NP is based on our experience in designing algorithms for hard problems. We are aware of more than 3000 problems in NP, many of them have been investigated for 40 years, for which no deterministic polynomial-time algorithm is known. It is not very probable that this is only the consequence of our disability to find efficient algorithms for them. Even if this had been the case, for the current practice, the classes P and NP are different[16] because we do not have polynomial-time algorithms for numerous problems from NP.

Note that assuming P $\subsetneq$ NP is not the same as assuming the validity of the Church–Turing thesis. While the Church–Turing thesis is an axiom that cannot be proved, the truth or the untruth of P $\subsetneq$ NP may be provable and hence the assumption P $\subsetneq$ NP cannot be considered to be an axiom.

We assume P $\subsetneq$ NP for the rest of this section. How does this assumption help to prove results of the kind $L \notin$ P for concrete languages L? The idea is to specify a subset of the hardest problems in NP. This specification of hardness has to be done in such a way that the membership of a hard problem in P would directly imply P $=$ NP. Since we assume P $\subsetneq$ NP, none of the hardest problems in NP may be in P.

Similarly as in the complexity theory, we use the classical mathematical concept of reduction in order to specify the subclass of hard problems with respect to polynomial-time solvability. A language (decision problem) L is hard if the recognition of any language from NP (solving any decision problem from NP) can be efficiently reduced to recognizing L.

Definition 6.49. *Let $L_1 \subseteq \Sigma_1^*$ and $L_2 \subseteq \Sigma_2^*$ be some languages. We say that*

$$L_1 \textit{ is } \textbf{polynomial-time reducible} \textit{ to } L_2,\ \boldsymbol{L_1 \leq_p L_2},$$

if there exists a polynomial-time TM *(algorithm) A (Fig. 6.6) that computes a mapping from Σ_1^* to Σ_2^* such that, for every $x \in \Sigma_1^*$,*

[16] Thus, if one proves the nonexistence of a polynomial-time algorithm for a problem under the assumption P $\subsetneq$ NP, then it really means that this problem is hard for the current practice.

$$x \in L_1 \Leftrightarrow A(x) \in L_2.$$

A is called the **polynomial-time reduction** *from* L_1 *to* L_2.

We observe that the reduction $\leq_p$ can be obtained from the reduction $\leq_{\mathrm{m}}$ by an additional requirement on the efficiency of the reduction (Fig. 6.6). Again,

$$L_1 \leq_p L_2$$

means that L_2 is at least as hard as L_1, but the hardness is considered with respect to polynomial-time solvability instead of with respect to pure solvability.

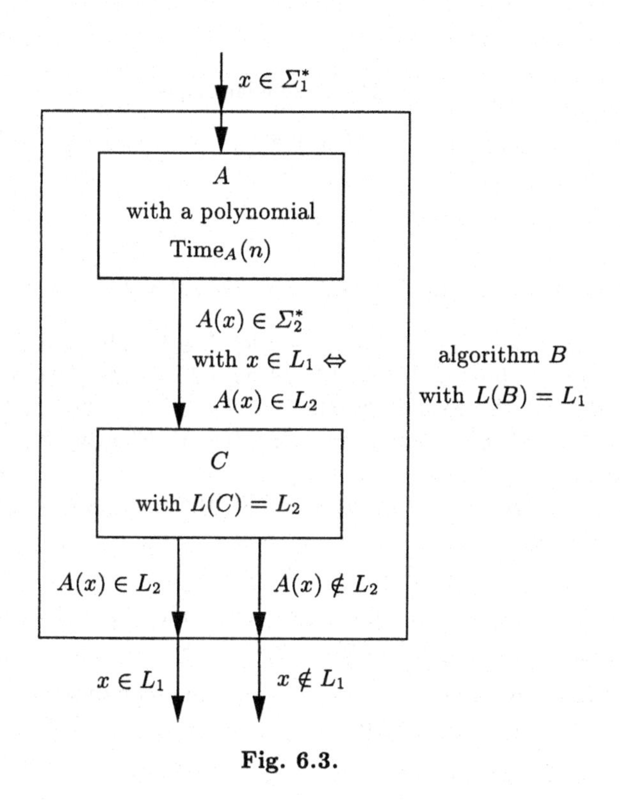

Fig. 6.3.

Definition 6.50. *A language L is* **NP-hard**, *if , for every* $L' \in \mathrm{NP}$,

$$L' \leq_p L.$$

A language L is **NP-complete**, *if*

(i) $L \in \mathrm{NP}$, *and*
(ii) L *is* NP*-hard.*

Now, the set of NP-complete languages is considered to be the subclass of the hardest problems in NP we have been searching for. The following lemma shows that our specification of hardness satisfies the aimed property of hard problems, namely that the intersection of P and the set of hard problems is empty (Fig. 6.6) if P $\neq$ NP.

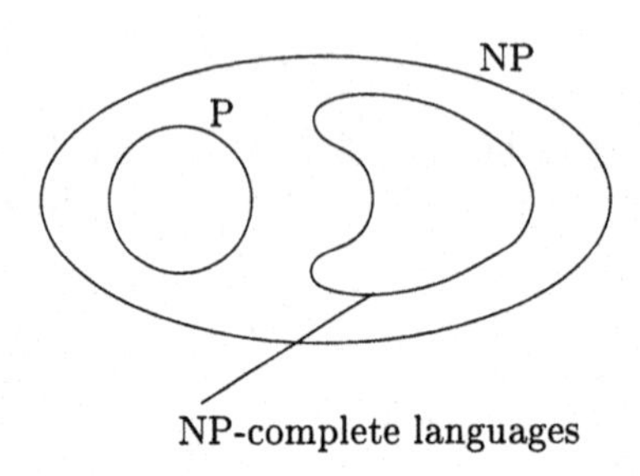

Fig. 6.4.

Lemma 6.51. *If* $L \in$ P *and* L *is* NP*-hard, then* P = NP.

Proof. Let L be a NP-hard language and let $L \in$ P. The fact $L \in$ P implies the existence of a polynomial-time TM M with $L = L(M)$. We prove that for every $U \in$ NP, $U \subseteq \Sigma^*$ for an alphabet Σ, there is a polynomial-time MTM (algorithm) A_U with $L(A_U) = U$, i.e., that $U \in$ P. Clearly, this would imply P = NP.

Since $U \leq_p L$ for each language $U \in$ NP, there exists a polynomial-time TM B_U, such that

$$x \in U \Leftrightarrow B_U(x) \in L.$$

Now, we describe the work of a polynomial-time MTM A_U with $L(A_U) = U$. For any input $x \in \Sigma^*$, $A_U$computes as follows.

1. A_U simulates the computation of B_U on x and computes $B_U(x)$.
2. A_U simulates the work of M on $B_U(x)$. A_U accepts x iff M accepts $B_U(x)$.

Since $x \in U \Leftrightarrow B_U(x) \in L$, we have $L(A_U) = U$. Since

(i) $\text{Time}_{A_U}(x) = \text{Time}_{B_U}(x) + \text{Time}_M(B_U(x))$,
(ii) $|B_U(x)|$ is polynomial in $|x|$, and
(iii) both B_U and M work in polynomial time,

we obtain that A_U also works in polynomial time. □

Now, we have the desired specification of the hardest problems in NP, provided that the set of the NP-complete languages is not empty. The next theorem shows this danger does not exist, because SAT is NP-complete[17] and so the concept of NP-completeness is reasonable.

The NP-hardness of SAT says that the language of Boolean formulae is powerful enough to express any decision problem in NP. More precisely, it means that, for any language $L \in \mathrm{NP}$, one can express the question

Is x in L?

as the question

Is a specific Boolean formula $\Phi_{x,L}$ satisfiable?

The expressional power of the language of Boolean formulae should not come as a surprise because one can use formulae to describe any text and texts can be used to represent arbitrary objects such as theorems, proofs, computations, graphs, etc. We present a simple example to guide this intuition.

We want to construct a Boolean formula in CNF that can be used to describe matrices of the size 3×3 over $\{-1, 0, 1\}$. A formula is satisfiable if and only if the corresponding matrix has exactly one element 1 in every row and every column. To do that, we consider 27 variables $x_{i,j,k}$ for all $i, j \in \{1, 2, 3\}$ and all $k \in \{-1, 0, 1\}$. The following meaning of the assignments of values to these variables enables us to describe any 3×3 matrix $A = (a_{ij})_{i,j=1,2,3}$.

$$\begin{aligned} x_{i,j,1} &= 1 \Leftrightarrow a_{ij} = 1 \\ x_{i,j,0} &= 1 \Leftrightarrow a_{ij} = 0 \\ x_{i,j,-1} &= 1 \Leftrightarrow a_{ij} = -1. \end{aligned}$$

Using the above interpretation, one can unambiguously describe the matrix

$$\begin{pmatrix} 1 & 0 & 0 \\ -1 & 0 & 1 \\ 0 & 1 & 0 \end{pmatrix}$$

by the following assignment of values to the 27 variables.

$$\begin{aligned} &x_{1,1,1} = 1, x_{1,1,0} = 0, x_{1,1,-1} = 0, \\ &x_{1,2,1} = 0, x_{1,2,0} = 1, x_{1,2,-1} = 0, \\ &x_{1,3,1} = 0, x_{1,3,0} = 1, x_{1,3,-1} = 0, \\ &x_{2,1,1} = 0, x_{2,1,0} = 0, x_{2,1,-1} = 1, \\ &x_{2,2,1} = 0, x_{2,2,0} = 1, x_{2,2,-1} = 0, \\ &x_{2,3,1} = 1, x_{2,3,0} = 0, x_{2,3,-1} = 0, \\ &x_{3,1,1} = 0, x_{3,1,0} = 1, x_{3,1,-1} = 0, \\ &x_{3,2,1} = 1, x_{3,2,0} = 0, x_{3,2,-1} = 0, \\ &x_{3,3,1} = 0, x_{3,3,0} = 1, x_{3,3,-1} = 0. \end{aligned}$$

[17] Remember that we have already proved that SAT $\in$ NP and so it remains only to prove that SAT is NP-hard.

Note, that there exist assignments that do not represent any matrix. For instance, the assignment

$$x_{1,1,1} = 1 = x_{1,1,0}$$

has the interpretation that the element a_{11} of a matrix takes both values 1 and 0, which is impossible. To exclude such possibilities, we start by constructing a Boolean formula that can be satisfied only by assignments that correspond to a representation of a 3×3 matrix over $\{-1, 0, 1\}$ (i.e., only by assignments that determine exactly one value from $\{-1, 0, 1\}$ to every position of the matrix). For all $i, j \in \{1, 2, 3\}$, the formula

$$F_{i,j} = (x_{i,j,1} \vee x_{i,j,0} \vee x_{i,j,-1}) \wedge \\ (\bar{x}_{i,j,1} \vee \bar{x}_{i,j,0}) \wedge (\bar{x}_{i,j,1} \vee \bar{x}_{i,j,-1}) \wedge (\bar{x}_{i,j,0} \vee \bar{x}_{i,j,-1})$$

guarantees that exactly one of the variables

$$x_{i,j,1}, x_{i,j,0}, x_{i,j,-1}$$

takes[18] the value 1, so the content of the position (i, j) of the matrix is unambiguously determined. Using this approach, every assignment satisfying the formula

$$\Phi = \bigwedge_{1 \leq i,j \leq 3} F_{i,j}$$

unambiguously determines a 3×3 matrix over $\{-1, 1, 0\}$.

For $i = 1, 2, 3$ the formula

$$Z_i = (x_{i,1,1} \vee x_{i,2,1} \vee x_{i,3,1}) \wedge \\ (\bar{x}_{i,1,1} \vee \bar{x}_{i,2,1}) \wedge (\bar{x}_{i,1,1} \vee \bar{x}_{i,3,1}) \wedge (\bar{x}_{i,2,1} \vee \bar{x}_{i,3,1})$$

guarantees that the i-th row contains exactly one 1. Analogously, the formula

$$S_j = (x_{1,j,1} \vee x_{2,j,1} \vee x_{3,j,1}) \wedge \\ (\bar{x}_{1,j,1} \vee \bar{x}_{2,j,1}) \wedge (\bar{x}_{1,j,1} \vee \bar{x}_{3,j,1}) \wedge (\bar{x}_{2,j,1} \vee \bar{x}_{3,j,1})$$

guarantees that the j-th column contains exactly one 1 for $j = 1, 2, 3$. Hence,

$$\Phi \wedge \bigwedge_{i=1,2,3} Z_i \wedge \bigwedge_{j=1,2,3} S_j$$

is the Boolean formula in CNF we are searching for.

Exercise 6.52. Give a set of Boolean variables that is sufficient for describing any state (situation) in chess. Describe the construction of a formula over these variables that takes the value 1 if and only if there are no chessmen on the chessboard, other than 8 queens and no queen attacks any other.

[18] The elementary disjunction $x_{i,j,1} \vee x_{i,j,0} \vee x_{i,j,-1}$ assures that at least one of the forthcoming variables is true. The elementary disjunction $(\bar{x}_{i,j,1} \vee \bar{x}_{i,j,0})$ assures that at least one of the variables $x_{i,j,1}$ and $x_{i,j,0}$ takes the value 0.

Using the strategy introduced above one can describe texts on a sheet. The sheet can be partitioned into $n \times m$ squares that form a matrix (a two-dimensional field) over symbols from Σ_{keyboard}. Then, $n \cdot m \cdot |\Sigma_{\text{keyboard}}|$ Boolean variables are sufficient to express any text on this sheet.

Exercise 6.53. Give a set of Boolean variables that suffices to describe any text on the sheet of the size 33×77. Describe the construction of a formula that has exactly one satisfiable assignment and this assignment corresponds to the text of the second page of the textbook "The Design and Analysis of Computer Algorithms" by Aho, Hopcroft and Ullman.

We see that one can describe (express) arbitrary texts by formulae. Since a configuration of a TM is also a text (word) over an alphabet, one can use formulae to describe configurations. The kernel of the proof of the following theorem is that it is even possible to use the satisfiability of Boolean formulae for describing semantic relationships of different parts of a text such as a configuration is reachable from another configuration in one computation step. The ability to express the relation of a computation step allows one to describe any consistent computation. Another crucial point of the following proof is that such formulae can be constructed by an algorithm and that their length is polynomial in the length of the described computation.[19]

Theorem 6.54.* Cook's Theorem
SAT *is* NP-*complete.*

Proof. We have already used Example 6.1 (see also Figure 6.5) to show that SAT belongs to the class VP, which is equal to the class NP.

It remains to show that all languages in NP are polynomial-time reducible to SAT. Following the definition of the class NP, for every language $L \in \text{NP}$, there is an NTM M with

$$L(M) = L \text{ and } \text{Time}_M(n) \in O(n^c)$$

for a $c \in N$. We view M as a finite representation of L and hence consider M as an input part of the polynomial-time reduction to be designed. Thus, we aim to show

"*For any polynomial-time* NTM M, $L(M) \leq_p \text{SAT}$."

Let $M = (Q, \Sigma, \Gamma, \delta, q_0, q_{\text{accept}}, q_{\text{reject}})$ be an arbitrary NTM with

$$\text{Time}_M(n) \leq p(n)$$

for a polynomial p. Let $Q = \{q_0, q_1, \ldots, q_{s-1}, q_s\}$, where $q_{s-1} = q_{\text{reject}}$ and $q_s = q_{\text{accept}}$, and let $\Gamma = \{X_1, \ldots, X_m\}$, with $X_m = \sqcup$.

[19] I.e., if the computation is polynomial in the length of the input, then the length of the formula is polynomial in the length of the input too.

We design a polynomial reduction $B_M : \Sigma^* \to (\Sigma_{\text{logic}})^*$, such that for all $x \in \Sigma^*$:

$$x \in L(M) \Leftrightarrow B_M(x) \in \text{SAT}$$

Let w be an arbitrary word from Σ^*. B_M has to construct a formula $B_M(w)$ such that

$$w \in L(M) \Leftrightarrow B_M(w) \text{ is satisfiable.}$$

We construct the formula $B_M(w)$ in such a way that one can use different value assignments to the variables of $B_M(w)$ in order to describe all possibilities (computations) of how M can act on w. The basic idea is to choose the meaning of the Boolean variables in such a way that one can describe any configuration of M at any time (after any number of computation steps when starting from the initial configuration on w). Remember that

$$\text{Space}_M(|w|) \leq \text{Time}_M(|w|) \leq p(|w|),$$

so any configuration can be described by a word of a length[20] of at most $p(|w|)+1$. Note, that for a fixed input w, $p(|w|)+2$ is a fixed positive integer. To simplify the description of configurations we represent any configuration

$$(\text{¢}Y_1Y_2\ldots Y_{i-1}qY_i\ldots Y_d)$$

for a $d \leq p(|w|)$ as

$$(\text{¢}Y_1Y_2\ldots Y_{i-1}qY_i\ldots Y_dY_{d+1}\ldots Y_{p(|w|)}),$$

where $Y_{d+1} = Y_{d+2} = \ldots = Y_{p(|w|)} = \sqcup$.

In this way the representations of all configurations have the same length $p(|w|)+1$. To simplify searching for an accepting configuration, we extend the definition of the transition function δ by

$$\delta(q_{\text{accept}}, X) = (q_{\text{accept}}, X, N)$$

for any $X \in \Gamma$, which does not change $L(M)$. Thus, when M has reached the state q_{accept}, it remains in q_{accept} forever without changing any part of the current configuration. Now, to recognize whether w is in $L(M)$, it suffices to search for q_{accept} only in configurations reachable in exactly $p(|w|)$ steps of M.

The formula $B_M(w)$ will be constructed from the following variable classes.

- $\boldsymbol{C\langle i,j,t\rangle}$ for $0 \leq i \leq p(|w|)$, $1 \leq j \leq m$, $0 \leq t \leq p(|w|)$.
  The meaning of $C\langle i,j,t\rangle$ is as follows:

  $C\langle i,j,t\rangle = 1 \Leftrightarrow$ the i-th cell of the tape of M contains the symbol $X_j \in \Gamma$ in time t (after t computation steps from the initial configuration of M on w).

[20] The representation of a configuration involves the content of tape cells on the positions $0, 1, \ldots, p(|w|)$ and the state.

Note, that the number of such variables is exactly

$$m \cdot ((p(|w|) + 1)^2 \in O((p(|w|))^2).$$

- $\boldsymbol{S\langle k, t\rangle}$ for $0 \le k \le s$, $0 \le t \le p(|w|)$.
  The meaning of the Boolean variable $S\langle k, t\rangle$ is:
  $S\langle k, t\rangle = 1 \Leftrightarrow$ The NTM M is in the state q_k in time t.
  The number of such variables is

$$(s + 1) \cdot (p(|w|) + 1) \in O(p(|w|)).$$

- $\boldsymbol{H\langle i, t\rangle}$ for $0 \le i \le p(|w|)$, $0 \le t \le p(|w|)$.
  The meaning of $H\langle i, t\rangle$ is:

  $H\langle i, t\rangle = 1 \Leftrightarrow$ The head of M is on the i-th position of the tape in time t.

  There are exactly

$$(p(|w|) + 1)^2 \in O((p(|w|))^2)$$

  such variables.

We observe that one can describe an arbitrary configuration by choosing an appropriate value for all variables with a fixed parameter t. For instance, the following configuration

$$(X_{j_0} X_{j_1} \ldots X_{j_{i-1}} q_r X_{j_i} \ldots X_{j_{p(|w|)}})$$

can be described by

- $C\langle 0, j_0, t\rangle = C\langle 1, j_1, t\rangle = \ldots = C\langle p(|w|), j_{p(|w|)}, t\rangle = 1$ and $C\langle k, l, t\rangle = 0$ for all remaining variables from this class,
- $H\langle i, t\rangle = 1$ and $H\langle j, t\rangle = 0$ for all $j \in \{0, 1, \ldots, p(|w|)\}, j \neq i$, and
- $S\langle r, t\rangle = 1$ and $S\langle l, t\rangle = 0$ for all $l \in \{0, 1, \ldots, s\}, l \neq r$.

Thus, by fixing one configuration for any possible time t, one can describe any computation of M on w of the length $p(|w|)$ by assigning appropriate Boolean values to the variables. It is also important to observe that there are assignments to the variables that do not have any interpretation on the level of computations of M on w. For instance,

$$S\langle 1, 3\rangle = S\langle 3, 3\rangle = S\langle 7, 3\rangle = 1 \text{ and } C\langle 2, 1, 3\rangle = C\langle 2, 2, 3\rangle = 1$$

would mean that after 3 computation steps M is at once in the states q_1, q_3, and q_7 and the second position of the tape contains the symbol X_1 as well as the symbol X_2. Clearly, this assignment to variables for $t = 3$ cannot correspond to any configuration.

To explain the situation in a transparent way, consider a sheet of the size $(p(|w|) + 2) \times (p(|w|) + 1)$. An assignment to variables can be used to fix a symbol for any position (i, j), $0 \le i, j \le p(|w|)$. The result of this assignment may be nonsense, but there are also assignments such that every row of the sheet corresponds to a configuration and the sequence of the rows corresponds to a computation. This transparent view of what we are aiming for is outlined in Fig. 6.6.

Now, our task is to construct a formula $B_M(w)$ in CNF over the variables $C\langle i,j,t\rangle$, $S\langle k,t\rangle$ and $H\langle i,t\rangle$, such that

$$B_M(w) \textit{ is satisfiable} \Leftrightarrow \text{there exists an accepting computation of } M \text{ on } w.$$

The algorithm B_M constructs the formula

$$B_M(x) = A \wedge B \wedge C \wedge D \wedge E \wedge F \wedge G$$

in CNF in the following seven steps (phases). To explain the concept in a transparent way, we first describe the meaning of the particular formulae A, B, C, D, E, F, and G.

A: A has to guarantee that, at any time t, the head is adjusted exactly on one cell (position) of the tape.
More precisely, the meaning is that A should be satisfied by an assignment if and only if the value 1 is assigned to exactly one of the variables $H\langle i,t\rangle$ for any fixed t.

B: B takes the value 1 if, for any time t, there is exactly one symbol from Γ on any position of the tape.

C: C takes the value 1 if, for any time t, M is exactly in one state.

D: D assures that only the symbol visited by the head may be changed in any step from a configuration to the next.

E: E assures that for any computation step, the change of the state, the movement of the head and the exchange of a symbol on the tape agree with a possible activity of M in the given configuration (i.e., agree with δ).

F: F assures that the variables with $t = 0$ determine the initial configuration of M on w.

G: G assures that the last $(((p(|w|) + 1)$-th) configuration is an accepting configuration.

We see that the satisfiability of $A \wedge B \wedge C$ guarantees that every row of our sheet in Fig. 6.6 contains a configuration. The part $D \wedge E$ of the formula $B_M(x)$ has additionally to assure that the context of the sheet corresponds to a computation of M. F guarantees that this computation is a computation of M on w and G guarantees that this computation reaches the state q_{accept}.

In the following construction we often need a formula over several variables that takes the value 1 if and only if exactly one of its variables takes the value 1. Let $x_1, x_2, \ldots, x_n$ be Boolean variables. The following formula has the above-formulated property with respect to $x_1, x_2, \ldots, x_n$.

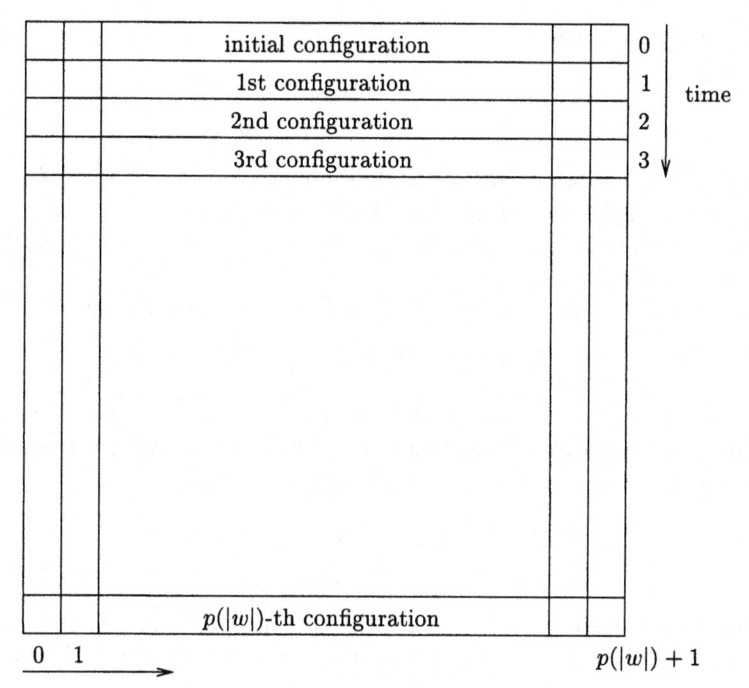

Fig. 6.5.

$$U(x_1, x_2, \dots, x_n) = (x_1 \vee x_2 \vee \dots \vee x_n) \wedge (\bigwedge_{\substack{1 \leq i,j \leq n \\ i \neq j}} (\overline{x_i} \vee \overline{x_j})).$$

The first part $x_1 \vee x_2 \vee \dots \vee x_n$ of $U(x_1, x_2, \dots, x_n)$ guarantees that the value 1 has to be assigned to at least one of the variables $x_1, x_2, \dots x_n$ in order to satisfy $U(x_1, x_2, \dots, x_n)$. Since the second part of $U(x_1, x_2, \dots, x_n)$ contains elementary clauses $\overline{x_i} \vee \overline{x_j}$ for all $i, j \in \{1, \dots, n\}, i \neq j$, one may not assign the value 1 to both variables x_i and x_j. Hence, any assignment satisfying the second part of the formula $U(x_1, x_2, \dots, x_n)$ contains at most one value 1. Note, that the length of $U(x_1, x_2, \dots, x_n)$ is quadratic in the number n of variables.

Now we construct the formulae A, B, C, D, E, F, and G.

(a) For each $t \in \{0, 1, 2, \dots, p(|w|)\}$ we construct

$$A_t = U(H\langle 0, t\rangle, H\langle 1, t\rangle, \dots, H\langle p(|w|), t\rangle).$$

The formula A_t is satisfied only if the head is adjusted on exactly one position $i \in \{0, 1, \dots, p(|w|)\}$ of the tape in time t.

To satisfy the formula

$$A = A_0 \wedge A_1 \wedge \ldots \wedge A_{p(|w|)} = \bigwedge_{0 \le i \le p(|w|)} A_i$$

any assignment has to guarantee that, for any time $t \in \{0,1,\ldots,p(|w|)\}$, the head is adjusted on exactly one position of the tape.
The number of literals in A is in $O((p(|w|))^3)$ because the number of literals in A_t is quadratic in $p(|w|)+1$.

(b) For all $i \in \{0,1,2,\ldots,p(|w|)\}$, $t \in \{0,1,\ldots,p(|w|)\}$ we define

$$B_{i,t} = U(C\langle i,1,t\rangle, C\langle i,2,t\rangle, \ldots, C\langle i,m,t\rangle).$$

The formula $B_{i,t}$ is satisfied if the i-th tape cell contains exactly one symbol after t computation steps. Since $|\Gamma| = m$ is a constant, the number of literals in $B_{i,t}$ is in $O(1)$, too. Satisfying the formula

$$B = \bigwedge_{0 \le i,t \le p(|w|)} B_{i,t}$$

one gets the assurance that all cells of the tape contain exactly one symbol at any time $t \in \{0,1,\ldots,p(|w|)\}$. Obviously, the number of literals in B is in $O((p(|w|))^2)$.

(c) For all $t \in \{0,1,\ldots,p(|w|)\}$ we define

$$C_t = U(S\langle 0,t\rangle, S\langle 1,t\rangle, \ldots, S\langle s,t\rangle).$$

If an assignment to the variables $S\langle 0,t\rangle, \ldots, S\langle s,t\rangle$ satisfies the formula C_t, then M is exactly in one state in time t. Since $|Q| = s+1$ is a constant, the number of literals in C_t in in $O(1)$. Clearly

$$C = \bigwedge_{0 \le t \le p(|w|)} C_t,$$

assures that M is in one state at each time. The number of literals in C is in $O(p(|w|))$.

(d) The formula

$$D_{i,j,t} = (C\langle i,j,t\rangle \leftrightarrow C\langle i,j,t+1\rangle) \vee H\langle i,t\rangle$$

for $0 \le i \le p(|w|), 1 \le j \le m, 0 \le t \le p(|w|)-1$ says that the only symbol that may be changed on the tape in the computation step $t+1$ is the symbol on which the head is adjusted (more precisely, if $H\langle i,t\rangle = 0$, the symbol on the i-th tape cell may not be rewritten in the next computation step). Obviously, $D_{i,j,t}$ can be transformed[21] in CNF in such a way that its length remains in $O(1)$.

[21] The formula $x \leftrightarrow y$ is equivalent to the formula $(\overline{x} \vee y) \wedge (x \vee \overline{y})$. Hence, $D_{i,j,t} \leftrightarrow (\overline{C\langle i,j,t\rangle} \vee C\langle i,j,t+1\rangle \vee H\langle i,t\rangle) \wedge (C\langle i,j,t\rangle \vee \overline{C\langle i,j,t+1\rangle} \vee H\langle i,t\rangle)$.

Hence, the formula we are searching for is

$$D = \bigwedge_{\substack{0 \le i \le p(|w|) \\ 1 \le j \le m \\ 0 \le t \le p(|w|)-1}} D_{i,j,t}$$

and D contains $O((p(|w|))^2)$ literals.[22]

(e) For all $i \in \{0, 1, 2, \ldots, p(|w|)\}$, $j \in \{1, \ldots, m\}$, $t \in \{0, 1, \ldots, p(|w|)\}$ and $k \in \{0, 1, \ldots, s\}$, we consider the formula

$$\begin{aligned} E_{i,j,k,t} = {} & \overline{C\langle i,j,t\rangle} \vee \overline{H\langle i,t\rangle} \vee \overline{S\langle k,t\rangle} \vee \\ & \bigvee_l (C\langle i, j_l, t+1\rangle \wedge S\langle k_l, t+1\rangle \wedge H\langle i_l, t+1\rangle) \end{aligned}$$

where l runs over all possible actions of the NTM M for the argument (q_k, X_j) with

$$(q_{k_l}, X_{j_l}, z_l) \in \delta(q_k, X_j), z_l \in \{L, R, N\} \text{ and}$$

$$i_l = i + \varphi(z_l), \varphi(L) = -1, \varphi(R) = 1, \varphi(N) = 0.$$

$E_{i,j,k,t}$ can be treated as a disjunction of the following four conditions:

- $\overline{C\langle i,j,t\rangle}$, i.e., the i-th position of the tape does not contain X_j in time t.
- $\overline{H\langle i,t\rangle}$, i.e., the head is not on the i-th position in time t.
- $\overline{S\langle k,t\rangle}$, i.e., M is not in the state q_k in time t.
- The change of the t-th configuration corresponds to a possible action for the argument (q_k, X_j) and the head position i.

Now, the idea of the construction of $E_{i,j,k,t}$ is obvious. If none of the first three constraints is fulfilled, then (q_k, X_j) is the actual argument for the $(t+1)$-th step and the head is adjusted on the i-th position of the tape. In such a case the changes have to follow the transition function δ for the argument (q_k, X_j). If one chooses the l-th possible action (q_{k_l}, X_{j_l}, z_l) for the argument (q_k, X_j), then X_j has to be exchanged by X_{j_l} in the i-th cell. M has to move to the state q_{k_l} and the head has to move with respect to z_l.

Since l is a constant, $E_{i,j,k,t}$ contains $O(1)$ literals, even if transformed to CNF. Hence,

$$E = \bigwedge_{\substack{0 \le i,t \le p(|w|) \\ 1 \le j \le m \\ 0 \le k \le s}} E_{i,j,k,t}$$

and E consists of $O((p(|w|))^2)$ literals.

(f) The initial configuration of M on w contains ¢w on the tape, the head has to be adjusted on the position 0, and M has to be in the state q_0. Let $w = X_{j_1} X_{j_2} \ldots X_{j_n}$ for some $j_r \in \{1, 2, \ldots, m\}$, $n \in \mathbb{N}$ and let $X_1 = ¢$. Then,

[22] Note that m is a constant.

the requirement to start the computation with the initial configuration of M on w in time $t = 0$ can be formulated by the following formula:

$$F = S\langle 0,0\rangle \wedge H\langle 0,0\rangle \wedge C\langle 0,1,0\rangle \wedge \bigwedge_{1\le r\le n} C\langle r, j_r, 0\rangle \wedge \bigwedge_{n+1\le d\le p(|w|)} C\langle d, m, 0\rangle$$

The number of literals in F is in $O(p(|w|))$ and F is in CNF.

(g) The simple formula

$$G = S\langle s, p(|w|)\rangle$$

assures that the last ($p(|w|)$-th) configuration of the computation contains the state q_{accept}.

Following the construction of the formula $B_M(w)$ we see that

(i) every assignment satisfying $B_M(w)$ corresponds to an accepting computation of M on w and so $B_M(w)$ is satisfiable iff there exists an accepting computation of M on w, and
(ii) the formula $B_M(w)$ can be algorithmically created from the data M, w and $p(|w|)$ in a time that is linear in the length of $B_M(w)$.

The number of literals in $B_M(w)$ is $O((p(|w|))^3)$. If one represents $B_M(w)$ over the alphabet Σ_{logic}, then every variable has to be represented in binary form. Since the number of variables is in $O((p(|w|))^2)$, each variable can be coded by $O(\log_2(|w|))$ bits. Hence, the length of $B_M(w)$ and so the time complexity of the reduction B_M are in

$$O((p(|w|))^3 \cdot \log_2(|w|)).$$

Thus B_M is a polynomial-time reduction from $L(M)$ to SAT. □

The proof of Theorem 6.54 shows that all languages (decision problems) from NP are polynomial-time reducible to SAT. The main consequence is that every instance of any decision problem in NP can be represented as the problem of satisfiability of a Boolean formula. A possible interpretation of this fact is that the language of Boolean formulae is powerful enough to describe any problem in NP.

The NP-completeness of SAT is the starting point[23] for classifying decision problems with respect to their membership to P. To prove the NP-completeness of other languages we use the method of reduction that is based on the following observation.

Lemma 6.55. *Let L_1 and L_2 be two languages. If $L_1 \le_p L_2$ and L_1 is* NP-*hard, then L_2 is* NP-*hard too.*

[23] SAT has in the complexity theory the same role as L_{diag} in the computability theory.

Exercise 6.56. Prove Lemma 6.55.

We use Lemma 6.55 for proving NP-completeness of some other languages from NP in the following. Our first aim is to show that the language of graphs (the language of relations) is also powerful enough to describe any problem from NP. To present the polynomial-time reductions in a transparent way we shall work directly with the terms graph and formula instead of using the words representing graph and formula. Thus,

$$\begin{aligned}
\text{SAT} &= \{\Phi \mid \Phi \text{ is a satisfiable formula in CNF}\},\\
\text{3SAT} &= \{\Phi \mid \Phi \text{ is a satisfiable formula in 3CNF}\},\\
\text{CLIQUE} &= \{(G,k) \mid G \text{ is a graph that contains,}\\
&\qquad \text{a } k\text{-clique}\}, \text{ and}\\
\text{VC} &= \{(G,k) \mid G \text{ is a graph that has a}\\
&\qquad \text{vertex-cover of size } k\},
\end{aligned}$$

where a vertex-cover of a graph $G = (V, E)$ is any set $U \subseteq V$ such that every edge in E has at least one endpoint in U.

Let Φ be a formula and let φ be an assignment to the variables of Φ. In the following we denote the truth value of Φ for the assignment φ by $\varphi(\Phi)$. Hence, Φ is satisfiable iff there exists as assignment φ with $\varphi(\Phi) = 1$.

Lemma 6.57.

$$\text{SAT} \leq_p \text{CLIQUE}.$$

Proof. Let

$$\Phi = F_1 \wedge F_2 \wedge \ \ldots \ \wedge F_m$$

be a formula in CNF, where

$$F_i = (l_{i1} \vee l_{i2} \vee \cdots \vee l_{ik_i}), k_i \in \mathbb{N} - \{0\}$$

for $i = 1, 2, \ldots, m$.

We construct an instance (G, k) of the clique problem, such that

$$\Phi \in \text{SAT} \Leftrightarrow (G,k) \in \text{CLIQUE}.$$

We set

$k = m$
$G = (V, E)$, where
$V = \{[i,j] \mid 1 \leq i \leq m, 1 \leq j \leq k_i\}$, i.e., we take a vertex for each occurrence of a literal in Φ,
$E = \{\{[i,j],[r,s]\} \mid$ for all $[i,j],[r,s] \in V$, with $i \neq r$ and $l_{ij} \neq \bar{l}_{rs}\}$, i.e., an edge $\{u, v\}$ connects vertices that correspond to literals from different clauses, if the literal of u is not the negation of the literal of v.

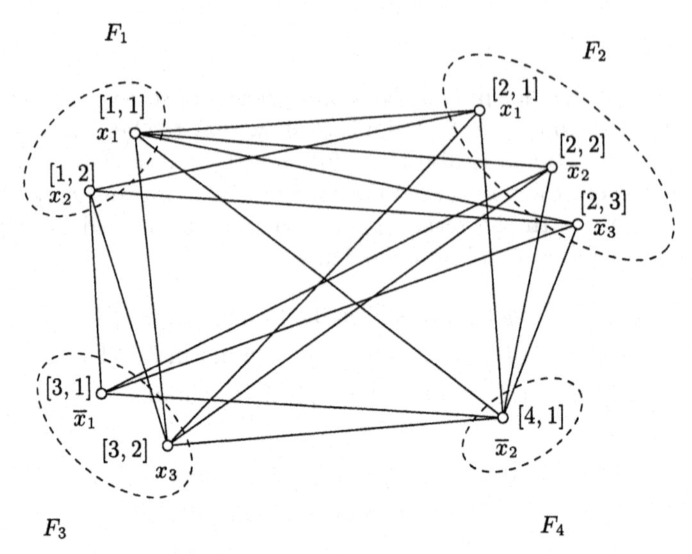

Fig. 6.6.

For instance, consider the formula

$$\Phi = (x_1 \vee x_2) \wedge (x_1 \vee \overline{x}_2 \vee \overline{x}_3) \wedge (\overline{x}_1 \vee x_3) \wedge \overline{x}_2.$$

Then, $k = 4$ and the corresponding graph G is depicted in Figure 6.6.

Obviously, (G, k) can be constructed from Φ in polynomial time.

Now, we show that

$$\Phi \textit{ is satisfiable } \Leftrightarrow G \textit{ contains a clique of size } k = m. \tag{6.1}$$

The proof idea is the following. Two literals l_{ij} and l_{rs} are "connected" in G iff they are from different clauses ($i \neq r$) and both can obtain the value 1 simultaneously.[24] Therefore, a clique in G corresponds to assignments to variables of Φ, that assign the value 1 to all literals of the vertices of a clique. For instance, the clique

$$\{[1,1], [2,1], [3,2], [4,1]\}$$

in Figure 6.6 determines the assignment $x_1 = 1$, $x_2 = 0$ ($\overline{x_2} = 1$), and $x_3 = 1$.

We prove the equivalence (6.1) by proving the corresponding implications separately.

(i) "$\Rightarrow$": Let Φ be a satisfiable formula. Then, there exists as assignment φ to the variables of Φ, such that $\varphi(\Phi) = 1$. Hence, $\varphi(F_i) = 1$ for all $i \in \{1, \ldots, m\}$. So, for any $i \in \{1, \ldots, m\}$ there exists an index $\alpha_i \in \{1, \ldots, k_i\}$, such that $\varphi(l_{i\alpha_i}) = 1$. We claim that the set

[24] There exists an assignment φ such that $\varphi(l_{ij}) = \varphi(l_{rs}) = 1$.

$$\{[i, \alpha_i] \mid 1 \leq i \leq m\}$$

is a clique in G.
Clearly, the vertices $[1, \alpha_1], [2, \alpha_2], \ldots, [m, \alpha_m]$ correspond to literals from different clauses.
For any i, j with $i \neq j$, the equality $l_{i\alpha_i} = \bar{l}_{j\alpha_j}$ implies $\omega(l_{i\alpha_i}) \neq \omega(l_{j\alpha_j})$ for every assignment w to the variables of Φ. Since $\varphi(l_{i\alpha_i}) = \varphi(l_{j\alpha_j}) = 1$ for all $i, j \in \{1, \ldots, m\}$, $l_{i\alpha_i} \neq \bar{l}_{j\alpha_j}$ must be true for all pairs (i, j) with $i \neq j$, thus

$$\{[i, \alpha_i], [j, \alpha_j]\} \in E$$

for all $i, j \in \{1, \ldots, m\}$. Hence, $\{[i, \alpha_i] \mid 1 \leq i \leq m\}$ is a clique of size m.

(ii) "$\Leftarrow$": Let Q be a clique of G with $k = m$ vertices. Since two vertices are connected in G only if they correspond to literals from different clauses, there exist $\alpha_1, \alpha_2, \ldots, \alpha_m$, $\alpha_p \in \{1, 2, \ldots, k_p\}$ for $p = 1, \ldots, m$, such that

$$Q = \{[1, \alpha_1], [2, \alpha_2], \ldots, [m, \alpha_m]\}.$$

Following the construction of G, there exists an assignment φ to the variables of Φ, such that

$$\varphi(l_{1\alpha_1}) = \varphi(l_{2\alpha_2}) = \cdots = \varphi(l_{m\alpha_m}) = 1.$$

The direct consequence is

$$\varphi(F_1) = \varphi(F_2) = \cdots = \varphi(F_m) = 1,$$

so φ satisfies the formula Φ. Hence, Φ is satisfiable.

□

Lemma 6.58.

$$\text{CLIQUE} \leq_p \text{VC}.$$

Proof. Let $G = (V, E)$ and k be an input instance of the clique problem. We construct an input instance of the vertex-cover problem as follows:

$m := |V| - k$
$\overline{G} = (V, \overline{E})$, where $\overline{E} = \{\{v, u\} \mid u, v \in V, v \neq u, \{u, v\} \notin E\}$.

Figure 6.7 illustrates the construction of the graph $\overline{G}$ from a given graph G. Since $\overline{G}$ can be obtained from G by converting 1s to 0s and 0s to 1s in the adjacency matrix of G, it is obvious that this construction can be executed in linear time.

To prove

$$\text{“}(G, k) \in \text{CLIQUE} \Leftrightarrow (\overline{G}, |V| - k) \in \text{VC''}$$

it is sufficient to show that

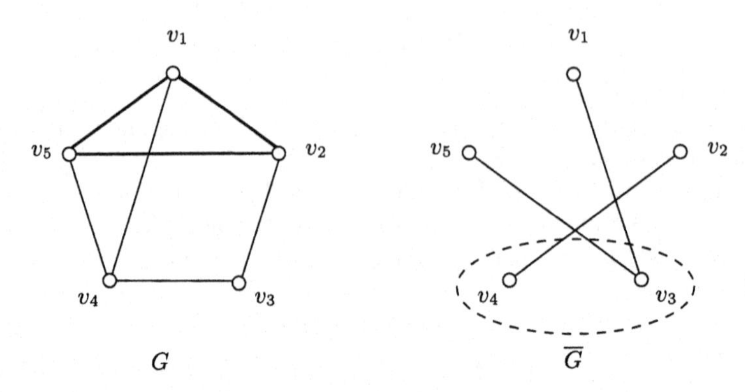

Fig. 6.7.

$$S \subseteq V \textit{ is a clique in } G \Leftrightarrow V - S \textit{ is a vertex-cover } \overline{G}.$$

In Figure 6.7 we see that the clique $\{v_1, v_2, v_5\}$ of G determines the vertex-cover $\{v_3, v_4\}$ in $\overline{G}$. Similarly, the clique $\{v_1, v_4, v_6\}$ determines the vertex-cover $\{v_2, v_3\}$, and the clique $\{v_2, v_3\}$ determines the vertex-cover $\{v_1, v_4, v_5\}$.

We prove this equivalence by proving the corresponding implications separately.

(i) "$\Rightarrow$": Let S be a clique in G. Hence, there is no edge between any pair of vertices from S in $\overline{G}$, so every edge in $\overline{G}$ is adjacent to at least one vertex in $V - S$. Thus, $V - S$ is a vertex-cover in $\overline{G}$.
(ii) "$\Leftarrow$": Let $C \subseteq V$ be a vertex-cover in $\overline{G}$. Following the definition of vertex-cover, every edge in $\overline{G}$ is adjacent to at least one vertex in C. Hence, there is no edge $\{u, v\}$ in $\overline{E}$ with both $u, v \in V - C$. Therefore $\{u, v\} \in E$ for all $u, v \in V - C$, $u \neq v$, and so $V - C$ is a clique in G.

□

The following result shows that the satisfiability problem remains hard even when restricting the set of problem instances to a special subclass of formulae. We say that a Boolean formula is in 3CNF, if it is in CNF and each clause consists of at most three literals. The 3CNF decision problem is to decide whether a given formula in 3CNF is satisfiable.

In the following we consider an assignment φ to the Boolean variables in a set $X = \{x_1, \ldots, x_n\}$ as a mapping $\varphi : X \to \{0, 1\}$. Let $Y = \{y_1, \ldots, y_r\}$ be a set of Boolean variables, $X \cap Y = \emptyset$. We say that $\omega : X \cup Y \to \{0, 1\}$ is an **extension** of $\varphi : X \to \{0, 1\}$, if

$$\omega(z) = \varphi(z) \text{ for all } z \in X.$$

Lemma 6.59.

$$\text{SAT} \leq_p \text{3SAT}.$$

Proof. Let $F = F_1 \wedge F_2 \wedge \ldots \wedge F_m$ be a formula in CNF over a set of Boolean variables $\{x_1, \ldots, x_n\}$. We construct a formula C in 3CNF such that

$$F \text{ is satisfiable } (F \in \text{SAT}) \Leftrightarrow C \text{ is satisfiable } (C \in \text{3SAT}).$$

The polynomial-time reduction from F to C is executed for each of the clauses $F_1, \ldots, F_m$ in the same way.

If F_i contains at most three literals, then one sets $C_i = F_i$.

Let

$$F_i = z_1 \vee z_2 \vee \cdots \vee z_k$$

with $k \geq 4, z_i \in \{x_1, \overline{x}_1, \cdots, x_n, \overline{x}_n\}$.

We construct C_i over the set of variables

$$\{x_1, \ldots, x_n, y_{i,1}, y_{i,2}, \ldots, y_{i,k-3}\},$$

where $y_{i,1}, y_{i,2}, \ldots, y_{i,k-3}$ are new variables that are not used in the construction of any C_j with $j \neq i$. We set

$$C_i = (z_1 \vee z_2 \vee y_{i,1}) \wedge (\overline{y}_{i,1} \vee z_3 \vee y_{i,2}) \wedge (\overline{y}_{i,2} \vee z_4 \vee y_{i_3})$$
$$\wedge \cdots \wedge (\overline{y}_{i,k-4} \vee z_{k-2} \vee y_{i,k-3}) \wedge (\overline{y}_{i,k-3} \vee z_{k-1} \vee z_k).$$

For instance, for $F_i = \overline{x}_1 \vee x_3 \vee \overline{x}_2 \vee x_7 \vee \overline{x}_9$, one constructs

$$C_i = (\overline{x}_1 \vee x_3 \vee y_{i,1}) \wedge (\overline{y}_{i,1} \vee \overline{x}_2 \vee y_{i,2}) \wedge (\overline{y}_{i,2} \vee x_7 \vee \overline{x}_9).$$

To show that

"*$F = F_1 \wedge \cdots \wedge F_m$ is satisfiable $\Leftrightarrow C = C_1 \wedge \cdots \wedge C_m$ is satisfiable*"

it is sufficient to prove the following assertion:

> "*An assignment φ to the variables $\{x_1, \ldots, x_n\}$ satisfies F_i iff there exists an extension φ' of φ to $\{x_1, \ldots, x_n, y_{i,1}, \ldots, y_{i,k-3}\}$ that satisfies C_i.*"

(i) "$\Rightarrow$": Let φ be an assignment to the variables $\{x_1, x_2, \ldots, x_n\}$, such that $\varphi(F_i) = 1$. Hence, there exists a $j \in \{1, \ldots, k\}$ with $\varphi(z_j) = 1$. We choose

$$\varphi' : \{x_1, \ldots, x_n, y_{i,1}, \ldots, y_{i,k-3}\} \to \{0, 1\},$$

such that

a) $\varphi(x_l) = \varphi'(x_l)$ for $l = 1, \ldots, n$,
b) $\varphi'(y_{i,1}) = \cdots = \varphi'(y_{i,j-2}) = 1$, and
c) $\varphi'(y_{i,j-1}) = \cdots = \varphi'(y_{i,k-3}) = 0$.

Since $\varphi'(z_j) = 1$, the $(j-1)$-th clause of C_i is satisfied. $\varphi'(y_{i,r}) = 1$ assures that the r-th clause of C_i is satisfied for $r = 1, \ldots, j-2$. $\varphi'(y_{i,s}) = 0$ (i.e. $\overline{y}_{i,s} = 1$) assures that the $(s+1)$-th clause of C_i is satisfied for $s = j-1, j, \ldots, k-3$. Hence, φ' satisfies all $k-2$ clauses if C_i and hence $\varphi'(C_i) = 1$.

(ii) "$\Leftarrow$": We prove this implication by the indirect proof. Let φ be an assignment such that $\varphi(F_i) = 0$. We have to show that no extension φ' of φ exists such that $\varphi'(C_i) = 1$. $\varphi(F_i) = 0$ implies

$$\varphi(z_1) = \varphi(z_2) = \cdots \varphi(z_k) = 0.$$

To satisfy the first clause of C_i, one has to assign the value 1 to the variable $y_{i,1}$. Then $\varphi'(\overline{y}_{i,1}) = 0$ and one has to set $\varphi'(y_{i,2}) = 1$ in order to satisfy the second clause of C_i. Using the same argument

$$\varphi'(y_{i,1}) = \varphi'(y_{i,2}) = \cdots = \varphi'(y_{i,k-3}) = 1$$

in order to satisfy the first $k-3$ clauses of C_i Then, $\varphi'(\overline{y}_{i,k-3}) = 0$ and since $\varphi(z_{k-1}) = \varphi(z_k) = 0$, the last clause $\overline{y}_{i,k-3} \vee z_{k-1} \vee z_k$ of C_i remains unsatisfied.

□

Exercise 6.60. Prove the following polynomial-time reducibilities.

(i) VC $\leq_p$ CLIQUE
(ii) 3SAT $\leq_p$ VC

The concept of NP-completeness became the basic instrument for classifying algorithmic problems with respect to their hardness. Currently, we are aware of more than 3000 NP-complete problems. Above we showed how to apply this concept for decision problems. In the following we show that we can modify this approach for classifying optimization problems too. In order to do this we first need some counterparts for the classes P and NP in the world of optimization problems.

Definition 6.61. **NPO** *is defined as the class of optimization problems, where*

$$U = (\Sigma_I, \Sigma_O, L, \mathcal{M}, cost, goal) \in \text{NPO},$$

if the following conditions are met:

(i) $L \in P$.
{One can efficiently verify whether a word $x \in \Sigma_I^*$ is a feasible input (the representation of an instance of the problem U).}

(ii) There exists a polynomial p_U, such that
(a) for every $x \in L$ and every $y \in \mathcal{M}(x)$, $|y| \leq p_U(|x|)$
{The size[25] of any feasible solution is polynomial in the input length.}, and

[25] The representation length

(b) there exists a polynomial-time algorithm A that, for any $y \in \Sigma_O^*$ and any $x \in L$ with $|y| \leq p_U(|x|)$, decides whether $y \in \mathcal{M}(x)$ or $y \notin \mathcal{M}(x)$.

(iii) The function cost can be computed in polynomial time.

We see that an optimization problem U is in NPO, if:

1. One can efficiently check whether a given word is an instance of the problem U (i.e., the hardness of U does not rely on solving the decision problem (Σ_I, L)).
2. The size of all feasible solutions to an input is polynomially bounded in the input size and one can efficiently verify whether a candidate for a solution is really a feasible solution (i.e., the hardness of U does not rely on solving the decision problem $(\Sigma_O, \mathcal{M}(x))$ for any x).
3. The cost of any feasible solution can be efficiently computed (i.e., the estimation of the "quality" of a feasible solution does not influence the hardness if U).

Observe, that the condition (ii.b) is the main analogy between NPO and VP. But the crucial point is that conditions (i), (ii), and (iii) are natural because they reduce the hardness of solving an optimization problem U to the pure optimization process (i.e., to the search for the best solutions among the feasible solutions) and make the hardness of U independent of decision problems such as checking whether an input represents a problem instance of U and whether a solution candidate is a feasible solution for the input.

The following considerations show that MAX-SAT belongs to NPO.

1. One can efficiently decide whether an $x \in \Sigma_{\text{logic}}^*$ codes a Boolean formula in CNF.
2. For every input x, each assignment $\alpha \in \{0,1\}^*$ to the variables of the formula Φ_x has the property $|\alpha| < |x|$ and one can verify whether $|\alpha|$ is the number of variables in Φ_x in linear time.
3. For any given assignment α to the variables of Φ_x, one can compute the number of satisfied clauses of Φ_x in time linear in $|x|$ and hence efficiently estimate the cost of α.

Now, let us consider the following optimization problems. The **maximum cut problem, MAX-CUT**, is to find for a given graph $G = (V, E)$, a cut of the maximal cardinality in G. A **cut** of $G = (V, E)$ is any pair (V_1, V_2) with

$$V_1 \cup V_2 = V \text{ and } V_1 \cap V_2 = \emptyset.$$

The cost of a cut (V_1, V_2) of G is the number of edges between V_1 and V_2, i.e.,

$$cost(V_1, V_2) = |E \cap \{\{v, u\} \mid v \in V_1, u \in V_2\}|.$$

The **minimum vertex cover problem, MIN-VC**, is to find a vertex cover of minimal cardinality for any given graph.

Exercise 6.62. Give formal definitions of the above-defined optimization problems MAX-CUT and MIN-VC and show that both belong to NPO.

The following definition introduces the class PO of optimization problems, which is a natural analogy of the class P.

Definition 6.63. PO *is the class of optimization problems* $U = (\Sigma_I, \Sigma_O, L, \mathcal{M}, cost, goal)$, *such that*

(i) $U \in NPO$ *and*
(ii) there exists a polynomial-time algorithm A, *such that, for every* $x \in L$, $A(x)$ *is an optimal solution for* x.

Now, we obtain the concept of NP-hardness for optimization problems by considering a simple reduction from specific decision problems to optimization problems.

Definition 6.64. *Let* $U = (\Sigma_I, \Sigma_O, L, \mathcal{M}, cost, goal)$ *be an optimization problem from* NPO. *We define the* **threshold language of** $\boldsymbol{U}$ *as*

$$\boldsymbol{Lang_U} = \{(x, a) \in L \times \Sigma_{\text{bool}}^* \mid Opt_U(x) \leq Number(a)\},$$

if $goal = minimum$, *and*

$$\boldsymbol{Lang_U} = \{(x, a) \in L \times \Sigma_{\text{bool}}^* \mid Opt_U(x) \geq Number(a)\},$$

if $goal = maximum$.

We say that U *is* **NP-hard** *if* $Lang_U$ *is* NP-*hard.*

First we show that Definition 6.64 provides a concept for proving hardness of optimization problems. More precisely, we show that proving the NP-hardness of $Lang_U$ is a way of showing $U \notin$ PO under the assumption P $\neq$ NP. An important point of this concept is that we do not need to make a new assumption like PO $\neq$ NPO for this purpose.

Lemma 6.65. *If an optimization problem* $U \in$ PO, *then* $Lang_U \in$ P.

Proof. If $U \in$ PO, then there is a polynomial-time algorithm A that, for every input instance x of U, computes an optimal solution for x and therefore the value $Opt_U(x)$. Hence, A can be used to decide $Lang_U$. □

Theorem 6.66. *Let* $U \in$ NPO. *If* $Lang_U$ *is* NP-*hard and* P $\neq$ NP, *then* $U \notin$ PO.

Proof. We give an indirect proof. Assume the contrary, i.e., $U \in$ PO. Applying Lemma 6.65, $Lang_U \in$ P. Since $Lang_U$ is NP-hard, $Lang_U \in$ P implies P = NP. □

To illustrate the simplicity of this method for proving the hardness of optimization problems with respect to the polynomial time, we present the following examples.

Lemma 6.67. MAX-SAT *is* NP-*hard.*

Proof. Following Definition 6.64 we have to prove that $Lang_{\text{MAX-SAT}}$ is NP-hard. Since we know that SAT is NP-hard, it is sufficient to prove

$$\text{SAT} \leq_p Lang_{\text{MAX-SAT}}.$$

The reduction is straightforward. Let x code a formula Φ_x of m clauses. Take (x, m) as the input instance of the decision problem $(Lang_{\text{MAX-SAT}}, \Sigma^*)$. Obviously,

$$(x, m) \in Lang_{\text{MAX-SAT}} \Leftrightarrow \Phi_x \text{ is satisfiable.}$$

□

The **maximum clique problem, MAX-CL**, is to find a clique of maximal size in a given graph G.

Lemma 6.68. MAX-CL *is* NP-*hard.*

Proof. Observe, that CLIQUE $= Lang_{\text{MAX-CL}}$. Since CLIQUE is NP-hard, the proof is completed. □

Exercise 6.69. Prove that the MAX-CUT and MIN-VC are NP-hard.

6.7 Summary

The main goal of complexity theory is the classification of algorithmic problems with respect to the amount of computer resources needed to solve them. The results of this investigation are quantitative laws of information processing and the limits of tractability (practical algorithmic solvability).

The most fundamental complexity measures are time complexity and space complexity. The multitape Turing machines is the basic computing model of the abstract complexity theory. The complexity of an MTM (an algorithm) is considered as a function f of the input length, where $f(n)$ is the maximum of the complexities of all computations on inputs of length n. This complexity measurement is called the worst-case complexity measurement.

There exist algorithmically solvable problems with an arbitrary large complexity. Algorithms of exponential complexity are usually not considered to be executable in practice.

The tractability of problems is connected with the polynomial-time complexity. The class P is the set of all decision problems that can be solved by polynomial-time algorithms. The definition of the class P is robust in the sense that it is independent of the choice of the computing model.

The time complexity of a nondeterministic MTM M on an input w is the length of the shortest correct computation of M on w. The typical work of a nondeterministic algorithm starts with nondeterministic guesses and continues with a deterministic verification of the guesses. The class NP is the class of all languages that can be accepted by polynomial-time nondeterministic algorithms. The question of whether P is a proper subset of NP is the most famous open problem of theoretical computer science. The class NP contains many interesting, practical problems and one does not know whether they are in P or not. One can show that the question of whether P is a proper subset of NP, is equivalent to the question of whether verifying the correctness of given mathematical proofs is easier than creating mathematical proofs for a given theorem.

There has been no essential success in developing methods for proving lower bounds on complexity of concrete problems and so we lack a methodology for classifying problems into tractable problems and intractable ones. The concept of NP-completeness enables us to prove results of the kind of "$L \notin$ P" (L is not tractable) under the assumption P $\neq$ NP. The basic idea of this concept is that the NP-complete problems are the hardest problems in NP in the sense that if a NP-complete problem would be in P, then P would be equal to NP.

The complexity hierarchies were proven by Hartmanis, Stearns and Lewis [24, 25]. The concepts of polynomial-time reducibility and of NP-completeness were introduced in the seminal works by Cook [13] and Karp [35]. The classical book by Garey and Johnson [19] provides an excellent, detailed presentation of the theory of the NP-completeness. A fascinating discussion about "practical solvability" can be found in Lewis and Papadimitriou [42] and Stockmayer and Chandra [66].

There are several good textbooks devoted to complexity theory. The textbooks by Hopcroft and Ullman [28] and Sipser [65] present an excellent introduction to complexity theory. An extensive presentation of the complexity theory is given, for instance, by Bovet and Crescenzi [7], Papadimitriou [49] and Balcázar, Díaz, and Gabarró [2, 3]. As for literature written in the German language we warmly recommend the textbook by Reischuk [56].

When a scientist says:
"This is the very end,
nobody can do anything more here,"
then he is no scientist.

L. Gould

7

Algorithmics for Hard Problems

7.1 Objectives

The complexity theory provides methods for the classification of algorithmic problems with respect to their hardness which is measured in the amount of computational resources (computational complexity) needed to solve them. The theory of algorithms is devoted to the design of efficient algorithms for solving concrete problems. In this chapter we restrict our attention to the design of algorithms for hard (for instance, NP-hard) problems. This may be slightly surprising, because Chapter 6 claims that solving NP-hard problems is beyond physical limits. For instance, the execution of an algorithm of time complexity 2^n for inputs of size $n = 100$ takes more time than the age of the known universe. However, there are many hard problems of enormous importance for everyday practice, which explains why computer scientists have been searching for at least some small progress for over the past 30 years. The central idea is to make a hard problem easier[1] by slightly modifying the problem specification or weakening some requirements. The true art of algorithmics consists of discovering the possibilities of obtaining an enormous gain from infeasible computations, reducing it to a matter of a few minutes on a common PC by an as small as possible change of the problem constraints. To achieve such effects enabling the practical solvability of hard problems, one (or a combination) of the following concepts can be used.

1. *Hard problem instances versus typical problem instances*
   We measure the time complexity as the worst-case complexity, which means that the time complexity Time(n) is the complexity of the hardest problem instance of size n. Often the hardest problem instances are far from being natural and hence they do not appear in applications. Therefore, it may be helpful to deepen the complexity analysis by classifying particular problem instances with respect to their hardness. A successful

[1] Solvable in polynomial time

classification can lead to a specification of a large subclass of efficiently solvable problem instances of a hard problem. If the typical problem instances of an application belong to such an easy subclass, then one can successfully solve the problem for this application.

2. *Exponential algorithms*
One does not try to design a polynomial algorithm for the hard problem. Instead one attempts to discover a practical exponential algorithm. The idea is that some exponential functions do not take excessively large values for realistic input sizes. Table 7.1 shows clearly, that the execution of $(1.2)^n$ computer intructions for $n = 100$, or of $10 \cdot 2^{\sqrt{n}}$ operations for $n = 300$ terminates within a matter of seconds.

Table 7.1.

Complexity	$n = 10$	$n = 50$	$n = 100$	$n = 300$
2^n	1024	16 digits	31 digits	91 digits
$2^{\frac{n}{2}}$	32	$\approx 33 \cdot 10^6$	16 digits	46 digits
$(1.2)^n$	≈ 6	≈ 9100	$\approx 83 \cdot 10^6$	24 digits
$10 \cdot 2^{\sqrt{n}}$	≈ 89	≈ 1345	10240	$\approx 1.64 \cdot 10^6$

3. *Weakening of requirements*
One can deviate from the requirement to guarantee the computation of the correct result in different ways. Typical representants of this concept are randomized algorithms and approximation algorithms. In the case of randomized algorithms, one exchanges the deterministic control for a randomized one. The cost for this exchange is the loss of the assurance of computing correct solutions, i.e., a positive probability of computing a wrong result. The gain of this exchange can be an essential decrease of the complexity. If the error probability is below 10^{-9} for every input, then there is no practical difference between the reliability of a deterministic algorithm and that of the randomized one. Thus one has solved the problem efficiently. Approximation algorithms are used for solving optimization problems. Instead of requiring an optimal solution, one accepts feasible solutions whose cost (quality) does not essentially differ[2] from the cost of the optimal solutions. The gain of this approach can also be an exponential decrease of computational complexity.

The aim of this chapter is to present some concrete applications of these concepts. In Section 7.2 we introduce the pseudopolynomial algorithms that represent an exemplary application of the first concept searching for a large class of easy problem instances of hard problems. The concept of approximation algorithms is explained and illustrated in Section 7.3. Section 7.4 is devoted to local algorithms that provide the possibility of applying all three

[2] Are not much worse than the optimal cost

concepts mentioned above. The local search is the base for several heuristics. In Section 7.5 we present the simulated annealing heuristic, which is based on an analogy to the optimization of physical systems. Because of the importance of the concept of randomization we devote the entire of Chapter 8 to it.

7.2 Pseudopolynomial Algorithms

In this section we consider a special class of problems, whose inputs can be viewed as sequences of natural numbers. Such problems are called **integer problems**. In fact, one can interpret any input from $\{0,1,\#\}^*$ with an unbounded number of symbols $\#$ as an integer problem. Let

$$x = x_1\#x_2\#\ldots\#x_n, \; x_i \in \{0,1\}^* \text{ for } i = 1,2,\ldots,n$$

be a word over $\{0,1,\#\}^*$. We interpret x as the following vector

$$\mathbf{Int}(x) = (\mathit{Number}(x_1), \mathit{Number}(x_2), \ldots, \mathit{Number}(x_n))$$

of n natural numbers. Each graph-theoretical problem[3] whose problem instances can be represented by adjacency matrices is an integer problem, because every adjacency matrix can be represented as a sequence of 0s and 1s. The TSP is an integer problem where the sequence of integers represents the costs of particular edges.

We define for any $x = x_1\#x_2\#\ldots\#x_n$ with $x_i \in \{0,1\}^*$ for $i = 1,2,\ldots,n$,

$$\mathbf{MaxInt}(x) = \max\{\mathit{Number}(x_i) \mid i = 1,2,\ldots,n\}.$$

The idea of the concept of pseudopolynomial algorithms is designing algorithms that are efficient for problem instances x whose MaxInt(x) is not substantially larger than $|x|$. Since an integer $\mathit{Number}(y)$ is exponentially larger than the length of its binary representation y, this requirement is a real restriction.

Definition 7.1. *Let $\mathcal{U}$ be an integer problem and let A be an algorithm that solves $\mathcal{U}$. We say that A is a* **pseudopolynomial algorithm for $\mathcal{U}$**, *if there exists a polynomial p of two variables, such that*

$$\text{Time}_A(x) \in O(p(|x|, \text{MaxInt}(x)))$$

for all problem instances x of $\mathcal{U}$.

We observe immediately that for problem instances with MaxInt$(x) \leq h(|x|)$ for a polynomial h, the time complexity Time$_A(x)$ is polynomial in $|x|$.

[3] Such as the vertex cover problem or the clique problem

Definition 7.2. *Let $\mathcal{U}$ be an integer problem and let h be a mapping from $\mathbb{N}$ to $\mathbb{N}$. The* **h-value-bounded subproblem of $\mathcal{U}$, Value(h)-$\mathcal{U}$,** *is the subproblem of $\mathcal{U}$ whose instances are all instances x of $\mathcal{U}$ satisfying*

$$\text{MaxInt}(x) \leq h(|x|).$$

The following theorem shows that one can specify large classes of easy instances of a hard problem in this way.

Theorem 7.3. *Let $\mathcal{U}$ be an integer problem and let A be a pseudopolynomial algorithm that solves $\mathcal{U}$. Then, for any polynomial h, there exists a polynomial-time algorithm for the problem*[4] Value(h)-$\mathcal{U}$.

Proof. If A is a pseudopolynomial algorithm for $\mathcal{U}$, then there exists a polynomial p of two variables, such that

$$\text{Time}_A(x) \in O(p(|x|, \text{MaxInt}(x)))$$

for any instance x of the problem $\mathcal{U}$. Since h is a polynomial function, there exists a constant c such that

$$\text{MaxInt}(x) \in O(|x|^c)$$

for all instances x of the problem Value(h)-$\mathcal{U}$.

Therefore,

$$\text{Time}_A(x) \in O(p(|x|, |x|^c)).$$

Hence, A is a polynomial-time algorithm for Value(h)-$\mathcal{U}$. □

We show an application of the concept of pseudopolynomial-time algorithms for solving the knapsack problem, an NP-hard optimization problem.

Knapsack problem

Input: $2n+1$ positive integers $w_1, w_2, \ldots, w_n, c_1, c_2, \ldots, c_n, b$ for an $n \in \mathbb{N} - \{0\}$. {These positive integers provide information about n objects, where w_i is the weight of the i-th object and c_i is the cost of the i-th object for $i = 1, 2, \ldots, n$. The number b is the limit[5] on the weight of the knapsack content, where the content of the knapsack may consist of a subset of these n objects only.}

Feasible solutions: For every $I = (w_1, w_2, \ldots, w_n, c_1, \ldots, c_n, b)$

$$\mathcal{M}(I) = \left\{ T \subseteq \{1, \ldots, n\} \;\middle|\; \sum_{i \in T} w_i \leq b \right\}$$

is the set of feasible solutions.

{A feasible solution for I is any subset of the n objects, whose total weight does not exceed b.}

[4] If $\mathcal{U}$ is a decision problem, then Value(h)-$\mathcal{U} \in$ P. If $\mathcal{U}$ is an optimization problem, then Value(h)-$\mathcal{U} \in$ PO.

[5] Upper bound

Cost: For all inputs I and all $T \in \mathcal{M}(I)$,

$$cost(T, I) = \sum_{i \in T} c_i.$$

{The cost of a feasible solution T is the total cost of all objects in the knapsack.}

Goal: maximum.

Now, we design a pseudopolynomial algorithm for the knapsack problem using the method of dynamic programming. To compute an optimal solution for the instance $I = (w_1, w_2, \ldots, w_n, c_1, \ldots, c_n, b)$, we start with the subinstance $I_1 = (w_1, c_1, b)$ and continue via the subinstances

$$I_i = (w_1, w_2, \ldots, w_i, c_1, \ldots, c_i, b)$$

for $i = 2, 3, \ldots, n$ to reach $I = I_n$. More precisely, for every I_i and every $k \in \{0, 1, 2, \ldots, \sum_{j=1}^{i} c_j\}$, we compute the triple

$$(k, W_{i,k}, T_{i,k}) \in \left\{0, 1, 2, \ldots, \sum_{j=1}^{i} c_j\right\} \times \{0, 1, 2, \ldots, b\} \times \mathcal{P}(\{1, \ldots, i\}),$$

where $W_{i,k}$ is the minimal weight with which one can reach the cost k for the subinstance I_i. The set $T_{i,k} \subseteq \{1, \ldots, i\}$ is a set of indices that determines the cost k with the weight $W_{i,k}$, i.e.,

$$\sum_{j \in T_{i,k}} c_j = k \quad \text{and} \quad \sum_{j \in T_{i,k}} w_j = W_{i,k}.$$

We note that several different sets of indices that determine the same cost k and the same weight $W_{i,k}$ may exist. In such cases we simply take an arbitrary one in order to confirm the reachability[6] of $(k, W_{i,k})$. On the other hand it can happen that a cost k is not reachable for the instance I_i. In this case, we do not have any triple with the first element k. In what follows, we denote the set of all triples of I_i as $\mathbf{TRIPLE}_i$. Observe that

$$|\text{TRIPLE}_i| \leq \sum_{j=1}^{i} c_j + 1.$$

An important point is that one can compute TRIPLE_{i+1} from TRIPLE_i in time $O(\text{TRIPLE}_i)$. To compute TRIPLE_{i+1}, we first compute

$$\begin{aligned}\text{SET}_{i+1} = \text{TRIPLE}_i \cup \{&(k + c_{i+1}, W_{i,k} + w_{i+1}, T_{i,k} \cup \{i+1\}) \mid \\ &(k, W_{i,k}, T_{i,k}) \in \text{TRIPLE}_i \text{ and } W_{i,k} + w_{i+1} \leq b\}\end{aligned}$$

[6] Of the cost k by the weight $W_{i,k}$

by adding the $(i+1)$-th object to any triple in TRIPLE_i, if this addition does not cause the total cost to exceed b. TRIPLE_{i+1} is then a subset of SET_{i+1} that is obtained by choosing a triple of minimal weight for every achievable cost k. Obviously the triple with the maximal cost in TRIPLE_n estimates an optimal solution for the problem instance $I = I_n$. Thus we can formally describe the designed algorithm as follows.

Algorithm DPR

Input: $I = (w_1, w_2, \ldots, w_n, c_1, \ldots, c_n, b) \in (\mathbb{N} - \{0\})^{2n+1}$ for a positive integer n.

Phase 1. $\mathrm{TRIPLE}(1) = \{(0, 0, \emptyset)\} \cup \{(c_1, w_1, \{1\})\} \mid \text{ if } w_1 \leq b\}$.

Phase 2.

```
for i = 1 to n − 1 do
  begin SET(i + 1) := TRIPLE(i);
    for every (k, w, T) ∈ TRIPLE(i) do
      begin if w + w_{i+1} ≤ b then
        SET(i + 1) := SET(i + 1)
                  ∪ {(k + c_{i+1}, w + w_{i+1}, T ∪ {i + 1})};
      end
    Determine TRIPLE(i + 1) as a subset of SET(i + 1),
    where, for every reachable cost k in SET(i + 1) contains
    exactly one triple with the cost k by taking a triple with
    the minimal weight for the cost k.
  end
```

Phase 3. Compute

$$c := \max\{k \in \{1, 2, \ldots, \sum_{i=1}^{n} c_i\} \mid (k, w, T) \in \mathrm{TRIPLE}(n)\}.$$

Output: The set of indices T, such that $(c, w, T) \in \mathrm{TRIPLE}(n)$.

We illustrate the work of the algorithm DPR for the problem instance $I = (w_1, w_2, \ldots, w_5, c_1, \ldots, c_5, b)$, where

$$\begin{aligned} &w_1 = 23,\ w_2 = 15,\ w_3 = 15,\ w_4 = 33,\ w_5 = 32, \\ &c_1 = 33,\ c_2 = 23,\ c_3 = 11,\ c_4 = 35,\ c_5 = 11, \text{ and} \\ &b = 65. \end{aligned}$$

Obviously $I_1 = (23, 33, 65)$ and the only reachable costs are 0 and 33. Therefore,

$$\mathrm{TRIPLE}_1 = \{(0, 0, \emptyset), (33, 23, \{1\})\}.$$

$I_2 = (23, 15, 33, 23, 65)$ and the only reachable costs for I_2 are 0, 23, 33, and 56. Hence we obtain

$$\mathrm{TRIPLE}_2 = \{(0, 0, \emptyset), (23, 15, \{2\}), (33, 23, \{1\}), (56, 38, \{1, 2\})\}.$$

The subinstance $I_3 = (23, 15, 15, 33, 23, 11, 65)$. The addition of a third object is possible for every triple in TRIPLE_2 and by doing so we always obtain a new cost. Thus, the cardinality of TRIPLE_3 is twice that of TRIPLE_2 (i.e., $\text{SET}_3 = \text{TRIPLE}_3$) and

$$\begin{aligned}\text{TRIPLE}_3 = \{&(0, 0, \emptyset), (11, 15, \{3\}), (23, 15, \{2\}),\\ &(33, 23, \{1\}), (34, 30, \{2, 3\}), (44, 38, \{1, 3\}),\\ &(56, 38, \{1, 2\}), (67, 53, \{1, 2, 3\})\}.\end{aligned}$$

For the triples $(44, 38, \{1, 3\})$, $(56, 38, \{1, 2\})$ and $(67, 53, \{1, 2, 3\})$ from TRIPLE_3, one cannot pack a fourth object into the knapsack. Hence,

$$\begin{aligned}\text{TRIPLE}_4 = \text{TRIPLE}_3 \cup \{&(35, 33, \{4\}), (46, 48, \{3, 4\}),\\ &(58, 48, \{2, 4\}), (68, 56, \{1, 4\}), (69, 63, \{2, 3, 4\})\}.\end{aligned}$$

Finally, we obtain for the instance $I = I_5$

$$\begin{aligned}\text{TRIPLE}_5 = \{&(0, 0, \emptyset), (11, 15, \{3\}), (22, 47, \{3, 5\}),\\ &(23, 15, \{2\}), (33, 23, \{1\}), (34, 30, \{2, 3\}),\\ &(35, 33, \{4\}), (44, 38, \{1, 3\}), (45, 62, \{2, 3, 5\}),\\ &(46, 48, \{3, 4\}), (56, 38, \{1, 2\}), (58, 48, \{2, 4\}),\\ &(67, 53, \{1, 2, 3\}), (68, 56, \{1, 4\}), (69, 63, \{2, 3, 4\})\}.\end{aligned}$$

Hence $\{2, 3, 4\}$ is the optimal solution for I, since $(69, 63, \{2, 3, 4\})$ is the triple with the maximal cost 69 in TRIPLE_5.

Exercise 7.4. Simulate the work of the algorithm DPR for the problem instance $(1, 3, 5, 6, 7, 4, 8, 5, 9)$ of the knapsack problem.

In what follows we analyze the time complexity of the algorithm DPR.

Theorem 7.5. *For every instance I of the knapsack problem,*

$$\text{Time}_{\text{DPR}}(I) \in O(|I|^2 \cdot \text{MaxInt}(I))$$

and hence DPR *is a pseudopolynomial algorithm for the knapsack problem.*

Proof. The time complexity of the first phase is in $O(1)$. For the problem instance $I = (w_1, w_2, \ldots, w_n, c_1, \ldots, c_n, b)$, DPR computes $n - 1$ sets $\text{TRIPLE}(i + 1)$. The computation of $\text{TRIPLE}(i + 1)$ from $\text{TRIPLE}(i)$ can be executed in time $O(|\text{TRIPLE}(i + 1)|)$. Since

$$|\text{TRIPLE}(i + 1)| \leq \sum_{j=1}^{n} c_j \leq n \cdot \text{MaxInt}(I)$$

for every $i \in \{0, 1, \ldots, n\}$, the time complexity of the second phase is in $O(n^2 \cdot \text{MaxInt}(I))$.

The time complexity of the third phase is in $O(n \cdot \text{MaxInt}(I))$, because one has to estimate a maximum of at most $n \cdot \text{MaxInt}(I)$ values. Since $n \leq |I|$,

$$\text{Time}_{\text{DPR}}(I) \in O(|I|^2 \cdot \text{MaxInt}(I)).$$

□

We observe that pseudopolynomial algorithms can be very successful in many applications. The weights and costs are usually from a fixed interval and thus independent of the number of parameters of problem instances. Hence, one can efficiently compute solutions for typical[7] instances of the knapsack problem.

The interest in designing pseudopolynomial algorithms also poses a classification question. For which NP-hard integer problems do pseudopolynomial algorithms exist? Which hard problems are so hard that they do not allow any pseudopolynomial algorithm solution? Now, we are searching for a method that enables us to prove that concrete problems do not possess pseudopolynomial algorithms if P $\neq$ NP. We will see that the concept of NP-completeness also works for this purpose.

Definition 7.6. *An integer problem $\mathcal{U}$ is called* **strongly NP-hard**, *if there exists a polynomial p such that its subproblem* Value(p)-$\mathcal{U}$ *is* NP-*hard.*

The following assertion shows that the strong NP-hardness provides the instrument for proving the nonexistence of pseudopolynomial algorithms for concrete problems.

Theorem 7.7. *Let $\mathcal{U}$ be a strongly* NP-*hard integer problem. If* P $\neq$ NP, *then there exists no pseudopolynomial algorithm for $\mathcal{U}$.*

Proof. Since $\mathcal{U}$ is strongly NP-hard, there exists a polynomial p such that the problem Value(p)-$\mathcal{U}$ is NP-hard. Assume that there is a pseudopolynomial algorithm for $\mathcal{U}$. Then Theorem 7.3 implies, for every polynomial h, the existence of a polynomial-time algorithm for the problem Value(h)-$\mathcal{U}$. This implies the existence of a polynomial-time algorithm for the NP-hard problem Value(p)-$\mathcal{U}$ and hence P = NP. This contradicts our assumption P $\neq$ NP.
□

One can again apply the reduction method to prove the nonexistence of a pseudopolynomial algorithm for a given problem. In what follows we present an application of the strong NP-hardness by showing that TSP is strongly NP-hard. To do this we use the fact that to decide whether a graph contains a Hamiltonian cycle[8] (the Hamiltonian cycle problem, HC) is NP-complete.

---

[7] Typical in the above sense

[8] Remember that a Hamiltonian cycle (tour) of a graph G is a cycle that visits every vertex of G exactly once.

Lemma 7.8. TSP *is strongly* NP*-hard.*

Proof. Since HC is NP-hard, it is sufficient to show that

$$\text{HC} \leq_p Lang_{\text{Value}(p)\text{-TSP}}$$

for a polynomial $p(n) = n$.

Let $G = (V, E)$ be an instance of HC. Let $|V| = n$ for a positive integer n. We construct a weighted complete graph (K_n, c) with $K_n = (V, E_{\text{complete}})$ as follows.

$$E_{\text{complete}} = \{\{u, v\} \mid u, v \in V, u \neq v\},$$

and the weight function $c : E_{\text{complete}} \to \{1, 2\}$ is defined by

$$c(e) = 1 \text{ if } c \in E, \text{ and}$$
$$c(e) = 2 \text{ if } c \notin E.$$

We observe that G contains a Hamiltonian cycle if and only if the cost of an optimal solution of the TSP instance (K_n, c) is exactly n, i.e., when

$$((K_n, c), n) \in Lang_{\text{Value}(p)\text{-TSP}}.$$

Hence, every algorithm that decides $Lang_{\text{Value}(p)\text{-TSP}}$ can be used to solve HC. □

Exercise 7.9. Consider the following generalization of the vertex cover problem. Let $G = (V, E)$ be a graph, and let $w : V \to \mathbb{N}$ be a mapping that assigns a weight to every vertex of G. For any vertex cover S, the cost of S is the sum of the weights of vertices in S. The minimum weighted vertex cover problem is to find a cheapest vertex cover in G. Prove that the minimum weighted vertex cover problem is strongly NP-hard.

7.3 Approximation Algorithms

Here, we introduce the concept of approximation algorithms for solving hard optimization problems. The idea is to jump from an exponential-time complexity to a polynomial-time complexity by weakening the requirements. Instead of forcing the computation of an optimal solution, we are satisfied with an "almost optimal" or "nearly optimal" solution. What the term "almost optimal" means is defined below.

Definition 7.10. *Let $\mathcal{U} = (\Sigma_I, \Sigma_O, L, \mathcal{M}, cost, goal)$ be an optimization problem.*

We say that A is a **consistent algorithm for $\mathcal{U}$** *if, for every $x \in L$, the output $A(x)$ of the computation of A on x is a feasible solution for x (i.e., $A(x) \in \mathcal{M}(x)$).*

Let A be a consistent algorithm for $\mathcal{U}$. For every $x \in L$, we define the **approximation ratio $\mathbf{R}_A(x)$ of A on x** *as*

$$\mathbf{R}_A(x) = \max\left\{\frac{cost(A(x))}{\mathrm{Opt}_{\mathcal{U}}(x)}, \frac{\mathrm{Opt}_{\mathcal{U}}(x)}{cost(A(x))}\right\},$$

where $\mathrm{Opt}_{\mathcal{U}}(x)$ is the cost of an optimal solution for the instance x of $\mathcal{U}$.

For any positive real number $\delta > 1$, we say that A is a **δ-approximation algorithm for $\mathcal{U}$** *if*

$$\mathrm{R}_A(x) \leq \delta$$

for every $x \in L$.

First we illustrate the concept of approximation algorithms for the minimum vertex cover problem. The idea is to efficiently find a matching[9] in a given graph G and then to take all vertices incident to the edges of the matching as a vertex cover.

Algorithm VCA

Input: A graph $G = (V, E)$.
Phase 1. $C := \emptyset$;
{During the computation, C is always a subset of V and at the end of the computation, C is a vertex cover of G.}
$A := \emptyset$;
{During the computation, A is always a subset of E (a matching in G) and when the computation has finished, then A is a maximal matching.}
$E' := E$;
{During the computation, the set $E' \subseteq E$ contains exactly those edges, which are not covered by the actual C. At the end of the computation $E' = \emptyset$.}
Phase 2.

```
while E' ≠ ∅ do
  begin take an arbitrary edge {u, v} from E';
    C := C ∪ {u, v};
    A := A ∪ {{u, v}};
    E' := E' − {all edges incident to u or v};
  end
```

Output: C

Consider a possible run of the algorithm VCA on the graph in Figure 7.1(a). Let $\{b, c\}$ be the first edge chosen by VCA. Then

$$C = \{b, c\},\ A = \{\{b, c\}\} \text{ and } E' = E - \{\{b, a\}, \{b, c\}, \{c, e\}, \{c, d\}\}$$

[9] A matching in $G = (V, E)$ is a set $M \subseteq E$ of edges such that there is no vertex incident to more than one edge from M. A matching is maximal, if for every $e \in E - M$, the set $M \cup \{e\}$ is not a matching in G.

as depicted in Figure 7.1(b). If the second choice of an edge from E' by VCA is the edge $\{e, f\}$ (Figure 7.1(c)), then

$$C = \{b, c, e, f\}, A = \{\{b, c\}, \{e, f\}\} \text{ and } E' = \{\{d, h\}, \{d, g\}, \{h, g\}\}.$$

If the last choice of VCA is the edge $\{d, g\}$ then

$$C = \{b, c, e, f, d, g\}, A = \{\{b, c\}, \{e, f\}, \{d, g\}\} \text{ and } E' = \emptyset.$$

Hence, C is a vertex cover of cost 6. We observe that $\{b, e, d, g\}$ is the optimal vertex cover and this optimal vertex cover cannot be achieved by any choice of edges by the algorithm VCA.

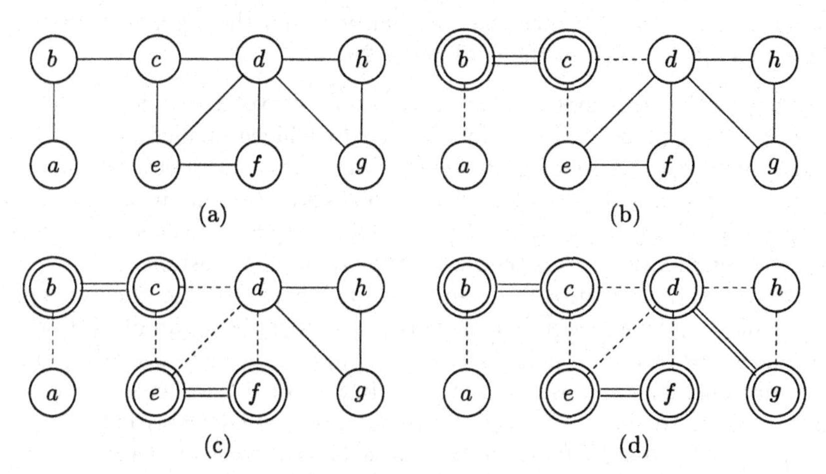

Fig. 7.1.

Exercise 7.11. Find a choice of edges in the second phase of VCA, such that the resulting vertex cover C contains all vertices of G in Figure 7.1(a).

Theorem 7.12. *The algorithm* VCA *is a 2-approximation algorithm for* MIN-VCP *and* $\text{Time}_{\text{VCA}}(G) \in O(|E|)$ *for any instance* $G = (V, E)$.

Proof. The claim

$$\text{Time}_{\text{VCA}}(G) \in O(|E|)$$

is obvious because every edge from E is manipulated exactly once in VCA.

Since $E' = \emptyset$ at the end of any computation, VCA computes a vertex cover in G (i.e., VCA is a consistent algorithm for MIN-VCP).

To prove

$$R_{\text{VCA}}(G) \leq 2$$

for every graph G, we observe that

1. $|C| = 2 \cdot |A|$, and
2. A is a matching in G.

To cover the $|A|$ edges of the matching A, one has to take at least $|A|$ vertices. Since $A \subseteq E$, the cardinality of any vertex cover of G is at least $|A|$, i.e.,

$$\mathrm{Opt}_{\mathrm{MIN\text{-}VCP}}(G) \geq |A|.$$

Hence

$$\frac{|C|}{\mathrm{Opt}_{\mathrm{MIN\text{-}VCP}}(G)} = \frac{2 \cdot |A|}{\mathrm{Opt}_{\mathrm{MIN\text{-}VCP}}(G)} \leq 2.$$

□

Exercise 7.13. Construct, for any positive integer n, a graph G_n, such that the optimal vertex cover has the cardinality n and the algorithm VCA can compute a vertex cover of the cardinality $2n$.

Whether the guarantee of an approximation ratio of 2 is sufficient, depends on particular applications. Usually one tries to achieve smaller approximation ratios, which requires much more demanding algorithmic ideas. On the other hand, one measures the approximation ratio of an algorithm in the worst-case manner, so a 2-approximation algorithm can provide solutions of essentially better approximation ratios than 2 for typical problem instances.

There are optimization problems that are hard also for the concept of approximation. Here, to be hard means that under the assumption P $\neq$ NP, there does not exist any polynomial-time d-approximation algorithm for the given problem for any positive integer d. In Section 7.2 we have showed that TSP is too hard for the concept of pseudopolynomial algorithms. In what follows, we show that TSP cannot be attacked by the concept of approximation algorithms either.

Lemma 7.14. *If* P $\neq$ NP, *then, for any positive integer d, there does not exist any polynomial-time d-approximation algorithm for* TSP.

Proof. We prove this by contradiction. Assume that we have a positive integer d, such that there exists a polynomial-time d-approximation algorithm A for TSP. Then, we show that there exists a polynomial-time algorithm B for the NP-complete HC problem, which contradicts the assumption P $\neq$ NP. The algorithm B for HC works for each input $G = (V, E)$ as follows.

1. B constructs an instance $(K_{|V|}, c)$ of TSP, where

$$\begin{aligned} K_{|V|} = (V, E'), \text{ with } E' &= \{\{u, v\} \mid u, v \in V, u \neq v\}, \\ c(e) &= 1, \quad \text{if} \quad e \in E, \text{and} \\ c(e) = (d-1) \cdot |V| + 2, &\quad \text{if} \quad e \notin E. \end{aligned}$$

2. B simulates the work of A on the input $(K_{|V|}, c)$. If the feasible solution computed by A is a Hamiltonian cycle of cost exactly $|V|$, then B accepts its input G. Otherwise, A rejects G.

The construction of the TSP instance $(K_{|V|}, c)$ can be executed in time $O(|V|^2)$. The second phase of B runs in polynomial time, because A works in polynomial time and the graphs G and $K_{|V|}$ are of the same size.

It remains to show that B really decides the HC problem. First, we observe the following facts.

(i) If G contains a Hamiltonian cycle, then $K_{|V|}$ has a Hamiltonian cycle of cost $|V|$, i.e.,
$$\mathrm{Opt}_{\mathrm{TSP}}(K_{|V|}, c) = |V|.$$
(ii) Every Hamiltonian cycle in $K_{|V|}$ containing at least an edge from $E' - E$, has cost of at least
$$|V| - 1 + (d-1) \cdot |V| + 2 = d \cdot |V| + 1 > d \cdot |V|.$$

Now we prove
$$G = (V, E) \in \mathrm{HC} \Leftrightarrow \text{ the output of } B \text{ is a solution of cost } |V|.$$

Let $G = (V, E)$ be in HC, i.e., G contains a Hamiltonian cycle C. Following the definition of the weight function c, the cost of C in $K_{|V|}$ is $|V|$, so
$$\mathrm{Opt}_{\mathrm{TSP}}(K_{|V|}, c) = |V|.$$
Fact (ii) implies that any Hamiltonian cycle in $K_{|V|}$ of cost larger than $|V|$ has cost at least
$$d \cdot |V| + 1 > d \cdot |V|.$$
Hence, the d-approximation algorithm A must compute a feasible solution of cost $|V|$, i.e., B accepts G.

Let $G = (V, E)$ not be in HC. Consequently, every feasible solution for $(K_{|V|}, c)$ has a cost larger than $|V|$, i.e., $cost(A(K_{|V|}, c)) > |V|$. Therefore B rejects G.

Thus, B is a polynomial-time algorithm that solves the NP-hard HC problem, which contradicts our assumption P $\neq$ NP. □

In order to attack the TSP problem, we will combine the approximation concept with the concept of searching for a set of easy TSP instances. In what follows we consider the metric TSP, Δ-TSP, that allows only TSP instances that satisfy the triangle inequality (Example 2.40). The triangle inequality is a natural restriction that is satisfied[10] in many applications. Next we show a polynomial-time 2-approximation algorithm for Δ-TSP.

Algorithm SB

Input: A complete graph $G = (V, E)$ with the weight function $c : E \to \mathbb{N}^+$, that satisfies the triangle inequality
$$c(\{u, v\}) \leq c(\{u, w\}) + c(\{w, v\})$$
for all pair-wise different vertices $u, v, w \in V$.

[10] Or almost satisfied

Phase 1. SB computes a minimal spanning tree[11] T of G with respect to c.
Phase 2. SB chooses an arbitrary vertex v from V and performs the depth-first search in T from v. Doing this, SB enumerates the vertices of T in the order in which they are visited. Let H be the resulting sequence of vertices that corresponds to this enumeration.
Output: The Hamiltonian cycle $\overline{H} = H, v$.

We illustrate the work of the algorithm SB for the problem instance B in Figure 7.2(a). A minimal spanning tree

$$T = (\{v_1, v_2, v_3, v_4, v_5\}, \{\{v_1, v_3\}, \{v_1, v_5\}, \{v_2, v_3\}, \{v_3, v_4\}\}$$

of G is depicted in Figure 7.2(b). Figure 7.2(c) shows the depth-first search in T from v_3. We observe that every edge in T is used exactly twice in the depth-first search. This depth-first search determines the sequence

$$H = v_3, v_4, v_1, v_5, v_2$$

of vertices, so

$$\overline{H} = v_3, v_4, v_1, v_5, v_2, v_3 = H, v_3$$

is the output of the algorithm SB (Figure 7.2(d)). The cost of $\overline{H}$ is $2 + 3 + 2 + 3 + 1 = 11$. An optimal solution is

$$v_3, v_1, v_5, v_4, v_2, v_3$$

with the cost $1 + 2 + 2 + 2 + 1 = 8$ (Figure 7.2(e)).

Theorem 7.15. *The algorithm* SB *is a polynomial-time 2-approximation algorithm for* Δ-TSP.

Proof. First we analyze the time complexity of SB. A minimal spanning tree of a graph $G = (V, E)$ can be found in time $O(|E|)$. The depth-first search in a tree $T = (V, E')$ runs in time $O(|V|)$. Hence,

$$\text{Time}_{\text{SB}}(G) \in O(|E|),$$

i.e., SB works in linear time.

Now we prove that the approximation ratio of SB is at most 2. Let H_{Opt} be an optimal Hamiltonian cycle with $cost(H_{\text{Opt}}) = \text{Opt}_{\Delta\text{-TSP}}(G)$ for a Δ-TSP instance $I = ((V, E), c)$. Let $\overline{H}$ be the output SB(I) of the algorithm SB for the input I. Let $T = (V, E')$ be the minimal spanning tree constructed in the first phase of SB. First, we observe that

$$cost(T) = \sum_{c \in E'} c(e) < cost(H_{\text{Opt}}), \tag{7.1}$$

[11] A spanning tree of a graph $G = (V, E)$ is a tree $T = (V, E')$ with $E' \subseteq E$. The cost of T is the sum of the costs of all edges in E'.

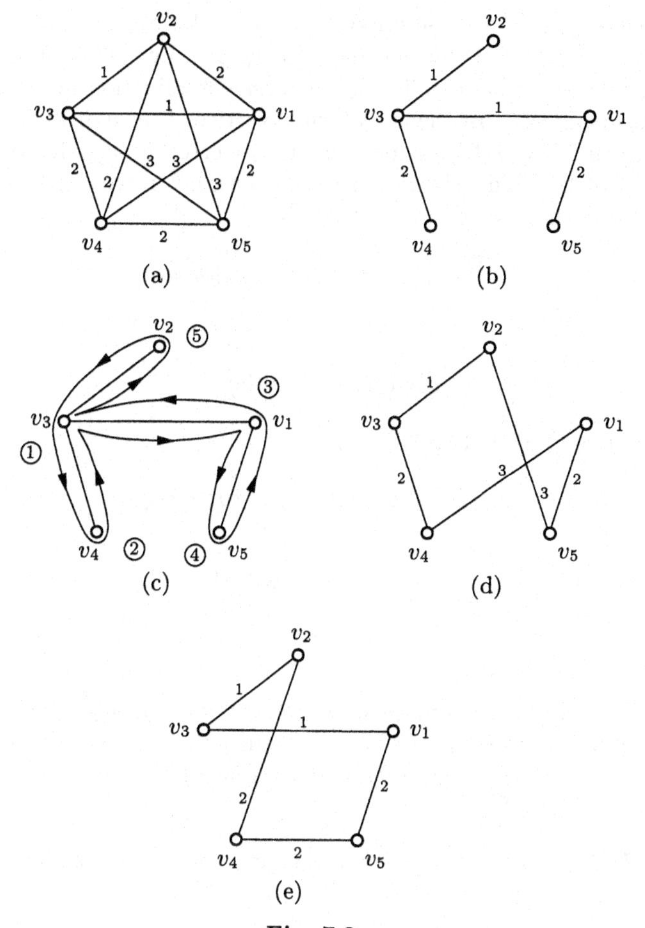

Fig. 7.2.

because by deleting an edge from H_{Opt}, one obtains a spanning tree of G and T is the optimal spanning tree in G.

Consider W to be the path that corresponds to the depth-first search of T. W goes exactly twice via every edge of T (once in every direction of every edge[12]). If $cost(W)$ is the sum of the costs of all edges of W, then

$$cost(W) = 2 \cdot cost(T). \tag{7.2}$$

Both inequality (7.1) and equality (7.2) imply

$$cost(W) < 2 \cdot cost(H_{\mathrm{Opt}}). \tag{7.3}$$

[12] Note that W can be considered as an Eulerian tour of a multigraph T_2 that is constructed from T by doubling every edge.

We observe that $\overline{H}$ can be obtained from W by exchanging[13] some subpaths $u, v_1, \ldots, v_k, v$ in W by the edges $\{u, v\}$ (i.e., by the direct connections u, v). In fact, this can be done by the successive application of a simple operation that exchanges subsequences of three vertices u, w, v of W by the direct connection u, v (this exchange happens if the vertex w has already occurred in the prefix of W). This simple operation does not increase the cost of the path because of the triangle inequality

$$c(\{u, v\}) \leq c(\{u, w\}) + c(\{w, v\}).$$

Hence

$$cost(\overline{H}) \leq cost(W). \tag{7.4}$$

The inequalities (7.3) and (7.4) provide

$$cost(\overline{H}) \leq cost(W) < 2 \cdot cost(H_{\mathrm{Opt}})$$

and hence

$$\frac{\mathrm{SB}(I)}{\mathrm{Opt}_{\Delta\text{-TSP}}(I)} = \frac{cost(\overline{H})}{cost(H_{\mathrm{Opt}})} < 2.$$

□

Exercise 7.16. Find, for any positive integer $n \geq 3$, a weight function c_n for the complete graph K_n of n vertices, such that there exist at least two different weights on the edges of K_n and the algorithm SB always computes an optimal solution.

Exercise 7.17.* For every positive integer $n \geq 4$, find an instance I_n of Δ-TSP with the property

$$\frac{\mathrm{SB}(I_n)}{\mathrm{Opt}_{\Delta\text{-TSP}}(I_n)} \geq \frac{2n-2}{n+1}.$$

7.4 Local Search

Local search is an algorithm design technique for optimization problems. The idea of this technique is to first compute a feasible solution α for a given input x, and then improve α by small (local) changes of α. What the term "small changes" means is defined by the following notion of neighborhood.

Definition 7.18. *Let $\mathcal{U} = (\Sigma_I, \Sigma_O, L, \mathcal{M}, cost, goal)$ be an optimization problem. For every $x \in L$, a* **neighborhood on $\mathcal{M}(x)$** *is any mapping f_x : $\mathcal{M}(x) \to \mathcal{P}(\mathcal{M}(x))$ such that*

[13] This happens when $v_1, \ldots, v_k$ were already visited before u has been visited, but v is visited for the first time in W.

(i) $\alpha \in f_x(\alpha)$ *for every* $\alpha \in \mathcal{M}(x)$
{The solution α *is always in the neighborhood of* α*.},*

(ii) if $\beta \in f_x(\alpha)$ *for* $\alpha, \beta \in \mathcal{M}(x)$*, then* $\alpha \in f_x(\beta)$
{If β *is in the neighborhood of* α*, the* α *is in the neighborhood of* β*.}, and*

(iii) for all $\alpha, \beta \in \mathcal{M}(x)$ *there exist a positive integer* k *and* $\gamma_1, \gamma_2, \ldots, \gamma_k \in \mathcal{M}(x)$*, such that*

$$\gamma_1 \in f_x(\alpha), \gamma_{i+1} \in f_x(\gamma_i) \text{ for } i = 1, \ldots, k-1, \text{ and } \beta \in f_x(\gamma_k).$$

{For all feasible solutions α *and* β*, it is possible to reach* β *from* α *by successively moving from a neighbor to another neighbor.}*

If $\alpha \in f_x(\beta)$ *for some* $\alpha, \beta \in \mathcal{M}(x)$*, we say that* **$\alpha$ and β are neighbors in $\mathcal{M}(x)$**. *The set* **$f_x(\alpha)$** *is called the* **neighborhood of α in $\mathcal{M}(x)$**.

A feasible solution $\alpha \in \mathcal{M}(x)$ *is called a* **local optimum for x with respect to the neighborhood f_x**, *if*

$$cost(\alpha) = goal\{cost(\beta) \mid \beta \in f_x(\alpha)\}.$$

For every $x \in L$*, let the function* f_x *be a neighborhood in* $\mathcal{M}(x)$*. The function*

$$f : \bigcup_{x \in L} (\{x\} \times \mathcal{M}(x)) \to \bigcup_{x \in L} \mathcal{P}(\mathcal{M}(x))$$

defined by

$$f(x, \alpha) = f_x(\alpha)$$

for all $x \in L$ *and all* $\alpha \in \mathcal{M}(x)$ *is a* **neighborhood for $\mathcal{U}$**.

Because of condition (ii) of Definition 7.18, any neighborhood on $\mathcal{M}(x)$ can be viewed as a symmetric relation on $\mathcal{M}(x)$. When one wants to define a neighborhood on $\mathcal{M}(x)$ in an application, then one usually does not work in the formalism of functions or relations. The common way to introduce a neighborhood on $\mathcal{M}(x)$ is to use the so-called **local transformations** on $\mathcal{M}(x)$. The term "local" is important in this context. The meaning of a local transformation of a solution α is that only a local part of the specification of α is changed in order to get another feasible solution. For instance, flipping the Boolean value assigned to a variable is a local transformation for MAX-SAT. Then, the neighborhood of a feasible solution α is the solution α itself and all feasible solutions that can be obtained by an application of chosen local transformations. Thus, for a formula of 5 variables,

$$\{01100, 11100, 00100, 01000, 01110, 01101\}$$

is the neighborhood of $\alpha = 01100$ with respect to the local transformation of flipping one Boolean value (bit) of the assignments.

Exercise 7.19. Prove that the local transformation of bit flipping for the problem MAX-SAT satisfies the definition of a neighborhood.

The most famous neighborhood for TSP is the so-called **2-Exchange** neighborhood (Figure 7.3). The simplest way to define it is to describe the corresponding local transformation. A 2-Exchange local transformation consists of

1. removing two edges $\{a, b\}$ and $\{c, d\}$ with $|\{a, b, c, d\}| = 4$ from a given Hamiltonian tour α that visits these 4 vertices in the order a, b, c, d, and
2. adding two edges $\{a, d\}$ and $\{b, c\}$ to α.

We observe (Figure 7.3) that the resulting object is again a Hamiltonian tour and that the edges $\{a, b\}$ and $\{c, d\}$ cannot be exchanged by edges other than $\{a, d\}$ and $\{b, c\}$ if one wants to get a new Hamiltonian tour.

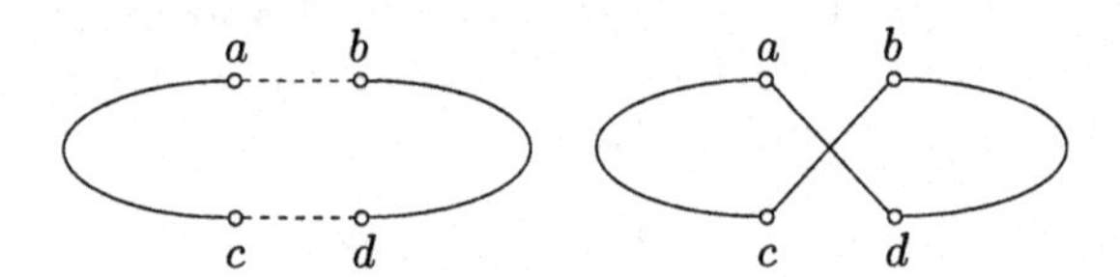

Fig. 7.3.

Exercise 7.20. Does 2-Exchange satisfy the conditions of Definition 7.18? Give formal arguments for your answer.

Exercise 7.21. Let H be a Hamiltonian tour in a graph G. Remove three edges $\{a, b\}$, $\{c, d\}$ and $\{e, f\}$ such that $|\{a, b, c, d, e, f\}| = 6$ and H visits these 6 vertices in the order a, b, c, d, e, f. Draw all possible triples of edges whose addition to H results again in a Hamiltonian tour. How many possibilities are there, if k edges forming a matching are removed for $k \geq 3$?

The local search with respect to a neighborhood is nothing other than an iterative movement in $\mathcal{M}(x)$ from a feasible solution to a better feasible solution. This iterative procedure stops when it reaches a feasible solution β whose neighbors are not better than β. Thus, one can present the scheme of the local search as follows.

Local Search Scheme according to a neighborhood Neigh
LS(Neigh)

Input: An input instance x of an optimization problem $\mathcal{U}$.
Phase 1. Find a feasible solution $\alpha \in \mathcal{M}(x)$
Phase 2.

```
while α is not a local optimum according to Neigh_x do
  begin
    find a β ∈ Neigh_x(α), such that
```

$cost(\beta) < cost(\alpha)$ if $\mathcal{U}$ is a minimization problem, and
$cost(\beta) > cost(\alpha)$ if $\mathcal{U}$ is a maximization problem;
$\alpha := \beta$
end

Output: α

We observe that LS(Neigh) always computes a local optimum with respect to the neighborhood Neigh. If all local optima are also global optima, then the local search guarantees solving the optimization problem. For instance, this is the case for the minimum spanning tree problem, where the neighborhood is determined by the local transformation of exchanging an edge.

If the costs of local optima do not differ too much from the cost of optimal solutions, then one can use the local search for designing approximation algorithms. This is the case for the maximum cut problem, MAX-CUT, defined as follows. The input is a graph $G = (V, E)$. Every pair (V_1, V_2) with

$$V_1 \cup V_2 = V \text{ and } V_1 \cap V_2 = \emptyset$$

is a cut of G. The cost of the cut (V_1, V_2) is the number of edges leading between the vertices from V_1 and V_2, i.e.,

$$cost((V_1, V_2), G) = |E \cap \{\{u, v\} \mid u \in V_1, v \in V_2\}|.$$

The goal is maximization. To define a neighborhood, we consider a local transformation that moves a vertex from one side to the other side. Applying the local-search schema for this neighborhood one obtains the following algorithm.

Algorithm LS-CUT

Input: A graph $G = (V, E)$.
Phase 1. $S = \emptyset$.
{During the computation we always consider the cut $(S, V - S)$. At the beginning the cut is $(\emptyset, V)$.}
Phase 2.

while there exists $v \in V$ such that
$cost(S \cup \{v\}, V - (S \cup \{v\})) > cost(S, V - S)$, or
$cost(S - \{v\}, (V - S) \cup \{v\}) > cost(S, V - S)$ **do**
begin
move v on the opposite side of the cut;
end

Output: $(S, V - S)$.

Theorem 7.22. LS-CUT *is a 2-approximation algorithm for* MAX-CUT.

Proof. Obviously, the algorithm LS-CUT computes a feasible solution for MAX-CUT.

It remains to show that

$$R_{\text{LS-CUT}}(G) \leq 2$$

for every graph $G = (V, E)$. Let (Y_1, Y_2) be the output of LS-CUT. Since (Y_1, Y_2) is a local minimum with respect to moving a vertex from one side to the other side, every vertex $v \in Y_1$ $[Y_2]$ has at least as many edges to the vertices in Y_2 $[Y_1]$ as the number of edges from v to vertices[14] in Y_1 $[Y_2]$. This simple counting argument assures that the cut (Y_1, Y_2) contains at least half of all edges in G. Since $\text{Opt}_{\text{MIN-CUT}}(G)$ cannot exceed $|E|$,

$$R_{\text{LS-CUT}}(G) = \frac{\text{Opt}_{\text{MIN-CUT}}(G)}{cost((Y_1, Y_2))} \leq \frac{|E|}{|E|/2} = 2.$$

□

Exercise 7.23. Prove that LS-CUT is a polynomial-time algorithm.

Exercise 7.24. Consider the so-called maximum weighted cut problem, MAX-WEIGHT-CUT, which is the following generalization of MAX-CUT. An input of MAX-WEIGHT-CUT is a graph $G = (V, E)$ and a weight function $c : E \to \mathbb{N}$, that assigns the weight $c(e)$ to every edge e. The cost of a cut is the sum of the weights of all edges in the cut. The goal is to find a cut with the maximal cost. Clearly, MAX-WEIGHT-CUT is an integer problem. Design a local-search algorithm for MAX-WEIGHT-CUT that would be a pseudopolynomial 2-approximation algorithm.

Algorithms based on the local search are also called local algorithms. Local algorithms are more or less determined by the choice of a neighborhood. The only free parameters of the local search scheme are

1. the strategy of searching for better neighbors, and
2. the decision whether to accept the first better neighbor found or to ultimately estimate one of the best solutions in the neighborhood as the successive feasible solution.

Assuming P $\neq$ NP, there are no polynomial-time local algorithms for NP-hard optimization problems. We observe that the time complexity of a local algorithm can be roughly estimated as

$$(\textit{the time of searching in the neighborhood}) \\ \times (\textit{the number of iterative improvements}).$$

Now, the following question is of interest.

> *For which* NP*-hard optimization problems does a neighborhood* Neigh *of a polynomial size exist such that* LS(Neigh) *always computes an optimal solution?*

[14] In the opposite case the vertex v would be moved to the other side.

This means that we are willing to accept an exponential number of iterative improvements in the worst case, if every iteration runs in polynomial time and the convergence to an optimal solution is guaranteed. The idea behind this is that increasing the size of a neighborhood on the one hand decreases the probability of getting stuck in a poor local optimum, but on the other hand, increases the time complexity of an iteration. The question is whether a neighborhood of a reasonable size exists such that every local optimum is a global optimum too. This question can be formalized as follows.

Definition 7.25. *Let* $\mathcal{U} = (\Sigma_I, \Sigma_O, L, \mathcal{M}, cost, goal)$ *be an optimization problem and let* Neigh *be a neighborhood for* $\mathcal{U}$*. The neighborhood* Neigh *is said to be* **exact** *if, for every* $x \in L$*, every local optimum for* x *with respect to* Neigh_x *is an optimal solution for* x*.*

A neighborhood Neigh *is called* **polynomial-time searchable**[15] *if there is a polynomial-time algorithm*[16] *that, for every* $x \in L$ *and every* $\alpha \in \mathcal{M}(x)$*, finds one of the best feasible solutions in* $\text{Neigh}_x(\alpha)$*.*

Thus for an optimization problem $\mathcal{U} \in$ NPO, our question can be reformulated in the terminology of Definition 7.25 as follows:

Does an exact polynomial-time searchable neighborhood exist for $\mathcal{U}$*?*

A positive answer to this question means that the hardness of the problem from the local search point of view is in the number of iterative improvements needed to reach an optimal solution. In many cases, it means that local search may be suitable for $\mathcal{U}$. For instance, if $\mathcal{U}$ is an integer-valued optimization problem, then the existence of a polynomial-time searchable exact neighborhood Neigh usually implies that LS(Neigh) is a pseudopolynomial algorithm[17] for $\mathcal{U}$. The famous positive example is the Simplex algorithm for linear programming. It is based on the existence of an exact, polynomial-time searchable neighborhood, but one cannot exclude the necessity of exponentially many iterative improvements.

A negative answer to this question implies that no polynomial-time searchable neighborhood can assure the success of local algorithms in searching for an optimal solution. Hence, one can at most try to obtain a local optimum in polynomial time in such a case.

---

[15] Note that if a neighborhood is polynomial-time searchable, this does not necessarily mean that this neighborhood is of a polynomial size.

[16] Note that we consider optimization problems from NPO only, and hence any algorithm working in polynomial time according to $|\alpha|$ is a polynomial-time algorithm according to $|x|$ too.

[17] For an integer-valued problem, any iterative step improves the solution cost at least by 1 and the cost of any solution is usually bounded by the sum of all input values.

Exercise 7.26. Let k-Exchange be the neighborhood for TSP, in which k edges may be exchanged when moving from a Hamiltonian tour to another one. What is the cardinality of the neighborhood $k\text{-Exchange}_G(H)$ of a Hamiltonian tour H when G has n vertices?

Exercise 7.27.* Prove that, for any positive integer k, k-Exchange is not an exact neighborhood for TSP.

Our next aim is to introduce a technique that can be successfully applied for proving the nonexistence of exact, polynomial-time searchable neighborhoods for concrete optimization problems. Again, we will see that the concept of NP-hardness works for this purpose.

Definition 7.28. *Let $\mathcal{U} = (\Sigma_I, \Sigma_O, L, \mathcal{M}, cost, goal)$ be an integer-valued optimization problem. We say that $\mathcal{U}$ is* **cost-bounded**, *if for every instance x with $\text{Int}(x) = (i_1, i_2, \ldots, i_n)$, $i_j \in \mathbb{N}$ for $j = 1, 2, \ldots, n$,*

$$cost(\alpha) \in \left\{1, 2, \ldots, \sum_{j=1}^{n} i_j\right\}$$

for every feasible solution $\alpha \in \mathcal{M}(x)$.

We observe that almost all known integer-valued optimization problems are cost-bounded and hence this requirement does not place any serious restriction on the applicability of the following method.

Theorem 7.29. *Let $\mathcal{U} \in$ NPO be a cost-bounded integer-valued optimization problem. If* P $\neq$ NP *and $\mathcal{U}$ is strongly* NP*-hard, then there does not exist any exact, polynomial-time searchable neighborhood for $\mathcal{U}$.*

Proof. We give an indirect proof of Theorem 7.29. Assume that $\mathcal{U}$ possesses an exact, polynomial-time searchable neighborhood Neigh. Then for any input x, one iteration step of $\text{LS}(\text{Neigh}_x)$ can be executed in a polynomial time of $p(|x|)$. Since $\mathcal{U} \in$ NPO, the initial feasible solution can also be computed in a polynomial time. In every iteration step of $\text{LS}(\text{Neigh}_x)$ the cost of the actual feasible solution is improved by at least 1, because the costs of feasible solutions are integers. Moreover, all possible costs are from the integer interval

$$\text{from 0 to } \sum_{j \in \text{Int}(x)} j \leq |x| \cdot \text{MaxInt}(x)$$

because $\mathcal{U}$ is a cost-bounded integer-valued optimization problem. Consequently, the number of iterative improvements is bounded by

$$|x| \cdot \text{MaxInt}(x).$$

Therefore, the overall time complexity of LS(Neigh) is in

$$O(p(|x|) \cdot |x| \cdot \text{MaxInt}(x)).$$

Since Neigh is an exact neighborhood, LS(Neigh) computes an optimal solution for x. Hence LS(Neigh) is a pseudopolynomial algorithm for $\mathcal{U}$. Since $\mathcal{U}$ is strongly NP-hard, the existence of a pseudopolynomial algorithm for $\mathcal{U}$ contradicts the assumption P $\neq$ NP (Theorem 7.7). □

In Chapter 6 we proved that TSP and MAX-CL are strongly NP-hard. We observe that both problems are cost-bounded, integer-valued optimization problems. Hence they do not possess any exact, polynomial-time searchable neighborhood. One can even prove that there is no exact neighborhood of size $2^{\sqrt[3]{n}}$ for TSP.

7.5 Simulated Annealing

In this section we introduce simulated annealing as a heuristic for solving hard optimization problems. The term heuristic in the area of combinatorial optimization is ambiguously specified and is used with different meanings. A heuristic algorithm in a very general sense is a consistent algorithm for an optimization problem that is based on some transparent, usually simple idea of searching in the set of all feasible solutions, and that does not guarantee finding any optimal solution. In this context people speak about local search heuristics, or a greedy heuristic, even if these heuristics provide approximation algorithms. In a narrow sense considered here, a heuristic is a technique providing a consistent algorithm for which nobody is able to prove that it provides feasible solutions of a reasonable quality in a reasonable time (for instance, polynomial time), but the idea of the heuristic seems to promise good behavior for typical instances of the optimization problem considered.

Despite its lack of any general assurance of a reasonable behavior, heuristics became very popular and widely used. Two main reasons for this are:

1. Heuristics are usually simple and hence easy to implement and test. Thus the investment for developing a heuristic algorithm is usually much lower than the cost of designing technical, very specialized algorithms for the given problem.
2. Heuristics are robust, which means that a heuristic can successfully work for a large class of problems, even if these problems have very different combinatorial structure. The consequence is that a change of the problem specification during the process of algorithm design does not cause any serious problems. Usually it is sufficient to change or readjust a few parameters of the heuristic. For the design of a very specific problem-oriented optimization algorithm, a change of the problem specification is often accompanied by an essential change of the combinatorial structure that requires to move back to the very beginning of the algorithm design.

If one applies local search for a hard problem and is unable to analyze the behavior of the resulting algorithm, then one can consider local search as a heuristic. We see that it has the property of robustness because it can be applied to almost any optimization problem and there is no doubt about its simplicity. The main drawback of local search is that it finishes the work in a local optimum, regardless of how good or bad it is. Our next aim is to improve the local search method by enabling the departure from a local optima in order to search for a better solution. The idea is to mimic the physical optimization based on the laws of thermodynamics.

In condensed matter physics, annealing is a process for obtaining low-energy states of a solid in a thermal bath. This process can be viewed as an optimization process in the following sense. At the beginning, one has a solid material with a number of imperfections in its crystal structure. The aim is to get the perfect structure of the solid, which corresponds to the state of the solid of minimal energy. The physical process of annealing consists of two steps.

1. The temperature of the thermal bath is increased to a maximum value at which the solid melts. This causes all the particles to arrange themselves randomly.
2. The temperature of the thermal bath is slowly decreased according to a given cooling schedule until a low-energy state of the solid (a perfect crystal structure) is achieved.

The crucial point is that this optimization process can be successfully modeled by the so-called Metropolisalgorithm that can be viewed as a randomized local search. In what follows, for a given state s of the solid, $\mathrm{E}(s)$ denotes the energy of this state. The Boltzmann constant is denoted by c_B.

Metropolis Algorithm

Input: A state s of the solid with energy $\mathrm{E}(s)$.
Phase 1. Choose the initial temperature T of the thermal bath.
Phase 2.

Generate a state q from s by applying a perturbation mechanism, which transfers s into q by a small random distortion (for instance, by a random displacement of a small particle).
`if` $\mathrm{E}(q) \leq \mathrm{E}(s)$ `then` $s := q$
{accept q as a new state}
`else` accept q as a new state with the probability
$$\mathrm{prob}(s \to q) = \mathrm{e}^{-\frac{\mathrm{E}(q)-\mathrm{E}(s)}{c_B \cdot T}};$$
{i.e., remain in the state s with the probability $1 - \mathrm{prob}(s \to q)$}

Phase 3.

Decrease T appropriately.
`if` T is not too close to 0 `then goto` Phase 2;
`else output`(s);

First of all, we observe a strong similarity between the Metropolis algorithm and the local search scheme. To move from the current state s, one considers only a small, local change in the description of s so as to generate q. If the generated state q is at least as good as s (or even better), then q is considered as the new state.

The main differences between the Metropolis algorithm and the local search scheme are:

(i) The Metropolis algorithm may accept a deterioration with the probability

$$\text{prob}(s \to q) = \mathrm{e}^{-\frac{\mathrm{E}(q)-\mathrm{E}(s)}{c_B \cdot T}}.$$

(ii) The value of the parameter T decides the termination of the Metropolis algorithm while the local optimality is the criterion for stopping in the local search scheme.

The probability $\text{prob}(s \to q)$ obeys the laws of thermodynamics that claim that at temperature T, the probability $\text{prob}(\Delta\mathrm{E})$ of an increase in energy of magnitude $\Delta\mathrm{E}$ is given by

$$\text{prob}(\Delta\mathrm{E}) = \mathrm{e}^{\frac{-\Delta\mathrm{E}}{c_B \cdot T}}.$$

This probability is essential for proving the convergence of the Metropolis algorithm to an optimal state. For our applications, the most important properties of $\text{prob}(s \to q)$ are:

(i) The probability $\text{prob}(s \to q)$ of the movement from the state s to the state q decreases with increasing $\mathrm{E}(q) - \mathrm{E}(s)$, i.e., large deteriorations are less probable than small deteriorations.
(ii) The probability $\text{prob}(s \to q)$ increases with T, i.e., large deteriorations are more probable at the beginning (when T is large) than later (when T becomes smaller and smaller).

The crucial point is that one allows deteriorations in order to get a possibility of leaving local optima. Without allowing deteriorations, one cannot guarantee the convergence to an optimum. More precisely, this means that at the beginning (when T is high), it is possible to overcome "high hills" around very deep local optima in order to reach some deep valleys, where later only some not too deep local optima are left. Intuitively, this optimization approach can be viewed as a recursive procedure in the following sense. First, one climbs to the top of a very high mountain and looks for the most promising areas (deep valleys). Then one goes to such an area and recursively continues to search for the minimum in this area only.

If one aims to use the strategy of the Metropolis algorithm in combinatorial optimization, then one has to use the following one-to-one correspondence between the terms of thermodynamic optimization and combinatorial optimization.

the set of system states $\widehat{=}$ the set of feasible solutions
the energy of a state $\widehat{=}$ the cost of a feasible solution
perturbation mechanism $\widehat{=}$ a random choice of a neighbor
an optimal state $\widehat{=}$ an optimal feasible solution
temperature $\widehat{=}$ a control parameter

The simulated annealing algorithm is a local-search algorithm that is based on the analogy to the Metropolis algorithm. If one fixes a neighborhood f for a minimization problem $\mathcal{U} = (\Sigma_I, \Sigma_O, L, \mathcal{M}, cost, Minimum)$, then the simulated annealing algorithm can be described as follows.

Simulated Annealing for $\mathcal{U}$ with respect to f
SA(f)

Input: A problem instance $x \in L$.
Phase 1. Compute a feasible solution $\alpha \in \mathcal{M}(x)$.
Select an initial temperature T.
Select a temperature-reduction function g as a mapping of two parameters T and the number of iterations I (time).
Phase 2.

```
I := 0;
  while T > 0 (or T is not too close to 0) do
    begin
      choose randomly a β from f_x(α);
      if cost(β) ≤ cost(α) then α := β;
      else
        begin
          generate a random number r uniformly in the range
          [0,1];
          if r < e^(-(cost(β) - cost(α))/T) then α := β;
        end
      I := I + 1;
      T := g(T, I);
    end
  end
```

Output: α.

Taking a "reasonable" neighborhood, and selecting an appropriate temperature T and temperature-reduction function g, one can prove that SA(f) converges to an optimum. The problem is that the number of iterations needed to reach an optimum cannot be bounded. Even attempts to give a guarantee of a good approximation ratio by SA(f) led to situations in which one needs more iterations of SA(f) than $|\mathcal{M}(x)|$ in order to assure a reasonable approximation ratio. Nevertheless, there are many applications where simulated annealing provides acceptable solutions and, due to its small implementation costs, it is preferred over other methods. Another positive property of simulated annealing is that the choice of the parameters T and g and of the termination criterion can be left to the user. Therefore the user alone can decide about the priorities with respect to the tradeoff between the running time and the solution quality.

7.6 Summary

Algorithm design techniques are crucial for the success in solving hard problems, because the quantitative jumps in the requirement on computer resources (for instance, from an exponential-time complexity to a polynomial one) are not achievable by any improvements in hardware technologies. To get efficient algorithms for hard problems, one has to pay on the level of requirements attached to the problem specification. Either one reduces the set of feasible inputs to a subclass to typical[18] inputs (i.e., one does not solve the problem in its general formal setting) or one solves the problem without the guarantee of always getting the correct or optimal solution. The art of algorithm design is in getting a big gain on the level of efficiency by paying a small "discount" on the level of problem setting (requirements).

Pseudopolynomial algorithms run in polynomial time on instances of integer problems, where the values of the integers are polynomial in the input length. One can design a pseudopolynomial algorithm for the knapsack problem. The concept of NP-completeness is also helpful here for proving the nonexistence of pseudopolynomial algorithms for concrete problems under the assumption $\mathrm{P} \neq \mathrm{NP}$.

Approximation algorithms are algorithms for optimization problems, that provide feasible solutions whose costs do not differ too much from the costs of optimal solutions. One can design approximation algorithms for the metric TSP, MAX-VC, and MAX-CUT. If $\mathrm{P} \neq \mathrm{NP}$, then there is no approximation algorithm for the general TSP. This negative result can be again obtained by a suitable application of the NP-hardness concept.

For an optimization problem, the local search scheme starts with an arbitrary feasible solution and tries to get a better solution by iteratively improving the current solution. One iteration consists of searching for a neighbor of the actual solution α that is better than α itself. The neighbors of a feasible solution α are defined by applying local transformations to α, where a local transformation may change only a local part of the specification of α. Local algorithms end always in a local optimum with respect to the neighborhood (allowed local transformations). Since the costs of local optima may essentially differ from the costs of optimal solutions, local search does not provide any guarantee for the quality of the solutions computed for many problems. Simulated annealing is a heuristic that is based on local search, but allows

[18] With respect to the application considered

the departure from local optima. Simulated annealing is robust and easy to implement and hence it is commonly used in many applications.

The concept of pseudopolynomial algorithms for integer problems and of strong NP-hardness stemmed from Garey and Johnson [19]. The pseudopolynomial algorithm for the knapsack problem was designed by Ibarra and Kim [34]. The first approximation algorithm was proposed by Graham [22]. Bock [6] and Coes [15] present the first local-search algorithms. The concept of exact, polynomial-time searchable neighborhood was introduced by Papadimitriou and Steiglitz [50]. The Metropolis algorithm for the simulation of the annealing process was discovered by Metropolis, A.W and M.N. Rosenbluth, A.M. and E. Teller [46]. The idea of applying the Metropolis algorithm in combinatorial optimization was independently presented by Černý [8] and Kirkpatrick, Gellat and Vecchi [37].

A systematic survey of methods for solving hard problems is given in [30]. For further reading on different algorithm design techniques, we strongly recommend Papadimitriou and Steiglitz [50], Cormen, Leiserson and Rivest [14] and Schöning [63]. Extensive sources about the theory of approximation algorithms are Ausiello, Crescenzi, Gambosi, Kann, Marchetti-Spaccamela and Protasi [1], Hochbaum [26], Mayr, Prömel and Steger [45] and Vazirani [69].

The tissue of the world
is built from necessities and randomness;
The intellect of men places itself between both
and can control them;
It considers the necessity
as the reason of its existence;
it knows how randomness can be
managed, controlled, and used ...

J. W. von Goethe

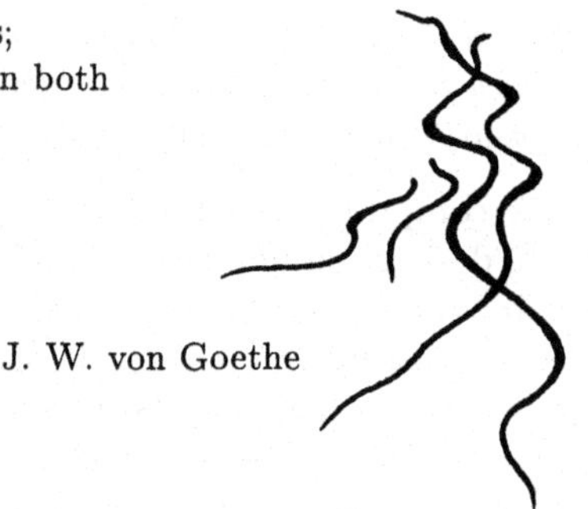

8 Randomization

8.1 Objectives

The notion of "randomness" is one of the most fundamental and most discussed terms in science. The fundamental question is whether randomness really exists or do we use this term only to model objects and events with unknown lawfulness. Philosophers and scientists dispute about the answer to this question since ancient times. The opinion of Demokrit was that

the randomness is the unknown,
and that the nature is determined in its fundamentals.

Thus, Demokrit asserted that order conquers the world and that this order is governed by unambiguous laws. In contrast to Demokrit, Epikur claimed that

the randomness is objective,
it is the proper nature of events.

Before the 20th century, the worldly view of people was based on causality and determinism, because of religions and later optimism caused by the success of natural sciences and mechanical engineering.

This belief in determinism also had emotional roots, because people related randomness to chaos and uncertainty, which were always connected with fear, so the possibility of random events was not accepted. Even Albert Einstein blamed randomness on insufficient knowledge and claimed that each randomized model of physical reality could be exchanged for a deterministic one, if the appropriate knowledge is discovered. The development of science (especially, physics and biology) in the 20th century returned the world to Epikur's view on randomness. Experimental physics confirmed the theory of quantum mechanics, whose core is based on random events.

Today, evolutionary biology considers the random mutation of DNA as a crucial instrument of the evolution. A nice, modern view on randomness is given by the Hungarian mathematician Alfréd Rényi.

Randomness and order do not contradict each other; more of less, both may be true at once The randomness controls the world and due to this in the world there are order and laws, which can be expressed in measures of random events that follow the laws of probability theory.

For us as computer scientists, it is important to realize that it can be very profitable to design and implement randomized algorithms and systems instead of completely deterministic ones. This realization is nothing other than the acceptance of nature as a teacher. It seems to be the case that nature always uses the most efficient and simplest way to achieve its goal, and that randomization of a part of the control is an essential concept of nature's strategy. Computer science practice confirms this viewpoint. In many everyday applications, simple randomized systems and algorithms do their work efficiently with a high degree of reliability and we do not know any deterministic algorithms that would do the same with a comparable efficiency. We even know of examples where the design and use of deterministic counterparts of some randomized algorithms is beyond physical limits. This is also the reason why currently one does not connect the tractability (the class of practically solvable problems) with deterministic polynomial time, but with randomized polynomial time.

In this chapter, we do not aim to present the fundamentals of the design and analysis of randomized algorithms and the complexity theory of randomized computations, because this would require too much knowledge in probability theory, complexity theory, algebra, and number theory. Instead, we prefer to present three simple examples of randomized algorithms that transparently explain the randomization concept and even help to build the intuition why randomization may be more powerful than any deterministic control.

This chapter is organized as follows. We present some elementary fundamentals of probability theory in Section 8.2. In Section 8.3 we design a randomized communication protocol for comparing the contents of two large databases, that is substantially more efficient than any deterministic communication protocol for this purpose. Section 8.4 uses the randomized protocol designed in Section 8.3 in order to introduce the method of abundance of witnesses as a paradigm of the design of randomized algorithms. We apply this method once again to show how an efficient randomized algorithm for primality testing can be developed. Note that primality testing is one of the most important decision problems of current practice and that we do not know of any efficient[1] deterministic algorithm for this task. In Section 8.5 the fingerprinting method as a special case of the method of abundance of witness is presented. We apply this method to efficiently decide the equivalence of two polynomials. As usual, we close this chapter with a short summary.

---

[1] Note that recently [44] a first deterministic polynomial-time algorithm for primality testing has been discovered. This result is of enormous theoretical importance, but the time complexity of the algorithm is in $O(n^{12})$ and so it is not practical.

8.2 Elementary Probability Theory

If an event is an inevitable consequence of another event, then one speaks of causality or determinism. As already mentioned in the introduction, there may exist events that are not completely determined. Probability theory was developed in order to model and analyze situations and experiments with ambiguous outcomes. Simple examples of such experiments are tossing (flipping) a coin or rolling a 6-sided dice. If there is no (apparent) possibility of predicting the outcome of such experiments, then one speaks of **random events**. When modeling a probabilistic experiment, one considers all possible outcomes of the experiment, called **elementary events**. From the philosophical point of view it is important that these elementary events are atomic. Atomic means that an elementary event cannot be viewed as a collection of other[2] events of the experiments and so one elementary event excludes any other elementary events. For the tossing of a coin, the elementary events are "head" and "tail". For the rolling of a 6-sided dice the elementary events are "1", "2", "3", "4", "5" and "6". An **event** is a set of elementary events (i.e., a subset of the set of elementary events). For instance, $\{2, 4, 6\}$ is an event of dice rolling, that corresponds to rolling an even number. Since elementary events can be also considered as events, we represent elementary events as one-element sets.

In the following we consider only experiments with finitely many elementary events, which increases the transparency of the next definition. Our aim now is to develop a reasonable theory that assigns a probability to every event. This aim was not easy to achieve. The probability theory took almost 300 years to advance from the works of Pascal, Fermat and Huygens in the middle of the 17th century to the currently accepted axiomatic definition of probability by Kolmogorov. Limiting the set S of elementary events to a finite set here is helpful in removing the technicalities connected with a possible uncountability of S in the general definition of Kolmogorov. The basic idea is to define the probability of an event E as

the ratio between the sum of probabilities of (favorable) elementary events involved in E to the sum of the probabilities of all possible elementary events. (8.1)

Fixing the probability of events in the above way one standardizes the probability values in the sense, that the probability 1 corresponds to a certain event[3] and the probability 0 corresponds to an impossible (empty[4]) event.

Another important point is that the probabilities of elementary events unambiguously determine the probabilities of all events. For symmetric experiments such as tossing a coin, one wants to assign the same probability to all elementary events.

---

[2] Even more elementary

[3] To the event S consisting of all elementary events

[4] Called also null event

Let $\text{Prob}(E)$ be the probability of an event E. In our model the result of the experiment must be one of the elementary events, hence we set $\text{Prob}(S) = 1$ for the set S of all elementary events. Then, for the rolling of a 6-sided dice, we have

$$\begin{aligned}\text{Prob}(\{2,4,6\}) &= \frac{\text{Prob}(\{2\}) + \text{Prob}(\{4\}) + \text{Prob}(\{6\})}{\text{Prob}(S)} \\ &= \text{Prob}(\{2\}) + \text{Prob}(\{4\}) + \text{Prob}(\{6\}) \\ &= \frac{1}{6} + \frac{1}{6} + \frac{1}{6} = \frac{1}{2},\end{aligned}$$

i.e., the probability of getting an even number is exactly $1/2$. Following the concept (8.1) of measuring probability, we obtain

$$\begin{aligned}\text{Prob}(X \cup Y) &= \frac{\text{Prob}(X) + \text{Prob}(Y)}{\text{Prob}(S)} \\ &= \text{Prob}(X) + \text{Prob}(Y)\end{aligned}$$

for all disjoint events X and Y. These considerations result in the following axiomatic definition of probability.

Definition 8.1. *Let S be the set of all elementary events of a probability experiment. A* **probability distribution on S** *is every function*

$$\text{Prob} : \mathcal{P}(S) \to [0, 1]$$

that satisfies the following conditions (probability axioms):

(i) $\text{Prob}(\{x\}) \geq 0$ *for every elementary event* x,
(ii) $\text{Prob}(S) = 1$, *and*
(iii) $\text{Prob}(X \cup Y) = \text{Prob}(X) + \text{Prob}(Y)$ *for all events* $X, Y \subseteq S$ *with* $X \cap Y = \emptyset$.

Prob(X) *is called the* **probability of the event X**. *The pair* (S, Prob) *is called a* **probability space**. *If*

$$\text{Prob}(\{x\}) = \text{Prob}(\{y\})$$

for all $x, y \in S$, Prob *is called the* **uniform probability distribution on S**.

Exercise 8.2. Prove that the following properties hold for every probability space (S, Prob):

(i) $\text{Prob}(\emptyset) = 0$.
(ii) $\text{Prob}(S - X) = 1 - \text{Prob}(X)$ for every $X \subseteq S$.
(iii) $\text{Prob}(X) \leq \text{Prob}(Y)$, for all $X, Y \subseteq S$ with $X \subseteq Y$.
(iv) $\text{Prob}(X \cup Y) = \text{Prob}(X) + \text{Prob}(Y) - \text{Prob}(X \cap Y)$
$\leq \text{Prob}(X) + \text{Prob}(Y)$ for all $X, Y \subseteq S$.
(v) $\text{Prob}(X) = \sum_{x \in X} \text{Prob}(x)$ for all $X \subseteq S$.

We observe that all properties from Exercise 8.2 correspond to our intuition, and hence to the informal concept (8.1) of probability. Thus the addition of probabilities corresponds to the idea that the probability of several pairwise exclusive (disjoint) events is the sum of the probabilities of these events.

To what does the multiplication of two probabilities correspond? Consider two probabilistic experiments that are independent in the sense that the result of an experiment has no influence on the result of the other experiment. An example of such a situation is the rolling of a dice twice. It does not matter, whether one rolls two dice at once or whether one uses the same dice twice, because the results do not influence each other. For instance, the elementary event "3" of the first roll does not have any influence on the result of the second roll. We know that $\text{Prob}(i) = \frac{1}{6}$ for both experiments and for all $i \in \{1, 2, \ldots, 6\}$. Consider now joining both probabilistic experiments into one probabilistic experiment. The set of elementary events of this joined experiment is

$$S_2 = \{(i, j) \mid i, j \in \{1, 2, \ldots, 6\}\},$$

where for an elementary event $\{(i, j)\}$ of S_2, i is the result of the first roll and j is the result of the second roll. What is the fair probability distribution Prob_2 on S_2, that can be determined from the basic experiment$(\{1, 2, \ldots, 6\}, \text{Prob})$? We consider our hypothesis that the probability of an event consisting of two fully independent events is equal to the product of the probabilities of these events, so

$$\text{Prob}_2(\{(i, j)\}) = \text{Prob}(\{i\}) \cdot \text{Prob}(\{j\}) = \frac{1}{6} \cdot \frac{1}{6} = \frac{1}{36}$$

for all $i, j \in \{1, 2, \ldots, 6\}$. We verify the correctness of this hypothesis. The set S_2 contains exactly 36 elementary events and each of these elementary events is equally probable. Hence

$$\text{Prob}_2(\{(i, j)\}) = \frac{1}{36}$$

for all $(i, j) \in S_2$.

Exercise 8.3. Let k be a positive integer. Let (S, Prob) be a probability space where Prob is a uniform probability distribution over $S = \{0, 1, 2, \ldots, 2^k - 1\}$. Create (S, Prob) from k coin tossing experiments.

It remains to explain how the probability theory can be used to model, design, and analyze randomized algorithms. One possibility is to start with the model of an NTM with finite computations and replace each nondeterministic branching (guess) by a random experiment. This means that for a nondeterministic choice from k possibilities one considers the k possibilities as k elementary events each of probability $\frac{1}{k}$. Then the probability of a computation is the product of probabilities of all random decisions of this computation. Let $S_{A,x}$ be the set of all computations of an NTM (a nondeterministic program) A on an input x. Assigning the probability Prob(C) to any computation C from $S_{A,x}$ in the above-described way, one obtains the probability space $(S_{A,x}, \text{Prob})$.

Exercise 8.4. Prove that $(S_{A,x}, \text{Prob})$ is a probability space.

The sum of the probabilities of the computations from $S_{A,x}$ with a wrong output for the input x is called the **error probability of the algorithm *A* on an input *x***, denoted $\mathbf{Error}_A(x)$. The **error probability of the algorithm *A*** is defined as a function $\text{Error}_A : \mathbb{N} \to \mathbb{N}$ with

$$\mathbf{Error}_A(n) = \max\{\text{Error}_A(x) \mid |x| = n\}.$$

Modeling randomized algorithms by probabilistic experiments can also be used for analyzing the probability[5] that a computation of A on x finishes in at most $t(n)$ steps for a given function t, hence judging the efficiency of a randomized algorithm.

Another simple possibility of modeling randomized algorithms is to consider a randomized algorithm as a probability distribution over a set of deterministic algorithms. This corresponds to the idea of giving a TM A (of a deterministic algorithm) an additional tape containing a sequence of random bits as an addition input. Each sequence of random bits determines unambiguously a (deterministic) computation of A on the given input x. Considering the random bit sequences as elementary events corresponds to considering the set $S_{A,x}$ of all computations of A on x as the set of elementary events. Usually, all random sequences have the same probability, so the randomized algorithm is a uniform probability distribution over all computations from $S_{A,x}$. The examples of randomized algorithms presented in the next sections are based on this simple modeling of randomized control. In these examples the sequences of random bits are viewed as random numbers (primes) that decide the choice of a deterministic strategy for solving a given problem.

8.3 A Randomized Communication Protocol

The main aim of this section is to show that randomized algorithms can be much more efficient than their most efficient deterministic counterparts. Consider the following task. We have two computers C_{I} and C_{II} that are very far apart (for instance, one in Europe and the other one in America). Originally, both computers contained the same database, but then they developed independently with the aim of getting the complete information about the database subject (for instance, genome sequences) in both locations. After some time, we want to check whether this process is successful, i.e., whether

[5] This probability is nothing other than the sum of the probabilities of all computations shorter than $t(|x|) + 1$.

C_{I} and C_{II} contain the same data. Let n be the size of the database in bits. For instance, n can be approximately 10^{16}, which is realistic for biological applications. Our goal is to design a communication protocol between C_{I} and C_{II} that is able to determine whether the data saved in both computers are the same or not. The complexity of the communication protocol is the number of bits that has to be exchanged between C_{I} and C_{II} in order to solve this decision problem, and obviously we try to minimize this complexity.

One can prove that every deterministic communication protocol solving this task must exchange at least n bits[6] between C_{I} and C_{II}, i.e., there exists no deterministic protocol that solves this task by communicating $n-1$ or fewer bits. Sending 10^{16} bits and additionally assuring that all arrive safely[7] at the other side is a practically nontrivial task, so one would probably not do it in this way.

A reasonable solution can be given by the following randomized protocol, which is based on the prime number theorem (Theorem 2.67).

$R = (C_{\mathrm{I}}, C_{\mathrm{II}})$ (Randomized Communication Protocol for Equality)

Initial situation: C_{I} has a sequence x of n bits, $x = x_1 \dots x_n$,
C_{II} has a sequence y of n bits $y = y_1 \dots y_n$.
Goal: Determine whether $x = y$ or $x \neq y$.
Phase 1. C_{I} chooses uniformly a prime p from the interval $[2, n^2]$ at random.
{Note, that there are approximately $Prim\,(n^2) \sim n^2/\ln n^2$ primes in this interval, and hence $\lceil \log_2 n^2 \rceil \leq 2 \cdot \lceil \log_2 n \rceil$ random bits are sufficient for this choice (representation).}
Phase 2. C_{I} computes the number

$$s = Number(x) \bmod p$$

and sends the binary representation of s and p to C_{II}.
{Since the binary representations of s and p consist of at most $\lceil \log_2 n^2 \rceil$ bits ($s \leq p < n^2$), the length of the message is at most $4 \cdot \lceil \log_2 n \rceil$. }
Phase 3. After reading s and p, C_{II} computes the number

$$q = Number(y) \bmod p.$$

If $q \neq s$, then C_{II} outputs "$x \neq y$".
If $q = s$, then C_{II} outputs "$x = y$".

Now, we analyze the work of $R = (C_{\mathrm{I}}, C_{\mathrm{II}})$. First we look at the complexity measured in the number of communication bits and then we analyze the reliability (error probability) of $R = (C_{\mathrm{I}}, C_{\mathrm{II}})$.

---

[6] This means that sending all data of C_{I} to C_{II} for the comparison is an optimal communication strategy.

[7] Without flipping a bit

The only communication of the protocol involves submitting the binary representations of the positive integers s and p. As we have already observed, $s \leq p < n^2$, hence the length of the message is at most

$$2 \cdot \lceil \log_2 n^2 \rceil \leq 4 \cdot \lceil \log_2 n \rceil.$$

For $n = 10^{16}$, the binary length of the message is at most $4 \cdot 16 \cdot \lceil \log_2 10 \rceil = 256$. This is a very short message that can be safely transferred.

Now we not only show for most inputs (initial situations) that this randomized strategy works, but also that the probability of providing the right answer is high for every input. Analyzing the error probability we distinguish two possibilities with respect to the real relation between x and y.

(i) Let $x = y$. Then

$$Number(x) \bmod p = Number(y) \bmod p$$

for all primes p. Therefore, C_{II} outputs "equal" with certainty, i.e., the error probability is equal to 0.

(ii) Let $x \neq y$. One obtains the wrong answer "equal" only if C_{I} has chosen a prime p such that

$$z = Number(x) \bmod p = Number(y) \bmod p.$$

This means that

$$Number(x) = x' \cdot p + z \text{ and } Number(y) = y' \cdot p + z$$

for some natural numbers x' and y'.
This implies that

$$Number(x) - Number(y) = x' \cdot p - y' \cdot p = (x' - y') \cdot p,$$

i.e., p divides the number

$$w = |\,Number(x) - Number(y)|.$$

Thus, the protocol $R = (C_{\mathrm{I}}, C_{\mathrm{II}})$ outputs a wrong answer only if the chosen prime p divides w. The prime p is randomly chosen from the $Prim\,(n^2)$ primes from $\{2, 3, \ldots, n^2\}$ with respect to the uniform probability distribution. Thus, to estimate the error probability it is sufficient to estimate how many primes from these

$$Prim\,(n^2) \sim n^2 / \ln n^2$$

primes can divide w. Since the length of the binary representations of x and y is equal to n,

$$w = |\,Number(x) - Number(y)| < 2^n.$$

Obviously,[8] we can factorize w to get

[8] We know from elementary number theory that every positive integer has a unique factorization.

$$w = p_1^{i_1} p_2^{i_2} \dots p_k^{i_k},$$

where $p_1 < p_2 < \cdots < p_k$ are primes and $i_1, i_2, \dots, i_k$ are positive integers. Our aim is to prove

$$k \leq n - 1.$$

We prove it by contradiction. Assume $k \geq n$. Then

$$w = p_1^{i_1} p_2^{i_2} \dots p_k^{i_k} \geq p_1 p_2 \dots p_n > 1 \cdot 2 \cdot 3 \cdot \dots \cdot n = n! > 2^n,$$

which contradicts the fact that $w < 2^n$. In this way we have proved that w has at most $n-1$ different prime factors. Since every prime in $\{2, 3, \dots, n^2\}$ has the same probability of being chosen, the probability of choosing a prime p dividing w is at most

$$\frac{n-1}{Prim\,(n^2)} \leq \frac{n-1}{n^2/\ln n^2} \leq \frac{\ln n^2}{n}$$

for sufficiently large n.
Thus the error probability of R for two different inputs x and y is at most

$$\frac{\ln n^2}{n}$$

which is at most $0.36892 \cdot 10^{-14}$ for $n = 10^{16}$.

An error probability of this size is no real risk, but let us assume that a pessimist is not satisfied with this error probability and wants to have an error probability below all physical limits. In such a case one can execute the work of the protocol $R = (C_{\rm I}, C_{\rm II})$ ten times, always with an independent, new choice of a prime.

Protocol $\mathbf{R_{10}}$

Initial situation: $C_{\rm I}$ has n bits $x = x_1 \dots x_n$ and $C_{\rm II}$ has n bits $y = y_1 \dots y_n$.
Phase 1. $C_{\rm I}$ chooses uniformly 10 random primes

$$p_1, p_2, \dots, p_{10}$$

from $\{2, 3, \dots, n^2\}$.
Phase 2. $C_{\rm I}$ computes

$$s_i = Number(x) \bmod p_i$$

for $i = 1, 2, \dots, 10$ and sends the binary representations of

$$p_1, p_2, \dots, p_{10}, s_1, s_2, \dots, s_{10}$$

to $C_{\rm II}$.

Phase 3. Upon receiving $p_1, p_2, \ldots, p_{10}, s_1, s_2, \ldots, s_{10}$, C_{II} computes

$$q_i = \mathit{Number}(y) \bmod p_i$$

for $i = 1, 2, \ldots, 10$.
If there exists an $i \in \{1, 2, \ldots, 10\}$ such that $q_i \neq s_i$, then C_{II} outputs "$x \neq y$". Else (if $q_j = s_j$ for all $j \in \{1, 2, \ldots, 10\}$), then C_{II} outputs "$x = y$".

We observe that the communication complexity of R_{10} is 10 times larger than those of R. But for $n = 10^{16}$, the message consists of at most 2560 bits, which is no issue for discussion.

What is the gain with respect to the error probability?

If $x = y$, then we have again the situation that the protocol R_{10} provides the right answer "$x = y$" with certainty, i.e., the error probability is 0.

However, if $x \neq y$, R_{10} outputs the wrong answer "$x = y$" only if all 10 chosen primes belong to the maximal $n - 1$ primes that divide the difference

$$w = |\,\mathit{Number}(x) - \mathit{Number}(y)|.$$

Since the 10 primes are chosen in ten independent experiments, the error probability is at most

$$\left(\frac{n-1}{\mathit{Prim}\,(n^2)}\right)^{10} \leq \left(\frac{\ln n^2}{n}\right)^{10} = \frac{2^{10} \cdot (\ln n)^{10}}{n^{10}}.$$

For $n = 10^{16}$, the error probability is smaller than

$$0.4717 \cdot 10^{-141}.$$

If one takes into account that the number of microseconds since the big bang is a number of 24 digits, and that the number of protons in the known universe is a number of 79 digits, an event with a probability below 10^{-141} is a real wonder. Note that also in the case when a deterministic protocol communication 10^{16} would be executable, the costs speak clearly in favor of the implementation of the above randomized protocol.

We can learn a lot from the construction of the protocol R_{10} that consists of independent repetitions of R. We see that the error probability of an algorithm A can be substantially pushed down by executing several independent runs of A. In cases such as the above communication protocol, even a few repetitions result in an enormous decrease in the error probability.

Exercise 8.5. Let k be a positive integer. Consider the protocol R_k that is based on the choice of k primes (i.e., on k independent runs of R). Estimate the error probability of R_k as a function of k.

Exercise 8.6. Another approach for decreasing the error probability is to change R for a protocol Q_r, $r \in \mathbb{N} - \{0,1,2\}$. The protocol Q_r works exactly as R, except it randomly chooses a prime from the set $\{2,3,\ldots,n^r\}$ instead of taking it from the set $\{2,3,\ldots,n^2\}$. Estimate the communication complexity and the error probability of Q_r for each positive integer $r \geq 2$.

Exercise 8.7. Let $\delta > 1$ be a positive integer. Design a randomized protocol for the comparison of two databases of size n that works with an error probability of at most $1/\delta$. Which of the two approaches from Exercises 8.5 and 8.6 is more efficient with respect to the communication complexity? Is it better to choose a few small primes or one large prime?

8.4 Abundance of Witnesses and Randomized Primality Testing

In this section we first aim to explain (or at least provide an intuition) why the randomized protocol R is exponentially more efficient than any deterministic protocol solving the same task. We show that we have designed the protocol R by a simple application of the method of abundance of witnesses. We now introduce this method of abundance of witnesses.

Consider an arbitrary decision problem, where the task is to decide whether a given input has a specific property or not. Assume there exists no efficient algorithm for this problem (or at least, no efficient algorithm for this decision task has yet been successfully designed). Applying the method of abundance of witnesses, one starts with the search for a suitable definition of a witness. A witness (see Definition 6.43) has to be an additional information to an input, which is helpful for efficiently proving that the input has the required property or that the input does not have the required property. In the design of the protocol R, a prime p is the witness of $x \neq y$ if

$$\mathit{Number}(x) \bmod\ p \neq \mathit{Number}(y) \bmod\ p.$$

If one has an oracle that provides such a p, then one can efficiently prove that "x is different from y". Obviously, one does not have such an oracle, but the main point is that one cannot efficiently compute such a witness because it would result in an efficient deterministic algorithm for the hard decision problem. To design an efficient randomized algorithm we need a set of witness candidates for any input and this set of candidates has to contain sufficiently many witnesses.

In the example of the protocol R, the witness candidates are the

$$\mathit{Prim}\left(n^2\right) \sim n^2/\ln n^2$$

many primes from the interval $[2,n^2]$. From these $\mathit{Prim}\left(n^2\right)$ candidates, there are at least $\mathit{Prim}\left(n^2\right)-(n-1)$ witnesses of "$x \neq y$" for any pair-wise different inputs x and y.

Hence, the probability of choosing a witness from the set of witness candidates is at least

$$\frac{\frac{n^2}{\ln n^2}-(n-1)}{\frac{n^2}{\ln n^2}} \geq 1-\frac{\ln n^2}{n}.$$

This is very favorable because this value is very close to 1. But even in a situation where the probability of choosing a witness is only 1/2, the abundance of witnesses is still more than sufficient. It is enough to simply choose a few candidates at random. In this way the probability of getting at least one witness grows quickly.

Now, we ask how is it possible that we are unable to find a witness efficiently in a deterministic way, when there are so many witnesses in the set of witness candidates. An approach would be to try to systematically check the set of candidates in such a way that a witness would be found in a short time. But the kernel of the problem is that, for each input, the distribution of witnesses among the candidates may be completely different. For each proposed search strategy, one can always find inputs where this strategy fails.[9]

Let us consider our previous example. Here, we can even prove that no strategy exists for C_{I} and C_{II} to efficiently find a witness for any input (x, y). We omit the presentation of the technical proof of this fact, but we can give a transparent idea by considering a simple search strategy that looks for primes in the order from the smallest one to the largest one. Clearly, after n probes, the strategy finds a witness because there are at most $n-1$ nonwitnesses. Unfortunately n probes cause the communication complexity $n \cdot 4 \cdot \log_2 n$, which is more than sending all bits of C_{I} to C_{II}. Why is there no assurance of finding a witness after a few probes of this strategy? Because for inputs (x, y) with

$$Number(x) - Number(y) = p_1 \cdot p_2 \cdot \dots \cdot p_k,$$

where $k = \frac{n}{2(\log n)^2}$ and $p_1 < p_2 \cdots < p_k$ are the smallest primes, the strategy needs $k+1$ probes to find a witness. One can easily imagine that for any other denumeration of primes one can find inputs that require many probes in order to find a witness.

The method of abundance of witnesses is a successful and powerful method for designing randomized algorithms. Efficient randomized primality testing[10] is one of the exemplary applications of this method. The best-known deterministic algorithms for primality testing would need billions of years for checking numbers of sizes typical for the current cryptographic applications. Explaining how to define suitable witnesses for primality testing is a too complex task for this introduction. Instead we will only solve a subproblem of primality testing. We show how to design a randomized algorithm that tests primality for all odd numbers with odd $\frac{n-1}{2}$.

---

[9] Is not efficient enough

[10] Note that primality testing belongs to the most fundamental algorithmic tasks of enormous practical importance.

First, let us explain what the term "efficient" means for number theoretical problems. For a number n, the input size of n is $\lceil \log_2 n \rceil$. This means that a polynomial-time algorithm for primality testing must have a time complexity that is polynomial in $\log_2 n$. In many applications, one needs to test numbers of several hundred digits (for instance, $\log_{10} n \approx 500$) and hence one can accept neither an exponential nor a high degree polynomial complexity.

The naive deterministic algorithm that checks whether a number from $\{2, 3, \ldots, \lfloor \sqrt{n} \rfloor\}$ divides the given number n is of exponential-time complexity in $\log_2 n$ (at least of $\sqrt{n} = 2^{\frac{\log_2 n}{2}}$). For this approach, a witness of the fact "p is composite" is every positive integer $m > 1, m \neq p$, that divides n. But such witnesses[11] are not sufficiently abundant in general. When $n = p \cdot q$ for two primes p and q, then there are only two witnesses p and q of the compositeness of n and the number of candidates is at least $\Omega(\sqrt{n})$. Therefore, one has to look for another definition of witnesses.

Theorem 8.8. Fermat's Theorem

For every prime p and every natural number a with $\gcd(a, p) = 1$,

$$a^{p-1} \mod p = 1.$$

Proof. We use the fact that every positive integer has a unique factorization. Since p is a prime, we have

$$c \cdot d \mod p = 0 \Leftrightarrow c \mod p = 0 \text{ or } d \mod p = 0 \tag{8.2}$$

for all natural numbers c and d.

Let a be an arbitrary positive integer with $\gcd(a, p) = 1$. Consider the numbers

$$m_1 = 1 \cdot a,\ m_2 = 2 \cdot a,\ \ldots,\ m_{p-1} = (p-1) \cdot a.$$

We claim that

$$m_u \mod p \neq m_v \mod p$$

for all $u, v \in \{1, \ldots, p-1\}$ with $u \neq v$. We prove this by contradiction.

Assume

$$m_u \mod p = m_v \mod p$$

for some $u, v \in \{1, \ldots, p-1\}$, $u > v$. Then p divides the number

$$m_u - m_v = u \cdot a - v \cdot a = (u - v) \cdot a.$$

But this is not possible because

(i) $u - v < p$, so p cannot divide $u - v$, and
(ii) $\gcd(a, p) = 1$, and hence p cannot divide a.

---

[11] That are based on the classical definition of primes

Therefore

$$|\{m_1 \mod p,\ m_2 \mod p,\ \ldots,\ m_{p-1} \mod p\}| = p - 1.$$

Now, we claim that each number $m_i \mod p$ is different from 0 (i.e., that p does not divide any m_i). Assume the opposite

$$m_u \mod p = (u \cdot a) \mod p = 0$$

for a u. Applying (8.2), we obtain

$$u \mod p = 0 \text{ or } a \mod p = 0.$$

But the prime p divides neither u nor a, because $u < p$ and $\gcd(a, p) = 1$. Hence

$$\{m_1 \mod p, m_2 \mod p, \ldots, m_{p-1} \mod p\} = \{1, 2, \ldots, p-1\}. \tag{8.3}$$

Now, consider the number

$$m = m_1 \cdot m_2 \cdot \cdots \cdot m_{p-1}.$$

¿From the definition of m_i, we get

$$m = 1 \cdot a \cdot 2 \cdot a \cdot \cdots \cdot (p-1) \cdot a = 1 \cdot 2 \cdot \cdots \cdot (p-1) \cdot a^{p-1}. \tag{8.4}$$

The set equality (8.3) implies

$$m \mod p = 1 \cdot 2 \cdot \cdots \cdot (p-1) \mod p. \tag{8.5}$$

The equalities (8.4) and (8.5) imply

$$1 \cdot 2 \cdot \cdots \cdot (p-1) \cdot a^{p-1} \mod p = 1 \cdot 2 \cdot \cdots \cdot (p-1) \mod p,$$

i.e.,

$$a^{p-1} \mod p = 1.$$

□

The following assertion strengthens Fermat's theorem.

$$\text{“}p \text{ is a prime”} \Leftrightarrow a^{\frac{p-1}{2}} \mod p \in \{1, p-1\}$$
$$\text{for all } a \in \{1, \ldots, p-1\}.$$

This assertion provides a new definition of primes. Following this definition, one can obtain the following new kind of witnesses of compositeness.

A number $a \in \{1, \ldots, n-1\}$ is a witness of the compositeness of n, if

$$a^{\frac{n-1}{2}} \mod n \notin \{1, n-1\}.$$

The following theorem shows that for odd composite numbers n with odd $\frac{n-1}{2}$, one has an abundance of witnesses of n's compositeness.

Theorem 8.9. *For every odd positive integer n with odd $\frac{n-1}{2}$ (i.e., with $n \mod 4 = 3$),*

(i) if n is a prime, then

$$a^{\frac{n-1}{2}} \mod n \in \{1, n-1\} \textit{ for all } a \in \{1, \ldots, n-1\} \textit{ and}$$

(ii) if n is composite, then

$$a^{\frac{n-1}{2}} \mod n \notin \{1, n-1\}$$

for at least half of the numbers a from $\{1, \ldots, n-1\}$.

Exercise 8.10.* Prove Theorem 8.9.

In this way, for any composite number m with $m \mod 4 = 3$, the probability of choosing a witness of m's compositeness is at least $\frac{1}{2}$. To be really satisfied with this concept of witnesses, one still has to fix that $a^{\frac{n-1}{2}} \mod n$ can be efficiently computed. Obviously, one cannot do it by $\frac{n-1}{2}$ multiplications with a because the time complexity would be exponential in $\lceil \log_2 n \rceil$. If one has to compute $a^b \mod p$ for $b = 2^k$, then one can do it easily with k multiplications by the following method of repeated squaring:

$$\begin{aligned}
a^2 \mod p &= a \cdot a \mod p, \\
a^4 \mod p &= (a^2 \mod p) \cdot (a^2 \mod p) \mod p, \\
a^8 \mod p &= (a^4 \mod p) \cdot (a^4 \mod p) \mod p, \\
&\vdots \\
a^{2^k} \mod p &= (a^{2^{k-1}} \mod p)^2 \mod p.
\end{aligned}$$

Let us consider the general case, where

$$b = \sum_{i=1}^{k} b_i \cdot 2^{i-1}$$

(i.e., $b = Number(b_k b_{k-1} \ldots b_1)$) for a $k \in \mathbb{N} - \{0\}$ and $b_i \in \{0, 1\}$ for $i = 1, \ldots, k$. Then, we can express a^b as

$$a^b = a^{b_1 \cdot 2^0} \cdot a^{b_2 \cdot 2^1} \cdot a^{b_3 \cdot 2^2} \cdot \ldots \cdot a^{b_k \cdot 2^{k-1}}.$$

Note, to compute $a^b \mod p$, one first computes all numbers

$$a_i = a^{2^i} \mod p$$

by repeated squaring. After that one computes the product $\mod p$ of all numbers a_i, for which $b_i = 1$. Applying this approach for computing $a^{\frac{n-1}{2}} \mod n$ for an $a \in \{1, \ldots, n-1\}$ has the advantage that during the whole computation one works with numbers from $\{0, 1, \ldots, n-1\}$ only, i.e., with numbers whose binary representation length is at most $\lceil \log_2 n \rceil$. The number of multiplications of such numbers in the computation of $a^{\frac{n-1}{2}} \mod n$ is smaller than

$$2 \cdot \lceil \log_2 \frac{n-1}{2} \rceil \in O(\log_2 n).$$

Taking the logarithmic cost measurement of time complexity, the complexity of the whole computation is in $O((\log_2 n)^2)$.

In this way we obtain the following efficient randomized algorithm for primality testing.

Solovay–Strassen algorithm

Input: An odd number n with odd $\frac{n-1}{2}$.
Phase 1. Choose uniformly at random an $a \in \{1, 2, \ldots, n-1\}$.
Phase 2. Compute $x := a^{\frac{n-1}{2}} \pmod n$.
Phase 3.

```
if x ∈ {1, n − 1} then output ("prime");
else output ("composite");
```

Let us now analyze the error probability of this randomized primality testing.

If n is a prime, then Theorem 8.9 (i) claims that

$$a^{\frac{n-1}{2}} \pmod n \in \{1, n-1\}$$

for all $a \in \{1, \ldots, n-1\}$ and hence the output of the algorithm is always "prime", i.e., the error probability is 0.

If n is composite, Theorem 8.9 (ii) asserts that a uniformly chosen $a \in \{1, 2, \ldots, n-1\}$ is a witness of n's compositeness with a probability of at least $\frac{1}{2}$. Hence, the error probability is at most $\frac{1}{2}$. This error probability is clearly too large. But independently choosing twenty numbers $a_1, \ldots, a_{20}$ from $\{1, \ldots, n-1\}$ at random instead of one, and giving the answer "prime" only if

$$a_i^{\frac{n-1}{2}} \mod n \in \{1, n-1\}$$

for all $i \in \{1, \ldots, 20\}$, one reduces the error probability to

$$(\frac{1}{2^{20}} < 10^{-6}.$$

Exercise 8.11. Let $k \geq 2, k \in \mathbb{N}$. How far can one reduce the error probability of the Solovay–Strassen algorithm by k independent runs? Present careful arguments for your answer.

8.5 Fingerprinting and Equivalence of Two Polynomials

In Section 8.3 we have applied the method of abundance of witnesses in order to design an efficient randomized communication protocol for the comparison of two large numbers *Number*(x) and *Number*(y). This special kind of abundance of witnesses application is also called fingerprinting and it can be generally described as follows.

Fingerprinting Method

Task: Decide the equivalence of two objects O_1 and O_2 whose complete representation is very large.

Phase 1. Let M be a "suitable" set of mappings from the full descriptions of considered objects to a partial representation (fingerprint) of these objects.
Choose a mapping h from M at random.

Phase 2. Compute $h(O_1)$ and $h(O_2)$.
$h(O_i)$ is called the **fingerprint of O_i** for $i = 1, 2$.

Phase 3.
if $h(O_1) = h(O_2)$ **then output** "O_1 and O_2 are equivalent";
else output "O_1 and O_2 are not equivalent";

For the designed randomized protocol in Section 8.3, O_1 and O_2 are two large numbers of n bits ($n = 10^{16}$). The set M was

$$\{h_p \mid h_p(m) = m \bmod p \text{ for all } m \in \mathbb{N},\ p \text{ is a prime, } p \leq n^2\}.$$

For a randomly chosen prime p,

$$h_p(O_1) = O_1 \bmod p \text{ and } h_p(O_2) = O_2 \bmod p$$

are the fingerprints of O_1 and O_2, respectively.

The main idea of this method is that $h_p(O_i)$ has a substantially shorter representation than O_i, so the comparison of $h_p(O_1)$ and $h_p(O_2)$ is significantly simpler than the direct comparison of O_1 and O_2. But this gain can be achieved only if $h_p(O_i)$ is not a full description of O_i, so one has to take into account the possibility of a wrong decision. The rest of the idea is based on the method of abundance of witnesses. The set M is the set of candidates for witnesses of the nonequivalence of O_1 and O_2. If, for any pair of objects (O_1, O_2), there are many[12] witnesses of "$O_1 \neq O_2$" in M, then one can suppress the error probability to an arbitrary small size.

The art of applying the fingerprinting method is based on a convenient choice of the set M. On the one hand, the fingerprints have to be as short as possible in order to make their comparison efficient. On the other hand, they

[12] With respect to $|M|$

have to contain as much information about the objects as possible[13] in order to suppress the probability of losing vital information that distinguishes two objects in their fingerprints. Therefore, an algorithm designer has to take care with the tradeoff between the compression degree from O to $h(O)$ and the error probability. In Section 8.3 we succeeded in getting an error probability tending towards 0 with growing input size, and gaining a logarithmic compression from O to $h(O)$.

Next, we consider an equivalence problem for which no deterministic polynomial-time algorithm is known and that can be solved efficiently by fingerprinting. The problem is to decide on the equivalence of two polynomials of several variables over a finite field $\mathbb{Z}_p$ for a prime p. Two polynomials $P_1(x_1, \ldots, x_n)$ and $P_2(x_1, \ldots, x_n)$ are said to be equivalent over $\mathbb{Z}_p$, if for all $(\alpha_1, \ldots, \alpha_n) \in (\mathbb{Z}_p)^n$,

$$P_1(\alpha_1, \ldots, \alpha_n) \equiv P_2(\alpha_1, \ldots, \alpha_n) \pmod{p}.$$

One does not have any polynomial-time algorithm for this problem. A naive observer may deny this by pointing out that the comparison of two polynomials is easy because it it simply done by comparing the coefficients of the corresponding terms. We know that two polynomials P_1 and P_2 are identical if and only if the coefficients of P_1 are the same as those of P_2. But, the real problem is that in order to perform this coefficient comparison, one has to first transform the polynomials in their normal form. The normal form of a polynomial of n variables $x_1, x_2, \ldots, x_n$ and a degree[14] d is

$$\sum_{i_1=0}^{d} \sum_{i_2=0}^{d} \cdots \sum_{i_n=0}^{d} c_{i_1, i_2, \ldots, i_n} \cdot x_1^{i_1} \cdot x_2^{i_2} \cdot \ldots \cdot x_n^{i_n}.$$

However the input polynomials for our equivalence test may be in an arbitrary form, for instance, as

$$P(x_1, x_2, x_3, x_4, x_5, x_6) = (x_1 + x_2)^{10} \cdot (x_3 - x_4)^7 \cdot (x_5 + x_6)^{20}.$$

Applying the binomial formula

$$(x_1 + x_2)^n = \sum_{k=0}^{n} \binom{n}{k} \cdot x_1^k \cdot x_2^{n-k}$$

it is obvious that $P(x_1, x_2, x_3, x_4, x_5, x_6)$ has exactly $10 \cdot 7 \cdot 20 = 1400$ terms with nonzero coefficients. Thus the normal form of a polynomial can be exponentially longer than its input representation and hence, in general, one

[13] This is the reason for the name of this method, because fingerprints are considered to be almost an unambiguous means of identification.

[14] The degree of a polynomial of several variables is the maximum of degrees of particular variables.

cannot compute the normal form from a given form in polynomial time. If one wants to be efficient, one has to find a way of comparing two polynomials without creating their normal forms. To do this, we apply the method of fingerprinting. Let $P_1(x_1, \ldots, x_n)$ and $P_2(x_1, \ldots, x_n)$ be polynomials over the field $\mathbb{Z}_p$ for a prime p. We say that an $\alpha = (\alpha_1, \ldots, \alpha_n) \in (\mathbb{Z}_p)^n$ is a witness of

$$\text{“}P_1(x_1, \ldots, x_n) \not\equiv P_2(x_1, \ldots, x_n)\text{”}$$

if

$$P_1(\alpha_1, \ldots, \alpha_n) \bmod p \neq P_2(\alpha_1, \ldots, \alpha_n) \bmod p.$$

In the language of fingerprints

$$h_\alpha(P_1) = P_1(\alpha_1, \ldots, \alpha_n) \bmod p$$

is the fingerprint of P_1. The above definition of a fingerprint (of a witness of nonequivalence) directly specifies the following randomized algorithm.

Algorithm AQP

Input: A prime p and two polynomials P_1 and P_2 over n variables $x_1, \ldots, x_n$ for a positive integer n and of a degree of at most d for a $d \in \mathbb{N}$.

Phase 1. Choose randomly[15] an $\alpha = (\alpha_1, \ldots, \alpha_n) \in (\mathbb{Z}_p)^n$.

Phase 2. Compute the fingerprints
$h_\alpha(P_1) = P_1(\alpha_1, \ldots, \alpha_n) \bmod p$, and
$h_\alpha(P_2) = P_2(\alpha_1, \ldots, \alpha_n) \bmod p$.

Phase 3.
if $h_2(P_1) = h_2(P_2)$ **then output** "$P_1 \equiv P_2$";
else output "$P_1 \not\equiv P_2$";

Now we analyze the error probability of the algorithm AQP. If P_1 and P_2 are equivalent over $\mathbb{Z}_p$, then

$$P_1(\alpha_1, \ldots, \alpha_n) \equiv P_2(\alpha_1, \ldots, \alpha_n) \pmod p$$

for all $(\alpha_1, \alpha_2, \ldots, \alpha_n) \in (\mathbb{Z}_p)^n$. Hence, the error probability for inputs P_1 and P_2 with $P_1 \equiv P_2$ is equal to 0.

Let P_1 and P_2 be two polynomials that are not equivalent over $\mathbb{Z}_p$. We show that if $p > 2nd$, then the error probability of AQP is smaller than $\frac{1}{2}$. The question whether

$$P_1(x_1, \ldots, x_n) \equiv P_2(x_1, \ldots, x_n)$$

is equivalent to the question whether

$$Q(x_1, \ldots, x_n) = P_1(x_1, \ldots, x_n) - P_2(x_1, \ldots, x_n) \equiv 0.$$

[15] With respect to the uniform probability distribution over $(\mathbb{Z}_p)^n$

This means that if P_1 and P_2 are not equivalent then the polynomial Q is not identical to 0 (zero polynomial). Now, we show that the number of roots of a polynomial $Q \not\equiv 0$ over n variables and of a degree d is bounded. This means that there are sufficiently many witnesses $\alpha \in (\mathbb{Z}_p)^n$ with

$$Q(\alpha) \not\equiv 0 \pmod p \text{ (i.e., } P_1(\alpha) \not\equiv P_2(\alpha) \pmod p).$$

We start with the well-known theorem about the number of roots of polynomials over one variable.

Theorem 8.12. *Let d be a nonnegative integer. Every polynomial $P(x)$ of a single variable x over any field and of degree d has either at most d roots or is equal to 0 everywhere.*[16]

Proof. We prove this claim by induction with respect to the degree d.

(i) Let $d = 0$. Then $P(x) = c$ for a constant c. If $c \neq 0$, then $P(x)$ does not have any root.
(ii) Assume that Theorem 8.12 holds for $d-1$, $d \geq 1$. Now we prove this for d. Let $P(x) \not\equiv 0$ and let a be a root of P. Then

$$P(x) = (x-a) \cdot P'(x),$$

where $P'(x) = \frac{P(x)}{(x-a)}$ is a polynomial of degree $d-1$. Following the induction hypothesis $P'(x)$ has at most $d-1$ roots. Therefore $P(x)$ has at most d roots.

□

Now we are ready to prove that there are sufficiently many witnesses[17] of the nonequivalence of different P_1 and P_2 over $\mathbb{Z}_p$ for a sufficiently large prime p.

Theorem 8.13. *Let p be a prime, and let d, and n be positive integers. Let $Q(x_1, \ldots, x_n) \not\equiv 0$ be a polynomial over $\mathbb{Z}_p$ in n variables $x_1, \ldots, x_n$, where each variable has degree of at most d in Q. Then, Q has at most*

$$n \cdot d \cdot p^{n-1}$$

roots.

Proof. We prove Theorem 8.13 by induction with respect to the number n of variables in Q.

(i) Let $n = 1$. Then Theorem 8.12 implies that $Q(x_1)$ has at most

$$d = n \cdot d \cdot p^{n-1} \text{ (for } n = 1)$$

roots.

[16] Is a zero polynomial
[17] Nonroots of $Q(x_1, \ldots, x_n) = P_1(x_1, \ldots, x_n) - P_2(x_1, \ldots, x_n)$

(ii) Assume that the assertion of Theorem 8.13 is true for $n-1$, $n \in \mathbb{N} - \{0\}$. We prove it for n. We can express Q as

$$Q(x_1, x_2, \ldots, x_n) = Q_0(x_2, \ldots x_n) + x_1 \cdot Q_1(x_2, \ldots, x_n) + \ldots + x_1^d \cdot Q_d(x_2, \ldots, x_n)$$
$$= \sum_{i=0}^{d} x_1^i \cdot Q_i(x_2, \ldots, x_n)$$

for some polynomials

$$Q_0(x_2, \ldots x_n),\ Q_1(x_2, \ldots, x_n),\ \ldots,\ Q_d(x_2, \ldots, x_n).$$

If $Q(\alpha_1, \alpha_2, \ldots, \alpha_n) \equiv 0 \pmod p$ for an $\alpha = (\alpha_1, \ldots, \alpha_n) \in (\mathbb{Z}_p)^n$, then either

(a) $Q_i(\alpha_2, \ldots, \alpha_n) \equiv 0 \pmod p$ for all $i = 0, 1, \ldots, d$, or
(b) there exists a $j \in \{0, 1, \ldots, d\}$ with $Q_i(\alpha_2, \ldots, \alpha_n) \not\equiv 0 \pmod p$ and α_1 is a root of the polynomial.

$$\overline{Q}(x_1) = Q_0(\alpha_2, \ldots \alpha_n) + x_1 \cdot Q_1(\alpha_2, \ldots, \alpha_n) + \ldots + x_1^d \cdot Q_d(\alpha_2, \ldots, \alpha_n)$$

in the variable x_1.

Now we count the number of roots in cases (a) and (b) separately.

(a) Since $Q(x_1, \ldots, x_n) \not\equiv 0$, there exists a $k \in \{0, 1, \ldots, d\}$, such that

$$Q_k(x_2, \ldots, x_n) \not\equiv 0.$$

The induction hypothesis implies that the number of roots of Q_k is at most

$$(n-1) \cdot d \cdot p^{n-2}.$$

Hence, there are at most $(n-1) \cdot d \cdot p^{n-2}$ elements $\overline{\alpha} = (\alpha_2, \ldots, \alpha_n) \in (\mathbb{Z}_p)^{n-1}$, such that

$$Q_i(\overline{\alpha}) \equiv 0 \pmod p$$

for all $i \in \{0, 1, 2, \ldots, d\}$. Since the value α_1 of x_1 does not have any influence on the condition (a), α_1 can be chosen arbitrarily from $\{0, 1, \ldots, p-1\}$. Thus there are at most

$$p \cdot (n-1) \cdot d \cdot p^{n-2} = (n-1) \cdot d \cdot p^{n-1}$$

elements $\alpha = (\alpha_1, \alpha_2, \ldots, \alpha_n) \in (\mathbb{Z}_p)^n$ that have the property (a).

(b) Since $\overline{Q}(x_1) \not\equiv 0$, the polynomial $\overline{Q}$ has at most d roots[18] (i.e., at most d values $\alpha_1 \in \mathbb{Z}_p$ with $\overline{Q}(\alpha_1) \equiv 0 \pmod p$). Therefore, there are at most

$$d \cdot p^{n-1}$$

values[19] $\alpha = (\alpha_1, \alpha_2, \ldots, \alpha_n) \in (\mathbb{Z}_p)^n$, that satisfy the property (b).

---

[18] Theorem 8.12

[19] Note that the values $\alpha_1, \alpha_2, \ldots, \alpha_n$ can be chosen arbitrarily.

Combining (a) and (b), $Q(x_1, \ldots, x_n)$ has at most

$$(n-1) \cdot d \cdot p^{n-1} + d \cdot p^{n-1} = n \cdot d \cdot p^{n-1}$$

roots.

□

Corollary 8.14. *Let p be a prime, and let n and d be positive integers. For every polynomial $Q(x_1, \ldots, x_n) \not\equiv 0$ over $\mathbb{Z}_p$ of degree at most d, the number of witnesses of "$Q \not\equiv 0$" is at least*

$$\left(1 - \frac{n \cdot d}{p}\right) \cdot p^n.$$

Proof. The number of elements in $(\mathbb{Z}_p)^n$ is exactly p^n and Theorem 8.13 implies that at most $n \cdot d \cdot p^{n-1}$ of them are not witnesses. Hence, the number of witnesses is at least

$$p^n - n \cdot d \cdot p^{n-1} = \left(1 - \frac{n \cdot d}{p}\right) \cdot p^n.$$

□

Thus the probability of choosing uniformly a witness of "$Q \not\equiv 0$" at random from p^n elements of $(\mathbb{Z}_p)^n$ is at least

$$\left(1 - \frac{n \cdot d}{p}\right).$$

For $p > 2nd$, the probability of choosing a witness is at least $1/2$. By executing several independent random choices from $(\mathbb{Z}_p)^n$, the probability of finding at least one witness of $Q \not\equiv 0$ (i.e., of $P_1(x_1, \ldots, x_n) \not\equiv P_2(x_1, \ldots, x_n)$) can be brought arbitrarily close to 1.

For some applications of the algorithm AQP, it is important that the prime p can be appropriately selected. This degree of freedom can be achieved in situations when one can reduce an equivalence problem to the comparison of two polynomials without giving any requirements on the field over which the polynomials are considered.

8.6 Summary

One can view a randomized algorithm as a nondeterministic algorithm with a probability distribution over every nondeterministic choice or as a probability distribution over a set of deterministic algorithms. Random control is a nature of physical and biological processes. Its main characteristics are simplicity and efficiency. This is the case also in algorithmics, where simple randomized algorithms are often much more efficient than their best deterministic counterparts.

Designing a communication protocol for the comparison of the contents of two databases we have an instructive example of the power of randomization. The complexity of the best possible deterministic protocol for this task is exponential in the complexity of the randomized protocol presented. The idea behind the design of this randomized protocol is based on the method of abundance of witnesses. A witness is an additional information to the input, such that a witness allows one to efficiently compute the result despite the fact that no efficient deterministic approach[20] (without the use of a witness) for solving the problem is known. For a successful application of this method it is important to be able to find a set of witness candidates, such that the set of witnesses is a nonnegligible portion of it.[21] Then, one can get a witness by a (repeated) random choice from the set of witnesses. The reason why one is unable to find a witness efficiently in a deterministic way, lies in the highly irregular (random) distribution of witnesses among the candidates. Because of this chaotic structure of the set of witness candidates, every deterministic search strategy risks a large number of false attempts. The art of using this method is in the search for a suitable definition of witnesses. We have shown how to define witnesses for a randomized primality testing algorithm that works for all odd numbers n with odd $\frac{(n-1)}{2}$. The presented definition of a witness for compositeness can be extended to work for any positive integer.

A special case of the method of abundance of witnesses is the method of fingerprinting for solving equivalence problems. The idea is to assign a fingerprint (a short, but involved/relevant partial description) to any complex[22] object and thus reduce the equivalence test to the efficient comparison of fingerprints. The randomly chosen mappings from complex objects to their fingerprints play the role of witnesses in this scenario. Using fingerprinting one can develop an efficient (polynomial-time) randomized test for the equivalence of two polynomials. This is of interest because one does not know of any polynomial-time deterministic algorithm for this equivalence problem and there are several other equivalence problems of practical importance that can be reduced to the comparison of two polynomials.

Motwani and Raghavan provide a most exhaustive overview of the design of randomized algorithms in their seminal work [48]. Unfortunately, this excellent monograph can barely be recommended for beginners because of a nontrivial degree of the hardness of the topic. An introduction to the design of randomized algorithm is given in Chapter 5 of [30, 32]. More information about randomized protocols is presented in [29, 31]. Sipser [65] provides a very transparent presentation of an application of the randomized algorithm testing the equivalence of two polynomials for solving the problem of semantic

[20] Or even no efficient approach exists

[21] I.e., such that there are many witnesses among the candidates

[22] Or to an object of a very long full representation

comparisons of two data structures used for the representations of Boolean formulae. An impressive survey on ideas and concepts related to the development of randomized algorithms is given by Karp [36].

The first randomized polynomial-time algorithms for primality testing were discovered by Solovay and Strassen [53], Miller [47] and Rabin [54, 55]. In summer 2002, a fascinating breakthrough was achieved by Agrawal, Kayal and Saxena [44], who designed a deterministic polynomial-time algorithm for primality testing. This algorithm works in time $O((\log_2 n)^{12})$ for any input[23] n, so it cannot be considered as a serious competitor to the randomized algorithms[24] for real applications.

---

[23] Recall that the input length for a positive integer n is $\lceil \log_2(n+1) \rceil$.

[24] Running in time $O((\log_2 n)^3)$

Your idea is really crazy.
The principal question is,
whether it is crazy enough
to could be truth.

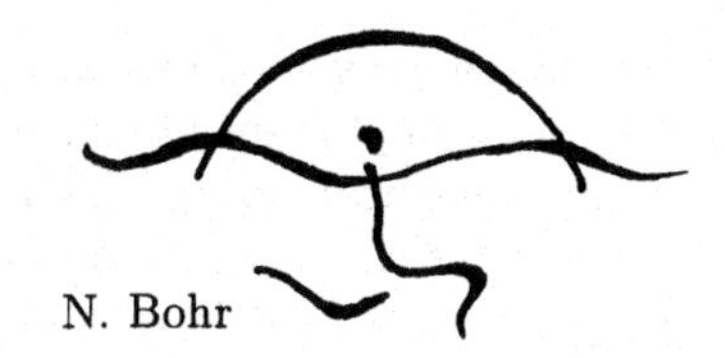

N. Bohr

9

Communication and Cryptography

9.1 Objectives

In the last century theoretical computer science mainly focused on the investigation of the sequential computing models that fit the framework given by von Neumann. What are the main problems of future computer science interest? Computer networks confront users not only with a particular computer, but also with an unclear, complex interconnected world full of asynchronous and unpredictable actions. The current understanding of computing in the interconnected world is not very deep and improving it is one of the main research tasks in computer science.

The manifoldness of research questions arising in the relation to distributive computing – cooperation and communication between computers, processes, and men – can hardly be presented in a short survey. For instance, the problems related to the design and analysis of communication algorithms (protocols) and to the design of reliable and capable networks depend strongly on the available technologies. These technologies rapidly developed from classical telephone networks to optical networks and every new technology opens up a new world of optimization and design problems. Since we do not see any possibility of giving a short, understandable overview of this topic, we restrict ourselves to presenting an instructive example of an interconnection network design in order to at least illustrate the problem formulations and techniques used in this area. The main focus of this chapter is on cryptography, which deals with the problems of secure communication in networks.

Our first objective is to provide the fundamental concepts of cryptography that aim to design cryptographic protocols (encryption methods) that assure that the messages exchanged over worldwide publicly accessible computer networks are

1. kept confidential (nobody can read them, except the right receiver), and
2. protected against manipulation.

We do not present these basic concepts of cryptography only because of their enormous importance in current practice.[1] The concept of modern cryptography is a natural continuation of concepts and ideas developed in the previous chapters of this book, especially the concepts of complexity theory, algorithmics for hard problems, and randomization. Moreover, cryptography is one of the fields that produces surprising, to some extent counterintuitive[2] results that open impressive possibilities considered before to be unrealistic dreams. Thus, cryptography is exactly the computer science discipline that can fascinate and consequently win the interest of young people for the study of theoretical computer science much more than any other area of computer science does.

This chapter is organized as follows. Section 9.2 is devoted to the introduction of the concept of classical cryptosystems. Section 9.3 presents the concept of the public-key cryptosystems and illustrates it with the famous RSA cryptosystem. In Section 9.4 we show how to use public-key cryptosystems for designing communication protocols for digital signatures. In Section 9.5 the concepts of interactive protocols and zero-knowledge proof systems are introduced in order to learn how one can verify mathematical proofs without reading them. Finally, Section 9.6 presents the design of a capable communication network as solving of a particular optimization task.

9.2 Classical Cryptosystems

Cryptology is the name of the science (teachings) of secret writing. In cryptology, we distinguish between cryptography and cryptoanalysis. Cryptography is devoted to the design of cryptosystems, while cryptoanalysis is devoted to the art of attacking cryptosystems (illegally intercepting messages). Here, we deal with cryptography only. The considered scenario is presented in Figure 9.1.

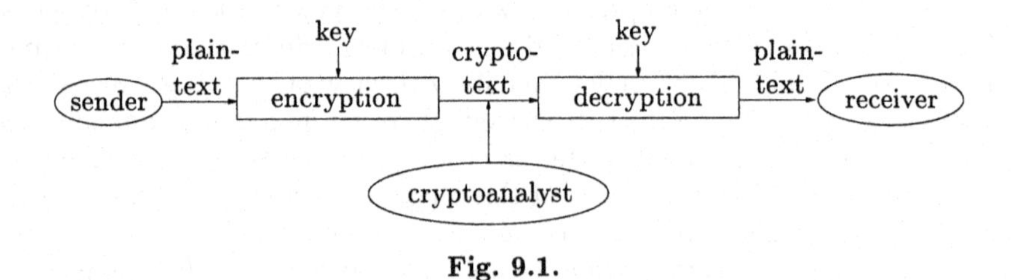

Fig. 9.1.

A person, called the **sender**, aims to send a secret message to another person, called the **receiver**. The secret message can be represented as a text,

[1] Digital signatures, E-commerce, electronic elections, etc.

[2] In the sense of being in contradiction with the present experience

called a **plaintext**. There is no other possibility besides communicating the secret through a publicly accessible network, where one cannot exclude the possibility that other persons can "listen" to the communication messages of the network. To exclude the possibility that an unauthorized person listening to the message learns about the secret, one sends the message in an encrypted form. The kind of encryption (decryption) is a common secret between the sender and the receiver, and the encryption is performed using the so-called **key**. The encrypted text is called the **cryptotext**. The cryptotext is submitted via the public network. Upon receiving the cryptotext, the receiver decrypts the cryptotext and obtains the original plaintext.

Formally, a cryptosystem is a triple $(\mathcal{K}, \mathcal{A}, \mathcal{S})$, where

(i) $\mathcal{K}$ is the set of all allowed (feasible) plaintexts,
(ii) $\mathcal{A}$ is the set of all possible cryptotexts, and
(iii) $\mathcal{S}$ is the set of keys.

Often, $\mathcal{K} = \Sigma^m$ for a positive integer m and an alphabet Σ. The meaning is that the plaintext has to be partitioned into blocks[3] of length m and every block is then encrypted separately. In such a case, we usually also have $\mathcal{A} = \Gamma^k$ for a positive integer k and an alphabet Γ.

Every key $\alpha \in \mathcal{S}$ unambiguously determines a one-to-one mapping E_α from $\mathcal{K}$ to $\mathcal{A}$. Thus, the encryption corresponds to computing $\mathrm{E}_\alpha(x)$ for a plaintext $x \in \mathcal{K}$ and the decryption corresponds to computing $\mathrm{E}_\alpha^{-1}(c)$ for a cryptotext $c \in \mathcal{A}$. Usually, one uses D_α to denote the function E_α^{-1} (inverse to E_α).

The common requirements on a cryptosystem are:

(i) The functions E_α and D_α have to be efficiently computable.
(ii) Without the knowledge of α, it should be "hard"[4] to compute the plaintext x from a given cryptotext $\mathrm{E}_\alpha(x)$.

CAESAR is probably the simplest cryptosystem. The set $\mathcal{K}$ of plaintexts as well as the set $\mathcal{A}$ of cryptotexts is the set of all words over the Latin alphabet with 26 symbols.[5] The set of keys $\mathcal{S}$ is $\{0, 1, 2, \ldots, 25\}$. For a given key $k \in \mathcal{S}$, the encryption consists of replacing every symbol of the plaintext for the symbol that is k steps to the right in the alphabetical order. If the end of the alphabet is reached, one continues cyclically from the beginning.

For instance, if $k = 3$, then given the plaintext

`CRYPTOGRAPHYISFASCINATING,`

one obtains the cryptotext

`FUBSWRJUDSKBLVIDVFLQDWLQJ.`

[3] Words of length m.
[4] At least in the sense of the complexity theory
[5] One can also consider $\mathcal{K}$ and $\mathcal{A}$ to be the Latin alphabet. Then the basic plaintext blocks have length 1, i.e., consist of 1 symbol.

Obviously the corresponding decryption is to go k symbols to the left in the alphabet. This cryptosystem is weak. If one knows that the cryptosystem used is CAESAR, then the cryptoanalyst simply tries all keys. This means that a cryptosystem with a small number of keys is useless in practice. The number of possible keys should be at least so large that trying all keys would be intractable. CAESAR can be improved by considering keys from $\{0, 1, \ldots, 26\}^m$ for a sufficiently large positive integer m. For a key

$$\alpha = \alpha_1, \alpha_2, \ldots, \alpha_m$$

one partitions the plaintext into blocks of length[6] m and replaces the i-th symbol of each block by the symbol being α_i positions after this symbol in the alphabetical order. If $\alpha = 3, 1, 6$, then one gets for the plaintext

$$\begin{matrix} C & R & Y & P & T & O & G & R & A & P & H & Y \\ 3 & 1 & 6 & 3 & 1 & 6 & 3 & 1 & 6 & 3 & 1 & 6 \end{matrix}$$

the cryptotext

$$F\,S\,E\,S\,U\,U\,J\,S\,G\,S\,I\,E.$$

This cryptosystem can be cracked too. For instance, the knowledge about the statistical density of the occurrence of particular symbols in texts of a natural language considered can be helpful for attacking such cryptosystems. But, there are also classical cryptosystems that are fast in encryption as well as in decryption, which nevertheless cannot be cracked given the current knowledge. The drawback of the classical cryptosystems is that they are usually based on a common secret between the sender and the receiver. Knowledge of the encryption mechanism directly implies knowledge of the decryption mechanism.[7] The consequence is that the sender and the receiver have to agree on a fixed key (secret) before using the cryptosystem, hence before having any secure channel for exchanging this secret. The main topic of the next section is how to solve this problem.

9.3 Public-Key Cryptosystems and RSA

The classical cryptosystems introduced in Section 9.2 are also known as symmetric cryptosystems, because the knowledge about the encryption [decryption] procedure provides the knowledge about the decryption [encryption] procedure. Due to this the sender and the receiver are equivalent and share a common secret. Besides the problem of having to agree on the common secret key without the help of any secure channel, symmetric cryptosystems have one other drawback. If one has a communication system with several

---

[6] That is, $\mathcal{K}$ and $\mathcal{A}$ are sets of words of length m over the alphabet.

[7] Often, the keys for the encryption and for the decryption are the same.

parties, then one traitor is sufficient to break the security of the entire secret communication exchange.

The revolutionary idea of the public-key cryptography overcoming the above-mentioned drawback is based on the following complexity-theoretical consideration. One searches for a **one-way function** f that has the following properties:

(i) f can be efficiently computed.
(ii) f^{-1} cannot be efficiently computed.
(iii) f^{-1} can be efficiently computed if one knows a special secret (called a **trapdoor**) which is analogous to the term witness (certificate) used in the previous chapters.

If the receiver has a one-way function f, then she/he can publicize f and each sender can use f to encrypt messages for this receiver. Through the publication of f (for instance, in a telephone list), property (ii) of f guarantees that nobody can decrypt the cryptotexts $f(x)$. Only the receiver, who has a trapdoor for f can compute f^{-1} and thus decrypt the messages. Obviously, such systems are not symmetric and because of the publication of f, we call them **public-key cryptosystems.**

The question is whether such one-way functions exist at all. One could even expect a negative answer because properties (i), (ii), and (iii) together seem somewhat contradictory. The following simple idea shows that the concept of one-way function is not so crazy.

Table 9.1.

	Name	Telephone number
C	Cook	00128143752946
R	Rivest	00173411020745
Y	Yao	00127345912233
P	Papadimitriou	00372453008122
T	Turing	00192417738429
O	Ogden	00012739226541
R	Rabin	00048327450028
A	Adleman	00173555248001
P	Papadimitriou	00372453008122
H	Hopcroft	00013782442358
Y	Yao	00127345912233

Let us consider the following encryption mechanism. Every symbol will be separately encrypted by a sequence of 14 digits. For every symbol of the plaintext, one nondeterministically chooses a name from a telephone book that begins with this symbol, and takes the corresponding telephone number[8] as the encryption. Table 9.1 provides an example of the encryption of the word "CRYPTOGRAPHY". Despite the fact that the encryption procedure is nondeterministic and many different cryptotexts may be assigned to a given plaintext, each cryptotext unambiguously determines the original plaintext. Now, the trapdoor of the receiver is a special telephone directory that is sorted with respect to the telephone numbers and hence the receiver can efficiently execute the decryption. Without the knowledge of the trapdoor (i.e., without possessing the special telephone directory sorted with respect to telephone numbers), the cost of decrypting one symbol corresponds to the exhaustive search in an unsorted list, i.e., to the complexity, which is proportional to the size of the telephone directory. Since the encryption method can be published, one can view this cryptosystem as a game on a public-key cryptosystem. We say "a game" only because anyone can sort the telephone directory with respect to the numbers, hence obtaining the trapdoor for this encryption function.

Thus, the above example cannot be seriously considered as a good candidate for a cryptosystem, but it shows that there may be a reasonable intuition behind the notions of a one-way function and its trapdoor. The following definition formalizes the notion of a one-way function in terms of computational complexity.

Definition 9.1. *Let Σ and Γ be alphabets. A function $f : \Sigma^* \to \Gamma^*$ is called a* **one-way function**, *if it satisfies the following properties:*

(i) There exist constants c and d from $\mathbb{N} - \{0\}$, such that for all $x \in \Sigma^*$,

$$\frac{1}{c} \cdot |x| \leq |f(x)| \leq d \cdot |x|.$$

{This means that $|x|$ and $|f(x)|$ are in a linear relation.}

(ii) The function f can be computed in polynomial time.

(iii) For every randomized polynomial-time algorithm A and every $k \in \mathbb{N} - \{0\}$, there exists a constant $n_{A,k}$, such that for all $n \geq n_{A,k}$ and a randomly chosen $w \in \Sigma^n$, the probability[9] that $A(f(w)) = w$ is smaller than n^{-k}. {This property assures that polynomially many independent runs of a randomized polynomial-time algorithm cannot provide the computation of f^{-1} with a constant error probability.}

Up till now, nobody has been able to prove for any concrete function that it is[10] a one-way function. This is related to the hardness of proving lower bounds on the computational resources necessary for solving a problem, i.e.,

---

[8] If the telephone number is shorter than 14 digits one adds leading 0s to get the required 14 digits.

[9] The probability is taken over the random choices of A as well as over the random choice of w.

[10] Even the existence of a one-way function has not been proved.

proving[11] property (iii) for f^{-1}. Nevertheless, we are aware of some plausible candidates for one-way functions. The most famous ones are:

(i) The multiplication of two numbers, i.e., the computation $f(x, y) = x \cdot y$. Obviously f can be computed efficiently. The inverse function to f is the factorization of a given number n. There is no known randomized polynomial-time algorithm for factorization of a given number and this problem is considered to be hard.[12]

(ii) One can compute $f(x) = a^x \mod n = b$ efficiently as already shown in Chapter 8 using the method of repeated squaring. The inverse function corresponds to solving the equality $a^x \equiv b \mod n$, for given a, b and n. This problem is called the discrete logarithm problem and it is considered to be hard (not solvable in randomized polynomial time).

Next, we present the famous RSA public-key cryptosystem[13] whose security is based on the assumption that factorization is hard. Let $\mathbf{gcd}(\boldsymbol{a}, \boldsymbol{b})$ denote the greatest common divisor of numbers a, $b \in \mathbb{N}$ in what follows.

The receiver determines the encryption procedure and the decryption procedure by the following calculation. She/He generates two large[14] random primes p and q and computes

$$n = p \cdot q \text{ and } \varphi(n) = (p-1) \cdot (q-1).$$

Here, φ is the so-called Eulerian function. For every positive integer n, $\varphi(n)$ is the number of integers a from $1, 2, \ldots, n-1$ with $\gcd(a, n) = 1$.

After that the receiver randomly chooses a large $d > 1$, such that

$$\gcd(d, \varphi(n)) = 1. \tag{9.1}$$

The equality 9.1 implies that d has a unique multiplicative inverse e. Thus the receiver computes[15] this e with the property

$$e \cdot d \mod \varphi(n) = 1. \tag{9.2}$$

The numbers n and e build the public key, and the numbers p, q, $\varphi(n)$, and d are the secret keys of the receiver. This is also why p, q, and d are randomly generated.

Now, we allow only numbers smaller than n as a plaintext. If the plaintext is larger or represented in another way, one first has to transform it into a sequence of digits of length $\lceil \log_{10} n \rceil - 1$, then encrypt these digits separately.

[11] Remember that we are even unable to prove a weaker requirement that f^{-1} cannot be computed in deterministic polynomial time.

[12] The factorization is not known to be NP-hard and one does not even believe that its hardness is so strong.

[13] The name RSA come from the names of its inventors River, Shamir, and Adleman.

[14] A few hundred digits long

[15] Computing e can be done efficiently by the Euclidean algorithm.

For a given number $w \in \{0, 1, \ldots, n-1\}$ the encryption is given by the encryption function

$$\mathrm{E}_{e,n}(w) = w^e \mod n.$$

For a given cryptotext c, the decryption function is

$$\mathrm{D}_{d,n}(c) = c^d \mod n.$$

As previously shown in Chapter 8, the functions $\mathrm{E}_{e,n}$ and $\mathrm{D}_{d,n}$ are efficiently computable by the method of repeated squaring. With the help of efficient randomized primality testing algorithms (see Section 8.4) one can efficiently generate large random numbers. The number d is randomly chosen. Then, one verifies whether $\gcd(d, \varphi(n)) = 1$, i.e., whether d and $\varphi(n)$ are coprimes. If not, one chooses another d and tests again. Since there is an abundance of coprimes to $\varphi(n)$, the number d can be efficiently found. As already mentioned, e can be efficiently determined by the Euclidean algorithm. Hence, the entire RSA public-key cryptosystem can be created efficiently.

We do not know of any efficient (randomized) algorithm capable of computing one of the numbers p, q, $\varphi(n)$, and d for a given public key (e, n). The knowledge of one of these four numbers p, q, $\varphi(n)$, and d is sufficient to crack the RSA cryptosystem. There is also no efficient algorithm known that can determine the plaintext x from the cryptotext $\mathrm{E}_{e,n}(x)$ and the public key (e, n).

Now, we want to show that RSA really works in the sense that

$$\mathrm{D}_{d,n}(\mathrm{E}_{e,n}(w)) = w$$

for all $w < n$. To prove that $\mathrm{D}_{d,n}$ is the inverse function to $\mathrm{E}_{e,n}$ for arguments from $\{0, 1, \ldots, n-1\}$, we need Euler's theorem that is a generalization of the Fermat's theorem.

Theorem 9.2. Euler's Theorem *Let w and n be two positive integers with* $\gcd(w, n) = 1$. *Let* $\varphi(n) = |\{a \in \{1, 2, \ldots, n\} \mid \gcd(a, n) = 1\}|$ *be the Eulerian number of n. Then*

$$w^{\varphi(n)} \mod n = 1.$$

Euler's theorem can be viewed as a consequence of the results of group theory that

(i) the order of every element of the group divides the order of the group, and
(ii) the cyclic multiplicative group $(Z/(n))^*$ has the order $\varphi(n)$.

Following the definition of the order of an element of a group,

$$w^k \mod n = 1$$

holds for every $w \in (Z/(n))^*$ with the order k.

Since (i) implies

$$\varphi(n) = k \cdot b$$

for a positive integer b, one obtains

$$\begin{aligned} w^{\varphi(n)} \mod n &= w^{k \cdot b} \mod n \\ &= (w^k \mod n)^b \mod n \\ &= (1)^b \mod n = 1. \end{aligned}$$

Exercise 9.3. Give an alternative proof of Euler's theorem by generalizing the proof of Fermat's theorem (Section 8.4) restated below.

Let $x_1, x_2, \ldots, x_{\varphi(n)} \in \{1, 2, \ldots, n-1\}$ be all numbers x_i with the property $\gcd(x_i, n) = 1$. Then, for every $a \in \{1, 2, \ldots, n-1\}$,

$$(ax_1 \bmod n, ax_2 \bmod n, \ldots, ax_{\varphi(n)} \bmod n)$$

is a permutation of $x_1, x_2, \ldots, x_{\varphi(n)}$.

Now we are ready to prove the correctness of RSA.

Theorem 9.4. *Let p, q, n, e, and d have the properties as described in the construction of* RSA. *Then for all $w < n$*

$$\mathrm{D}_{d,n}(\mathrm{E}_{e,n}(w)) = w^{ed} \mod n = w.$$

Proof. Following the choice of d and e with respect to equalities (9.1) and (9.2), we have

$$e \cdot d = j \cdot \varphi(n) + 1 \tag{9.3}$$

for a $j \in \mathbb{N} - \{0\}$. Thus we have to prove

$$w^{j \cdot \varphi(n)+1} \mod n = w \tag{9.4}$$

for all $w < n$. We distinguish three possibilities with respect to the relation between p, q and w.

(i) *Neither of the primes p, q divides w.*
If p and q do not divide w and $w < p \cdot q$, then

$$\gcd(p \cdot q, w) = 1.$$

Hence the assumptions of the Euler's theorem for $n = p \cdot q$ and w are fulfilled. Therefore

$$w^{\varphi(n)} \mod n = 1$$

and consequently

$$w^{j\varphi(n)} \mod n = 1. \tag{9.5}$$

If one multiplies both sides of equality (9.5) by w, one obtains the desired equality (9.4).

(ii) *Only one of the primes p and q divides w.*
Without loss of generality we assume that p divides w and q does not divide w. Fermat's theorem[16] implies

$$w^{q-1} \mod q = 1$$

and hence

$$w^{(q-1)\cdot(p-1)} \mod q = 1, \text{ i.e.,} w^{\varphi(n)} \mod q = 1.$$

Consequently,

$$w^{j\varphi(n)} \mod q = 1. \tag{9.6}$$

Since p divides w, the equality (9.6) holds for modulo $n = p \cdot q$ too, i.e.,

$$w^{j\varphi(n)} \mod n = 1.$$

Multiplying this equality by w, we obtain the claim of the theorem.

(iii) *Both primes p and q divide w.*
This case cannot arise because p and q are primes and $p \cdot q > w$.

□

Public-key cryptosystems have many advantages compared with symmetric cryptosystems. Besides those already mentioned, they are a basis for creating different communication protocols (for instance, digital signatures), that cannot be built by symmetric cryptosystems. On the other hand, classical symmetric cryptosystems also have an important advantage over public-key cryptosystems. Due to possible hardware implementations, the symmetric cryptosystems are often hundreds of times faster than public-key ones. Thus, it is common to use a public-key cryptosystem only for exchanging of the key of a symmetric cryptosystem. The rest of the communication[17] is then performed using the symmetric cryptosystem.

9.4 Digital Signatures

To show some transparent applications of the public-key cryptosystems, we present two simple protocols for digital (electronic) signatures. From the juridical point of view, the handwritten signature is a form of authenticity guarantee. Obviously, one cannot provide handwritten signatures by electronic communication. Moreover, we would like to have digital signatures that are harder to forge than handwritten ones.

---

[16] Note, that this is a special case of Euler's theorem because $\varphi(q) = q - 1$ for the prime q.

[17] That is, the main part of the communication

Consider the following scenario. A customer K wants to sign an electronic document for her/his bank B. For instance, K has to give the bank B an identity authentication for a money transfer from her/his account. One has the following natural requirements on communication protocols for such digital signatures.

(i) B must have a possibility of verifying the correctness of the digital signature of K, i.e., to authenticate K as the owner of the digital signature. This means that both K and B should be protected against attacks by a third party (a falsifier) F who pretends to be K in a communication with B.
(ii) K must be protected against messages forged by B, who claims to have received them properly signed from K. Furthermore, it means that B cannot be able to forge the signature of K.

Exercise 9.5. Design a communication protocol for digital signatures that is based on a symmetric cryptosystem and fulfills requirement (i).

Satisfying both requirements (i) and (ii) seems to be harder than satisfying the requirement (i) alone, because requirements (i) and (ii) are seemingly contradictory. On the one hand, property (i) requires that B has some nontrivial knowledge about K's signature for verification purposes. On the other hand, property (ii) requires that B should not know so much about K's signature (especially about the signature procedure) as to be able to forge it.

The following simple solution to our problem is provided by public-key cryptosystems.

The customer K has a public-key cryptosystem with an encryption function E_K and a decryption function D_K. The bank B knows the public encryption function E_K. Then K can sign a document as follows.

1. K computes $\mathrm{D}_K(w)$ for the document w and sends the pair $(w, \mathrm{D}_K(w))$ to B.
2. B checks whether $w = \mathrm{E}_K(\mathrm{D}_K(w))$ with the help of the public key E_K.

Let us verify that this communication protocol satisfies our requirements.

(i) Since nobody except K can compute $\mathrm{D}_K(w)$, B is certain that K has signed w. Moreover, since E_K is public, B also has the possibility of convincing anybody[18] that K has signed w with the pair $(w, \mathrm{D}_K(w))$.
(ii) Requirement (ii) is also satisfied because knowing E_K and the pair $(w, \mathrm{D}_K(w))$ is not helpful for signing another document u by $\mathrm{D}_K(u)$.

Note that it is important that this electronic signature changes the entire text w of the document, i.e., the signature is not only an additional text on the end of the document.

[18] Who knows E_K and is sure that E_K is the public key of K

Exercise 9.6. The protocol presented above does not try to work with w as a secret. Anybody listening to the communication can learn about w. The following additional requirement (iii) can be essential for many applications.

(iii) A third party may not learn about the contents of the signed document, even if it is able to listen to the entire communication.

Design a communication protocol that satisfies all three requirements (i), (ii) and (iii).

Now we consider a harder digital signature problem called the **authentication problem**. Here, one does not aim to sign a document, but rather to convince somebody about her/his identity. The requirements on a communication protocol for the authentication are as follows:

(i') as (i), and
(ii') K should be protected against the activities of B, where B attempts to convince[19] a third party that she/he is K.

The above-presented communication protocol is not satisfactory for authentication, because B learns the signature $(w, \mathrm{D}_K(w))$ in this protocol and can use it to convince a third party that she/he is K. Obviously, there are situations where this is undesirable. Moreover, anybody listening to the communication between K and B learns the signature $(w, \mathrm{D}_K(w))$ of K too. Then, the adversary, with the knowledge of E_K, can check the correctness of the signature $(w, \mathrm{D}_K(w))$ and use this to masquerade as K.

The authentication problem can be solved by taking an additional public-key cryptosystem. Exactly as before, K possesses a public-key cryptosystem $(\mathrm{D}_K, \mathrm{E}_K)$. Additionally, B possesses another public-key cryptosystem[20] $(\mathrm{D}_B, \mathrm{E}_B)$. Both encryption functions E_K and E_B are public, hence they are known to both B and K. The decryption function D_K is the secret of K and the decryption function D_B is the secret of B. Now, K signs in the following way.

1. B chooses a random number w and sends $\mathrm{E}_K(w)$ to K.
2. K computes $w = \mathrm{D}_K(\mathrm{E}_K(w))$.
   K computes $c = \mathrm{E}_B(\mathrm{D}_K(w))$ and sends it to B.
3. B checks whether

$$w = \mathrm{E}_K(\mathrm{D}_B(c)) = \mathrm{E}_K(\mathrm{D}_B(\mathrm{E}_B(\mathrm{D}_K(w)))).$$

Clearly, B is convinced about the identity of K in this way. K is the only person who knows D_K and can compute the number w from $\mathrm{E}_K(w)$ and the message $\mathrm{D}_K(w)$ from w.

[19] Thus (ii') is similar to (ii), because both are requirements restricting the possibility of forgery by B.

[20] Here, one requires that both cryptosystems are commutative, i.e., that $w = \mathrm{D}_K(\mathrm{E}_K(w)) = \mathrm{E}_K(\mathrm{D}_K(w))$ and $w = \mathrm{D}_B(\mathrm{E}_B(w)) = \mathrm{E}_B(\mathrm{D}_B(w))$.

The message $E_K(w)$ can be decrypted by K only and the message $E_B(D_K(w))$ can be decrypted by B only. Therefore a third party (a falsifier) cannot learn and check the signature $(w, E_B(D_K(w)))$, satisfying condition (i).

B learns the signature $(w, E_B(D_K(w)))$ of K in this communication. Through several executions of this authentication protocol between B and K, B can learn several such pairs. But this is not sufficient for B to convince a third party that she/he is K. If all parties in the interconnection network use this protocol, than each third party C sends the message $E_K(u)$ for a random u to K. If B as an active adversary[21] takes this message from the network,[22] then in general B cannot learn u because B does not know the secret D_K of K. The only attempt B can make is to compute $E_K(w)$ for all saved pairs $(w, E_B(D_K(w)))$ of K's signatures, and compare these with $E_K(u)$. If B gets lucky and finds $E_K(w) = E_K(u)$ for a saved u, then B can convince[23] C that she/he is K. But if u is a random number of several hundred digits, then the success probability of B is smaller than 1 over the number of protons in the known universe. This probability can be further reduced by a regular change of the keys of the public-key cryptosystems used.[24]

Exercise 9.7. Consider the authentication problem where one signs in order to convince everybody that she/he is K, but not to sign a document. The above-presented protocol for this purpose is reliable with a high probability. Introduce a small change to this protocol in such a way that the probability of a falsification attempt by B is reduced to 0. This has to hold independent of the size of B's list of saved signatures from K.

9.5 Interactive Proof Systems and Zero-Knowledge Proofs

In Chapter 6 we learned that every language from NP has a polynomial-time verifier. This means that all claims $x \in L$ for a $L \in$ NP have proofs of polynomial length in $|x|$ and the correctness of these proofs can be verified in polynomial time. Thus the languages from NP are easy with respect to proof verification. In Chapter 8 we argued that the practical solvability should be related to randomized polynomial time rather than to deterministic polynomial time. In this way, the following question arises

For which languages can one "practically" verify the proofs?

[21] Who wants to convince C that she/he is K
[22] That is, $E_K(u)$ never reaches K
[23] By sending the corresponding signature $E_C(D_K(w))$ to C
[24] There are also more elaborate protocols for the authentication problem, but their presentation is beyond the framework of this book.

To investigate this question we consider the following communication protocol model.

We have two parties – the **prover** and the **verifier**. The prover is an algorithm (a TM) with an unbounded computational power.[25] The verifier is a randomized polynomial-time algorithm (a randomized polynomial-time TM). Let L be a language. At the beginning both parties get the same input x. The verifier and the prover may communicate by exchanging messages of a polynomial length in $|x|$. The prover with an unbounded computational power aims to convince the verifier about the truth of the claim "$x \in L$". For this purpose, the prover is also allowed to lie (make false claims). The task of the verifier is to pose questions to the prover in order to estimate with high probability whether the prover has the proof of "$x \in L$" or not. The number of communication rounds[26] is at most polynomial in $|x|$ and the entire computation of the verifier runs in polynomial time. The verifier must finish the communication with the decision $x \in L$ or $x \notin L$. This communication protocol between the prover and the verifier is called an interactive proof system.

Definition 9.8. *Let $L \subseteq \Sigma^*$ for an alphabet Σ. We say that L possesses an* **interactive proof system,** *if there exists a verifier (a randomized polynomial-time algorithm) V such that for all $x \in \Sigma^*$ the following conditions hold:*

(i) If $x \in L$, then there exists a prover B such that V after communicating with B decides to accept x with probability greater than $\frac{2}{3}$.
{If $x \in L$, then there exists a prover that has a proof of the claim "$x \in L$" and V can efficiently[27] *check this proof in a communication with B.}*

(ii) If $x \notin L$, then for every prover B, the communication between V and B ends with the rejection of x with a probability of at least $\frac{2}{3}$.
{If $x \notin L$, then there is no proof of "$x \in L$", hence every strategy to convince V about the false claim "$x \in L$" has to be detected with a high probability.}

We define the class IP *by*

$$\mathbf{IP} = \{L \mid L \textit{ possesses an interactive proof system}\}.$$

The precise value of the probability bounds in Definition 9.8 is not essential. After $O(|x|)$ independent repetitions of the communication between the prover and the verifier, the error probability can be reduced to $2^{-|x|}$. Thus, forcing the probability $1 - 2^{-|x|}$ instead of $\frac{2}{3}$ for the correct decision does not change anything on the class IP.

The following result follows directly from Definition 9.8.

---

[25] Without any bound on complexity

[26] The number of messages exchanged

[27] It does not matter how long the proof is.

Lemma 9.9.

$$\mathrm{NP} \subseteq \mathrm{IP}.$$

Proof. Since NP = VP, there exists a polynomial-time verifier for each language $L \in \mathrm{NP}$. This means that for every language L and every $x \in L$, there exists a witness (a proof) c of the fact "$x \in L$" and c has a polynomial length in $|x|$. Hence, the prover that possesses c can send the entire c to the verifier and the verifier can deterministically verify whether c is the witness of "$x \in L$" in polynomial time. If $x \notin L$, there is no witness (proof) of "$x \in L$", so there is no possibility of convincing the deterministic verifier that "$x \in L$". □

The next question is whether there are languages outside NP that possess interactive proof systems. Consider the following problem. Given two graphs G_1 and G_2, one has to decide whether G_1 and G_2 are isomorphic.[28] The graph isomorphism problem is in NP, because the isomorphism can be nondeterministically guessed and deterministically verified in polynomial-time. The complementary problem of graph nonisomorphism is not known to be in NP. We conjecture that it is not in NP, because we do not see how nondeterministic guessing can help to prove the nonexistence of an isomorphism between two graphs in any way. Let

$$\mathrm{NONISO} = \{(G_1, G_2) \mid G_1 \text{ and } G_2 \text{ are not isomorphic}\}.$$

Now we describe an interactive proof system for NONISO. Let (G_1, G_2) be the input known to both V and B.

1. The verifier V checks whether G_1 and G_2 have the same number of edges and vertices. If not, V rejects the input with certainty.
2. If G_1 and G_2 have the same number of vertices and edges, V chooses an $i \in \{1,2\}$ and a permutation $(j_1, j_2, \ldots, j_n)$ of $(1, 2, \ldots, n)$ randomly,[29] where n is the number of vertices. Then V converts the Graph G_i to $G_i(j_1, j_2, \ldots, j_n)$ according to the permutation. After that V sends $G_i(j_1, j_2, \ldots, j_n)$ to the prover B and asks B to determine whether $G_i(j_1, j_2, \ldots, j_n)$ is isomorphic to G_1 or G_2.
3. If G_1 and G_2 are not isomorphic, then B can determine i (i.e., to compute which of the graphs G_1 and G_2 is isomorphic to $G_i(j_1, j_2, \ldots, j_n)$ and sends i to V.
   If G_1 and G_2 are isomorphic, then B does not have any possibility of computing i. B can only guess i, so B is forced to choose a random $k \in \{1,2\}$ and sends this to V.

[28] Two graphs are isomorphic if one can label the vertices of one of the graphs in such a way that both graphs become identical.

[29] Both i and the permutation are chosen with the uniform probability distribution.

4. If $k \neq i$, then V rejects the input (G_1, G_2).
   If $k = i$, then V repeats 2 sending an $s \in \{1, 2\}$ and a random permutation of G_s to B and B sends an $l \in \{1, 2\}$ to V according to 3.
   Now if $l \neq s$, then V rejects the input (G_1, G_2).
   If $k = s$, then V accepts the input (G_1, G_2).

We show that the above protocol is an interactive proof system for NONISO.

If $(G_1, G_2) \in$ NONISO, there is a prover that can distinguish between G_1 and G_2 and therefore this prover always gives the right answer. Hence, the verifier accepts (G_1, G_2) with certainty.

If $(G_1, G_2) \notin$ NONISO, there is no prover that can distinguish between G_1 and G_2 with respect to homomorphism. The probability that a prover correctly guesses the two randomly chosen values i and s from $\{1, 2\}$ is at most $\frac{1}{2} \cdot \frac{1}{2} = \frac{1}{4}$. Thus the error probability is at most $\frac{1}{4}$. One can reduce this to 2^{-k} by considering k independent questions of the verifier on the prover.

The next result shows the enormous power of randomized verification.

Theorem 9.10.* Shamir's Theorem

$$\text{IP} = \text{PSPACE}.$$

The most fascinating consequence of this theorem is that the shortest proofs of the claims "$x \in L$" for languages $L \in$ PSPACE may have an exponential length[30] in $|x|$, and hence the prover cannot send them (not even a nonnegligible portion of them) to the verifier. Thus one can verify proofs of exponential length without reading them with high probability in randomized polynomial time.

We close our elementary introduction to cryptography by introducing the **zero-knowledge proof systems**. We omit the formal definition. The idea is to place an additional requirement on a proof system, namely that the verifier may not learn any bit (anything) in the communication with the prover, except the information that the verifier can compute itself without communication with the prover. This "learning nothing" during the communication can be formalized in such a way that the entire communication between B and V is modeled as a probability space where the probability distribution is determined by the random choices of V. An interactive proof system is zero-knowledge, when there exists a randomized polynomial-time TM M that generates the entire communication with the same probability distribution. The consequence of this requirement is that the verifier does not learn any bit of the prover's proof in a zero-knowledge protocol.

Zero-knowledge proof systems have many cryptographic applications. An interesting application is related to the following scenario. One wants to design

[30] The fact $L \in$ PSPACE says only that the proofs of "$x \in L$" have a polynomial breadth, where the breadth of a proof is the length of the longest assertion occurring in a sequence of proof implications.

an access control for a computer that allows only users with a valid password to work, but at the same time, the users remain anonymous. This means the user takes the role of a prover that tries to convince the access control (the verifier) that she/he possesses a valid password without revealing any bit of the password to the access control.

Another important application of zero-knowledge proof systems is using private data of another person in a computation without learning them. Assume that the verifier V seeks to compute the value $f(x, y)$ and possesses the value y only. The prover B knows x and is willing to help V under the condition that V does not learn any bit of x.

To determine the class of functions and decision problems for which one has zero-knowledge proof systems is a challenging research problem. It is really surprising how many problems can be solved in this way. For instance, we know that all languages in NP possess a zero-knowledge proof system. Because of technical limitations, we do not present the proof of this result. To illustrate the work of zero-knowledge proof systems we only present an example of such systems for the graph isomorphism problem.

Input: A pair of graphs (G_1, G_2) for B and V. Let n be the number of vertices in G_1 and G_2.

B: The prover randomly chooses an $i \in \{1,2\}$ and a permutation $\pi = (j_1, j_2, \ldots, j_n)$ of $(1, 2, \ldots, n)$. Then B applies π on G_i and sends the resulting graph $G_i(\pi)$ to the verifier.

V: The verifier randomly chooses an $j \in \{1,2\}$ and sends it to the prover. This message j corresponds to the requirement of providing a proof of the homomorphism between $G_i(\pi)$ and G_j.

B: If G_1 and G_2 are isomorphic, the prover computes an isomorphism δ such that $G_j(\delta) = G_i(\pi)$ and sends δ to the verifier.
If G_1 and G_2 are not isomorphic and $i = j$, then the prover sends the random permutation $\delta = \pi$. Else (if $i \neq j$ and there is no isomorphism between G_j and $G_i(\pi)$), the prover sends π too.

V: The verifier accepts (G_1, G_2) iff $G_j(\delta) = G_i(\pi)$.

First we show that this protocol is an interactive proof system. If G_1 and G_2 are isomorphic, then the prover always finds a permutation δ such that $G_j(\delta) = G_i(\pi)$. Therefore, V accepts the input (G_1, G_2) with certainty. If G_1 and G_2 are not isomorphic, then B can send a δ with $G_j(\delta) = G_i(\pi)$ only when $i = j$. The probability that $i = j$ for two random numbers $i, j \in \{1,2\}$ is exactly $\frac{1}{2}$. By executing k independent repetitions of this protocol and accepting only if all answers are correct, one obtains an iterative proof system with an error probability of at most 2^{-k}.

Since we omitted the presentation of a formal definition of zero-knowledge proof systems, we cannot give a formal proof that our protocol has the zero-knowledge property. However, we can intuitively understand why this protocol is a zero-knowledge proof system. The communication between B and V does not provide any information about the isomorphism (if any) between G_1 and

G_2. The first message $G_i(\pi)$ can be considered to be a random event that is determined by the random choice of i and π. The second message is a random number j. The last message δ is also determined by the random permutation π (δ is either exactly π or a composition of π and the homomorphism between G_1 and G_2). Hence, for a fixed (G_1, G_2), the set of all communications $(G_i(\pi), j, \delta)$ between B and V is a probability space. One can show that a randomized polynomial-time TM can generate the triples $(G_i(\pi), j, \delta)$ with the same probability distribution.

9.6 Design of an Interconnection Network

The communication between different subjects such as men, computers, processors of a parallel computer, etc., can be executed by different kinds of communication media. Each communication technology provides different possibilities and has its limitations and hence one has to solve different tasks when designing telegraph networks, telephone networks, optical networks or architectures of parallel computers. The magnitude of problems appears also in designing communication strategies in given networks. The magnitude is so large that one cannot aim to present it systematically in one section. Therefore, we prefer to illustrate the problems of network communication by an exemplary design of a network only.

Consider the following task. We have $2n$ parties

$$x_0, x_1, \ldots, x_{n-1}, y_0, y_1, \ldots, y_{n-1},$$

that are represented as $2n$ nodes[31] of a network. We have to build a network between the n x-parties $x_0, x_1, \ldots, x_{n-1}$ and the n y-parties $y_0, y_1, \ldots, y_{n-1}$ in such a way that at any time each x-party can obtain a connection for a call to any y-party. The network is considered as a graph

$$G = (V, E),\ x_0, x_1, \ldots, x_{n-1}, y_0, y_1, \ldots, y_{n-1} \in V.$$

Connecting x_i and y_j means finding and reserving a path $x_i, v_1, \ldots, v_m, y_j$ between x_i and y_j in G in such a way that no edge of this path is simultaneously used for another communication between other pairs of nodes. This means that at any time, every edge of G can be used exclusively for only one call.

The simplest solution to this problem is to create an edge from each x-party to each y-party. In this way one obtains a complete bipartite graph as depicted in Figure 9.6 for $n = 4$. But this solution is not practical. First, one has n^2 edges for $2n$ parties. This means for $n = 10\ 000$, the number of wires (physical connections) increases to 100 million. Besides the high production

[31] When speaking about networks we prefer to use the term "node" instead of the term vertex used in graphs.

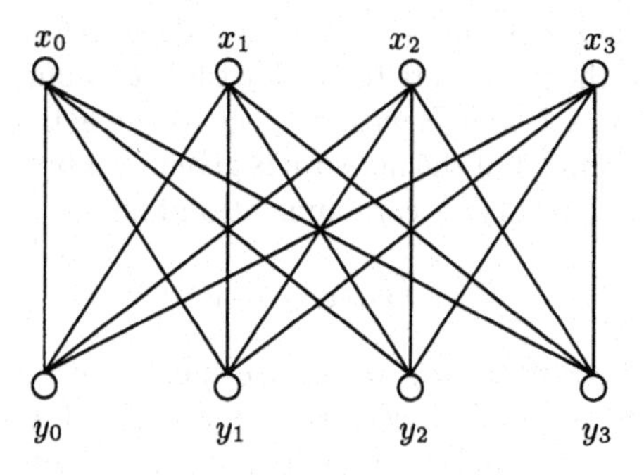

Fig. 9.2.

costs and the large number of connections rendering the network obscure, one cannot build such a network because of technological reasons. The number of connections to a node (the degree of a node) has to be bounded by a constant, i.e., the degree of the network may not grow with the number of parties.

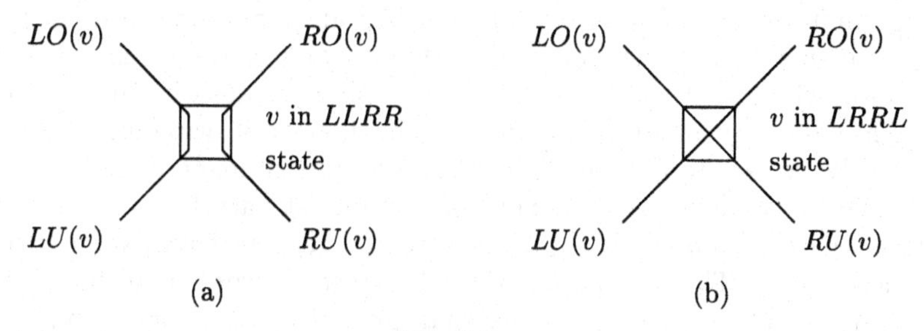

Fig. 9.3.

Now let us exactly formulate the requirements on an acceptable practical solution of our communication problem.

(i) Every node of the network has a degree bounded by 4, the nodes corresponding to parties have a degree of at most 2.

(ii) The nodes of the network that do not correspond to the parties are called the **switching nodes**. The structure of all switching nodes is the same (Figure 9.6). Every switching node v has exactly four incident edges $LO(v)$, $LU(v)$, $RO(v)$ and $RU(v)$ and it can be in one of the two (switching) states $LLRR$ or $LRRL$. If v is in the state $LLRR$ then one considers that there is a fixed connection between the edges $LO(v)$ and $LU(v)$ and a fixed connection between $RO(v)$ and $RU(v)$ (see Figure 9.6(a)). If v is in the state $LRRL$, then $LO(v)$ and $RU(v)$ lie on a common communication path and the connected edges $RO(v)$ and $LU(v)$ are a part of another communication path (see Figure 9.6(b)).

(iii) An x-party node u has only two incident edges $L(u)$ and $R(u)$ and it can decide which of them it wants to use for the communication. Hence, u has two possible states L and R with respect to the active edge (Figure 9.6).

(iv) A feasible requirement of the network is given by a permutation $(i_0, i_1, \ldots, i_{n-1})$ of $(0, 1, \ldots, n-1)$. A communication task

$$(i_0, i_1, \ldots, i_{n-1})$$

means that the x-party x_j wants to speak with the y-party y_{i_j} for $j = 0, 1, \ldots, n-1$. We assert that for each of the $n!$ permutations $(i_0, i_1, \ldots, i_{n-1})$ there exists an assignment of states to the nodes of the network that all the n calls

$$(x_0, y_{i_0}), (x_1, y_{i_1}), \ldots, (x_{n-1}, y_{i_{n-1}})$$

can be simultaneously performed. The guarantee of the simultaneous performance of the calls should be given by fixing (switching) n edge-disjoint paths between x_j and y_{i_j} for $j = 0, 1, \ldots, n-1$.

A network with $2n$ parties that can execute all communication tasks given by permutations of $(0, 1, \ldots, n-1)$ is called an **n-permutation network**. Observe that identifying (joining) vertices x_i and y_i results in a telephone network, where any party can at any time communicate with another party.

The cost of a permutation network is the number of switching nodes[32] and obviously we seek to minimize this cost. Another parameter, which we would also like to minimize, is the length of the longest path between an x-party and a y-party. It would also be nice to create a regular, transparent structure of the network. The modularity of the designed network could be of special importance. The modularity means that networks for $2n$ parties can be used as basic components for building networks of more (for instance, $4n$) parties. Clearly, modularity decreases the cost of the future expansion of the network, so it is important for the cost calculation.

The network in Figure 9.6 is a (reasonable) low priced solution for 8 parties $x_0, x_1, x_2, x_3, y_0, y_1, y_2, y_3$. The number of switching nodes is only 4 and all communication paths between an x-party and an y-party have the same length of exactly 2. The states of the modes of the 4-permutation network in Figure 9.6 determine the execution of the permutation $(3, 0, 2, 1)$.

Exercise 9.11. Prove that the network in Figure 9.6 is a 4-permutation network.

Now, we aim to design an asymptotically optimal network for the communication problem considered. First, we show that every network with $2n$ parties has to have at least $\Omega(n \log n)$ switching nodes.

[32] Note, that it does not matter whether one considers the number of switching nodes or the number of edges as the descriptional complexity measure.

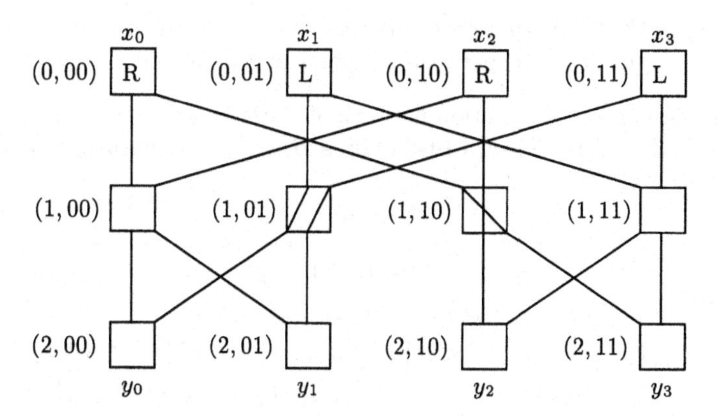

Fig. 9.4.

Theorem 9.12. *Let n be a positive integer. Every n-permutation network has at least $\Omega(n \log n)$ switching nodes.*

Proof. We prove the theorem by a simple counting argument. Let n be a positive integer and let Net_n be an n-permutation network. If one wants to perform a communication task in Net_n given by a permutation, then one has to find an assignment of states to the nodes of Net_n. Clearly, the different permutations force different assignments of states to the nodes. Thus, the number of possible assignments of states in Net_n must be at least

$$n!,$$

which is the number of all different communication tasks (permutations of n elements).

Let m be the number of switching nodes in Net_n. Hence, the number of different assignments of states to the nodes of Net_n is exactly

$$2^n \cdot 2^m.$$

Therefore, we obtain

$$2^m \cdot 2^n \geq n! \text{ (i.e., } 2^m \geq \frac{n!}{2^n}\text{)},$$

and hence[33]

$$m \geq \log_2(n!) - n \geq n \cdot \log n - n \cdot (\ln e + 1) \in \Omega(n \log n).$$

□

[33] With respect to Stirling's formula $n! \approx \frac{n^n}{e^n} \cdot \sqrt{2\pi n}$.

Exercise 9.13. Prove that each n-permutation network contains at least one path of a length of $\log_2 n$ between x-parties and y-parties.

To design an n-permutation network of a size in $O(n \log n)$, we start with the so-called **r-dimensional butterflies** But_r for any natural number n.

$But_r = (V_r, E_r)$, where

$$\begin{aligned} V_r &= \{(i,w) \mid i \in \{0,1,\ldots,r\},\ w \in \{0,1\}^r\}, \text{ and} \\ E_r &= \{\{(i,w),(i+1,w)\} \mid i \in \{0,1,\ldots,r-1\}\} \cup \\ &\quad \{\{(i,xay),(i+1,xby)\} \mid i \in \{0,1,\ldots,r-1\},\ x \in \{0,1\}^i, \\ &\qquad a,b \in \{0,1\}, a \neq b, y \in \{0,1\}^{r-i-1}\}. \end{aligned}$$

The 2-dimensional butterfly But_2 is shown in Figure 9.6. A transparent representation of But_r lay the $(r+1)\cdot 2^r$ nodes of But_r as a matrix of $r+1$ rows and 2^r columns. The position (i,j) of the matrix contains exactly the vertex (i,w) with $Number(w) = j$. An edge is between (i,x) and $(i+1,y)$ only if either

(i) $x = y$ (the vertical edges that run in every column from the first row to the last row), or
(ii) the only difference between x and y is the i-th bit of their representation (Figure 9.6).

If, for $j = 0,1,\ldots,2^r-1$, one assigns the x-party x_j to the node $(0,w)$ with $Number(w) = j$ and the y-party y_j to the node (r,w') with $Number(w') = j$, then, for every pair

$$(x_d, y_c),\ d,c \in \{0,1,\ldots,2^r-1\},$$

one can find the following path between x_d and y_c.

Let

$$d = Number(a_0a_1\ldots a_{r-1}) \text{ and } c = Number(b_0b_1\ldots b_{r-1})$$

for some $a_k, b_k \in \{0,1\}$ for $k = 0,1,\ldots,r-1$. Hence, the node $(0, a_0a_1\ldots a_{r-1})$ corresponds to the x-party x_d and the node $(r, b_0b_1\ldots b_{r-1})$ corresponds to the y-party y_c. If $a_0 = b_0$, then the path starts with the vertical edge

$$\{(0, a_0a_1\ldots a_{r-1}), (1, a_0a_1\ldots a_{r-1})\}.$$

If $a_0 \neq b = 0$, then the path starts with the "cross" edge

$$\{(0, a_0a_1\ldots a_{r-1}), (1, b_0a_1\ldots a_{r-1})\}.$$

Thus, one reaches in both cases the vertex

$$(1, b_0a_1\ \ldots\ a_{r-1}).$$

In general, the path from $(0, a_1 \ldots a_{r-1})$ to $(r, b_0 b_1 \ldots b_{r-1})$ reaches the node

$$(k, b_0 b_1 \ldots b_{k-1} a_k a_{k+1} a_{k+2} \ldots a_r)$$

after k edge choices and continues with the $(k+1)$-th edge

$$\{(k, b_0 b_1 \ldots b_{k-1} a_k a_{k+1} \ldots a_{r-1}), (k+1, b_0 b_1 \ldots b_k a_{k+1} \ldots a_{r-1})\}.$$

Thus, the s-th edge of the path is responsible for flipping the s-th bit a_s to b_s.

We showed that one can connect every x-party to an arbitrary y-party by a path of length r in But_r. Unfortunately, But_r is not able to perform each permutation by edge-disjoint paths. In our strategy, the node

$$(\lfloor \tfrac{r}{2} \rfloor, b_0 b_1 \ldots b_{\lfloor r/2 \rfloor} a_{\lfloor r/2 \rfloor + 1} \ldots a_{r-1})$$

is used in all paths with the endpoint in the set

$$\{(r, b_0 b_1 \ldots b_{\lfloor r/2 \rfloor} e_{\lfloor r/2 \rfloor + 1} \ldots e_{r-1}) \mid e_j \in \{0,1\} \text{ for } j = \lfloor \tfrac{r}{2} \rfloor + 1, \ldots, r-1\}$$

and with the starting point from the set

$$\{(0, f_0 f_1 \ldots f_{\lfloor r/2 \rfloor} a_{\lfloor r/2 \rfloor + 1} \ldots a_{r-1} \mid f_i \in \{0,1\} \text{ für } i = 0, 1, \ldots, \lfloor \tfrac{r}{2} \rfloor\}.$$

There are $2^{\frac{r}{2}-1}$ such paths, and we may use a switching node for at most 2 paths.

We use the r-dimensional butterflies as components for building a 2^n-permutation network, that is well known as the **Beneš-network**. One obtains the r-dimensional Beneš-network $Benes_r$ from two But_r networks A and B by pair-wise joining the corresponding nodes of the last (r-th) rows of A and B. Figure 9.6 shows $Benes_3$. We observe, that $Benes_r$ has $2r+1$ rows, where the first $r+1$ rows[34] build the r-dimensional butterfly A and the last $r+1$ rows build the r-dimensional butterfly B. Figure 9.6 shows a recursive definition of $Benes_r$ by building $Benes_r$ from two $Benes_{r-1}$ networks and some additional nodes and edges.

Exercise 9.14. Give, for every positive integer r, a formal description of $Benes_r$ as a graph.

Theorem 9.15. *For every positive integer r, $Benes_r$ is a 2^r-permutation network.*

Proof. We show that, for every permutation

$$(i_0, \ldots, i_{2^r-1}),$$

[34] With corresponding edges

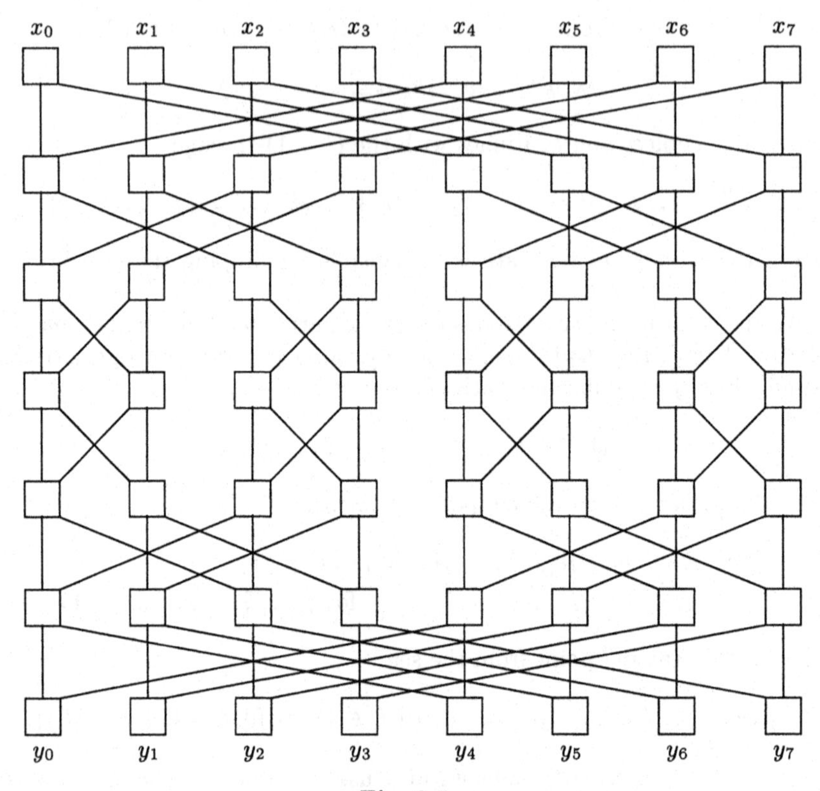

Fig. 9.5.

one can choose 2^r paths from x_j to y_{i_j} for $j = 0, 1, \ldots 2^r - 1$ in such a way that no edge is used for more than one path and every node lies exactly on one path.[35] Obviously, such a set of paths guarantees that one can find an assignment of the states to the nodes that exactly corresponds to these 2^r paths, solving the communication task $(i_0, \ldots, i_{2^r-1})$.

We prove this claim by induction with respect to r.

(i) Let $r = 1$.
Clearly (Figure 9.6), $Benes_1$ is a 2-permutation network.[36]

(ii) Let $r \geq 2$.
We assume,[37] that for every permutation of $(0, 1, \ldots, 2^{r-1} - 1)$ there exist 2^{r-1} node-disjoint paths between the 2^{r-1} x-parties and the 2^{r-1} y-parties in $Benes_{r-1}$. Now, we view $Benes_r$ as a network depicted in Figure 9.6,

[35] Such groups of paths are called pair-wise node-disjoint paths.
[36] Note, that for $r = 1$ even But_r is a 2-permutation network.
[37] With respect to our induction hypothesis

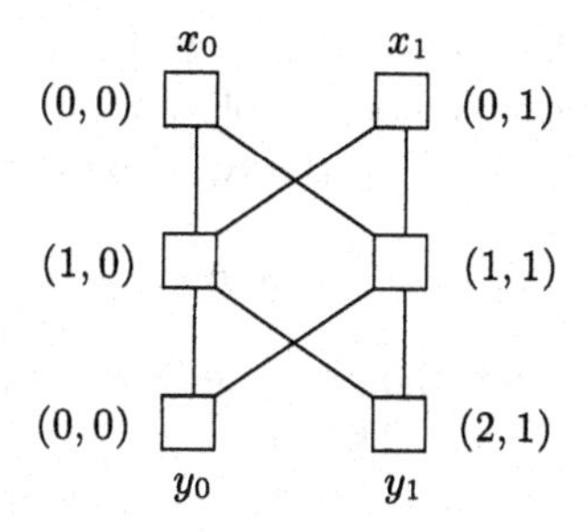

Fig. 9.6.

that consists of two $Benes_{r-1}$ networks A and B and two "first rows" of But_r.

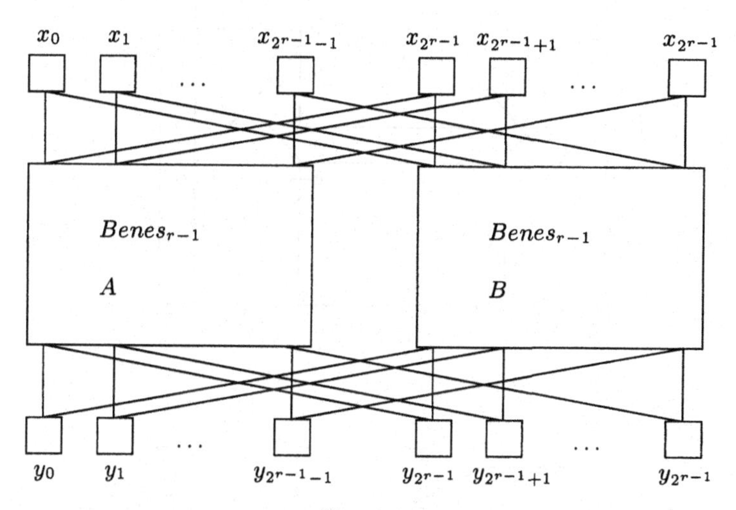

Fig. 9.7.

Due to the induction hypothesis it is sufficient to find a communication strategy that fulfils:

a) Exactly half of the paths go via the component $Benes_{r-1}$ A.
b) The communication tasks for A and B can be given by two permutation of $(0, 1, \ldots, 2^{r-1} - 1)$.

To assure this it is sufficient to choose the paths in the first row and the last row of $Benes_r$ in such a way that no node in the first row of A and B (i.e., in the second row of $Benes_r$) lies on two paths[38] and analogously

[38] That is, only one edge between v and the first row (row 0) of $Benes_r$ may by used.

no node u from the last row of A and B lies on two paths.[39] Formally, we can express these requirements as follows:

(1) For all $i \in \{0, 1, \ldots, 2^{r-1}\}$, the two paths starting in

$$x_i \text{ and } x_{i+2^{r-1}}$$

have to use different $Benes_{r-1}$-networks (a path must go via A and another via B) (Figure 9.6(a)).

(2) For all $j \in \{0, 1, \ldots, 2^{r-1}\}$, the two paths leading to

$$y_j \text{ and } y_{j+2^{r-1}}$$

must be routed via different components $Benes_{r-1}$ (Figure 9.6(a)).

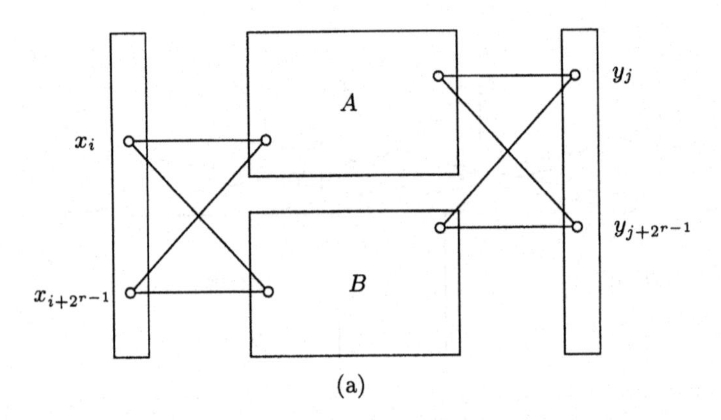

(a)

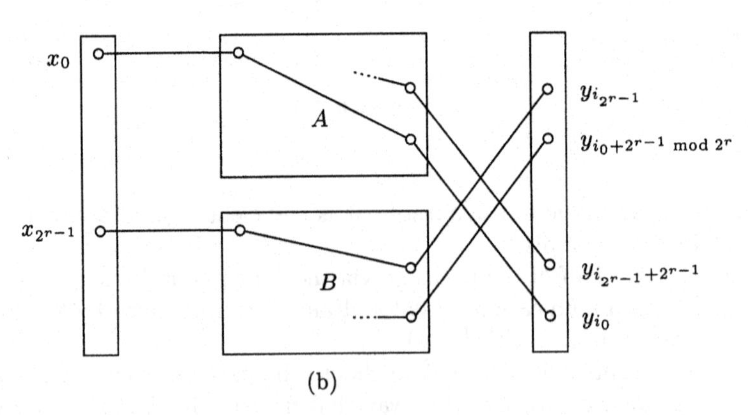

(b)

Fig. 9.8.

[39] That is, only one edge from u to the y-parties may be used.

In the following we show that the paths can be determined one after the other. Let

$$(i_0, i_1, \ldots, i_{2^r-1})$$

be an arbitrary permutation of $(0, 1, \ldots, 2^r - 1)$. We consider now the pairs

$$(x_0, y_{i_0}) \text{ and } (x_{2^r-1}, y_{i_{2^r-1}})$$

and determine the path from x_0 to y_{i_0} via A and the path from x_{2^r-1} to $y_{i_{2^r-1}}$ via B (Figure 9.6(b)).

If

$$|i_{2^r-1} - i_0| = 2^{r-1},$$

there are no consequences that would pose a requirement of routing other paths.

If

$$|i_{2^r-1} - i_0| \neq 2^{r-1}$$

we have to lay (because of (2)) the path to

$$y_{i_0+2^{r-1} \bmod 2^r}$$

via B and the path to

$$y_{i_{2^r-1}+2^{r-1} \bmod 2^r}$$

via A. Doing this, the consecutive requirements on the layout of two paths from the nodes

$$x_{q+2^{r-1} \bmod 2^r} \text{ and } x_{s+2^{r-1} \bmod 2^r}$$

for

$$i_q = (i_{2^r-1} + 2^{r-1}) \bmod 2^r \text{ and } i_s = (i_0 + 2^{r-1}) \bmod 2^r$$

can be posed.[40]

We see that continuing with this strategy, each layout of the paths can place at most one requirement on the layout of another pair of paths and this requirement can be easily satisfied. Thus, the proposed strategy assures the construction of node-disjoint paths for performing the given telephone calls. □

Exercise 9.16. Estimate the paths in *Benes*$_3$, for the permutations $(7, 2, 1, 3, 0, 5, 6, 4)$, $(0, 1, 2, 3, 7, 6, 5, 4)$, and $(0, 7, 1, 6, 2, 5, 3, 4)$.

Theorem 9.15 provides a solution to our problem of designing n-permutation networks. The network *Benes*$_r$ is a 2^r-permutation network with

$$(2r + 1) \cdot 2^r \text{ nodes and } r \cdot 2^{r+2} \text{ edges.}$$

Thus, the number of switching nodes is in $O(n \cdot \log_2 n)$ if the number of x-parties is 2^r. Moreover, the Beneš networks have a regular structure with a high degree of modularity (Figure 9.6). The distance between any x-party and any y-party is exactly $2r$ and hence logarithmic in the number of parties.

[40] If $|((q + 2^{r-1}) \bmod 2^r) - ((s + 2^{r-1}) \bmod 2^r)| \neq 2^{r-1}$.

Exercise 9.17.* Let $G = (V, E)$ be a graph. A **balanced cut** of G is a pair (V_1, V_2), such that

(i) $V = V_1 \cup V_2$, $V_1 \cap V_2 = \emptyset$, and
(ii) $-1 \leq |V_1| - |V_2| \leq 1$.

The cost of a cut (V_1, V_2) are

$$cost(V_1, V_2) = |E \cap \{\{v, u\} \mid v \in V_1, u \in V_2\}|,$$

i.e., the number of edges between V_1 and V_2. The **bisection width of G** is the minimum over the costs of all balanced cuts of G. In the area of network design, one searches for networks with a high bisection width. A large bisection width assures that, for any partition of the network into two almost equal-sided parts, there are sufficiently many communication channels between these two parts.

Prove that But_r and $Benes_r$ have a bisection width[41] in $\Omega(2^r)$.

9.7 Summary

Cryptography deals with the design of cryptosystems that enable a secret information exchange via public communication channels. For the classical (symmetric) cryptosystems, the key determines the encryption mechanism as well as the decryption mechanism, so the key is the common secret between the sender and the receiver. Public-key cryptosystems work with a public encryption key, because knowledge of it does not help to learn the decryption key. The idea of public-key cryptosystems is based on the concept of one-way functions. A one-way function is efficiently computable, but the corresponding inverse function is not efficiently computable without an additional knowledge (a secret, owned by the receiver only). Candidates for one-way functions are the multiplication, which inverse function is the factorization, and computing $a^b \pmod{n}$, which inverse function is the discrete logarithm.

Interactive proof systems enable an efficient verification of proof existence by randomization. All languages in PSPACE have interactive proof systems. Zero-knowledge proof systems can be used to verify whether another person owns a proof or a secret without revealing any bit of this secret. All languages in NP have zero-knowledge proof systems.

The capability of interconnection networks is measured by the amount of transferred data or by the amount of satisfied communication requests in a short time interval. The design of a network that should be optimal or almost optimal with respect to some quality parameters, usually leads to nontrivial

[41] The bisection width is approximately as high as the number of parties. Expressed in another way, the bisection width is in $\Omega(\frac{m}{\log m})$, where m is the number of nodes of these networks.

optimization problems. One tries to solve such problems using methods of discrete mathematics.

The revolutionary concept of public-key cryptosystems was proposed by Diffie and Hellman [17] in 1976. The famous RSA cryptosystems was discovered by Rivest, Shamir and Adleman [58]. The concepts of interactive and zero-knowledge proof systems are attributed to Goldwasser, Micali and Rackoff [21]. Shamir [64] was the first to prove IP = PSPACE.

The Beneš network was proposed by Beneš in his seminal papers [4, 5] about the design of telephone networks. The fact that Beneš networks are permutation networks has been proved by Waksman [70].

For a transparent, very well-written introduction to the cryptography we warmly recommend Salomaa [61]. Another excellent, modern textbook on cryptography is by Delfs and Knebl [16]. An extensive well-written survey about interactive systems is given by Bovet and Crescenzi [7] and Sipser [65]. Leighton [41] is the seminal textbook and monograph on network design and communication problems. An introduction to communication problems in interconnection networks is given by Hromkovič, Klasing, Monien and Peine [33].

References

1. G. Ausiello, P. Crescenzi, G. Gambosi, V. Kann, A. Marchetti-Spaccamela, and M. Protasi. *Complexity and Approximation. Combinatorial Optimization Problems and their Approximability Properties.* Springer-Verlag, 1999.
2. J.L. Balcázar, J. Díaz, and J. Gabarró. *Structural Complexity I.* Springer-Verlag, 1988.
3. J.L. Balcázar, J. Díaz, and J. Gabarró. *Structural Complexity II.* Springer-Verlag, 1990.
4. V. Beneš. Permutation groups, complexes and rearrangable multistage connecting networks. *Bell System Technical Journal*, 43:1619–1640, 1964.
5. V. Beneš. *Mathematical Theory of Connecting Networks and Telephone Traffic.* Academic Press, 1965.
6. F. Bock. An algorithm for solving "traveling-salesman" and related network optimization problems: abstract. *Bulletin 14th National Meeting of the Operations Research Society of America*, page 897, 1958.
7. D.P. Bovet and P. Crescenzi. *Introduction to the Theory of Complexity.* Prentice-Hall, 1994.
8. V. Černý. A thermodynamical approach to the traveling salesman problem: An efficient simulation algorithm. *Journal of Optimization Theory and Applications*, 45:41–55, 1985.
9. G.J. Chaitin. On the length of programs for computing finite binary sequences. *Journal of the ACM*, 13:407–412, 1966.
10. G.J. Chaitin. On the simplicity and speed of programs for computing definite sets of natural numbers. *Journal of the ACM*, 16:407–412, 1969.
11. G.J. Chaitin. Information-theoretic limitations of formal systems. *Journal of the ACM*, 13:403–424, 1974.
12. A. Church. An undecidable problem in elementary number theory. *American Journal of Mathematics*, 58:345–363, 1936.
13. S. Cook. The complexity of theorem-proving procedures. In *Proceedings of 3rd ACM STOC*, pages 151–157, 1971.
14. T. Cormen, C. Leiserson, and R. Rivest. *Introduction to Algorithms.* MIT Press and McGraw-Hill, 1990.
15. G.A. Croes. A method for solving traveling salesman problem. *Operations Research*, 6:791–812, 1958.

16. H. Delfs and H. Knebl. *Introduction to Cryptography. Priciples and Applications.* Springer-Verlag, 2002.
17. W. Diffie and M. Hellman. New directions in cryptographie. *IEEE Transitions Inform. Theory*, 26:644–654, 1976.
18. K. Erk and L. Priese. *Theoretische Informatik (Eine umfassende Einführung).* Springer-Verlag, 2000.
19. M. Garey and D. Johnson. *Computers and Intractability.* Freeman, 1979.
20. K. Gödel. Über formal unentscheidbare Sätze der Principia Mathematica und verwandter Systeme. *Monatshefte für Mathematik und Physik*, 38:173–198, 1931.
21. S. Goldwasser, S. Micali, and C. Rackoff. Knowledge complexity of interactive proofs. In *Proc. 17th ACM Symp. on Theory of Computation*, pages 291–304. ACM, 1985.
22. R. Graham. Bounds for certain multiprocessor anomalies. *Bell System Technical Journal*, 45:1563–1581, 1966.
23. D. Harel. *Algorithmics. The Spirit of Computing.* Addison Wesley, 1993.
24. J. Hartmanis, R. Stearns, and P. Lewis. Hierarchies of memory limited computations. In *Proceedings of 6th IEEE Symp. on Switching Circuit Theory and Logical Design*, pages 179–190, 1965.
25. J. Hartmanis and R.E. Stearns. On the computational complexity of algorithms. *Transactions of ASM*, 117:285–306, 1965.
26. D.S. Hochbaum. *Approximation Algorithms for NP-hard Problems.* PWS Publishing Company, Boston, 1997.
27. J.E. Hopcroft, R. Motwani, and J.D. Ullman. *Introduction to Automata Theory, Languages, and Computation.* Addison-Wesley, Reading, MA, 2001.
28. J.E. Hopcroft and J.D. Ullman. *Introduction to Automata Theory, Languages, and Computation.* Addison-Wesley, Reading, 1979.
29. J. Hromkovič. *Communication Complexity and Parallel Computing.* Springer-Verlag, 1997.
30. J. Hromkovič. *Algorithms for Hard Problems. Introduction to Combinatorial Optimization, Randomization, Approximation and Heuristics.* Springer-Verlag, 2001.
31. J. Hromkovič. Communication protocols: An exemplary study of the power of randomness. In J. Reif J. Rolim P. Pavdalos, S. Rajasekaran, editor, *Handbook of Randomized Computing*, volume 2, pages 533–596. Kluwer Academic Publishers, 2001.
32. J. Hromkovič. *Algorithmics for Hard Problems.* Springer-Verlag, 2003.
33. J. Hromkovič, R. Klasing, B. Monien, and R. Peine. Dissemination of information in interconnection networks. In *Combinational Network Theory*, pages 125–212, 1996.
34. O.H. Ibarra and C.E. Kim. Fast approximation algorithms for the knapsack and sum of subsets problem. *Journal of the ACM*, 21:294–303, 1974.
35. R. Karp. Reducibility among combinatorial problems. In R. Miller, editor, *Complexity of Computer Computation*, pages 85–104. Plenum Press, 1972.
36. R.M. Karp. An introduction to randomized algorithms. *Discrete Applied Mathematics*, 34:165–201, 1991.
37. S. Kirkpatrick, P.D. Gellat, and M.P. Vecchi. Optimization by simulated annealing. *Science*, 220:671–680, 1983.
38. S.C. Kleene. General recursive functions of natural numbers. *Mathematische Annalen*, 112:727–742, 1936.

39. A.N. Kolmogorov. Three approaches for defining the concept of information quantity. *Probl. Information Transmission*, 1:1–7, 1965.
40. A.N. Kolmogorov. Logical basis for information theory and probability theory. *IEEE Transactions on Information Theory*, 14:662–664, 1968.
41. F.T. Leighton. *Introduction to Parallel Algorithms and Architectures: Arrays, Trees, Hypercubes.* Morgan Kaufmann Publ. Inc., 1992.
42. H.R. Lewis and Ch. Papadimitriou. The efficiency of algorithms. *Scientific American*, 238(1), 1978.
43. M. Li and P.M.B. Vitányi. *An Introduction to Kolmogorov Complexity and Its Applications.* Springer-Verlag, 1993.
44. N. Saxona M. Agrawal, N. Kayal. Primes in p. Unpublished manuscript.
45. E.W. Mayr, H.J. Prömel, and A. Steger, editors. *Lectures on Proof Verification and Approximation Algorithms.* Number 1967 in Lecture Notes in Computer Science. Springer-Verlag, 1998.
46. N. Metropolis, A.W. Rosenbluth, M.N. Rosenbluth, A.H. Teller, and E. Teller. Equation of state calculation by fast computing machines. *Journal of Chemical Physics*, 21:1087–1091, 1953.
47. G. Miller. Rieman's hypothesis and test for primality. *Journal of Computer and System Sciences*, (13):300–317, 1976.
48. R. Motwani and P. Raghavan. *Randomized Algorithms.* Cambridge University Press, 1995.
49. Ch.H. Papadimitriou. *Computational Complexity.* Addison-Wesley, 1994.
50. Ch.H. Papadimitriou and K. Steiglitz. *Combinatorial Optimization: Algorithms and Complexity.* Prentice-Hall, 1982.
51. E. Post. Finite combinatory process-formulation. *Journal of Symbolic Logic*, 1:103–105, 1936.
52. E. Post. A variant of a recursively unsolvable problem. *Transactions of ASM*, 52:264–268, 1946.
53. V. Strassen R. Solovay. A fast monte carlo test for primality. *SIAM Journal on Computing 6*, pages 84–85, 1977.
54. M.O. Rabin. Probabilistic algorithms. In J.F. Traub, editor, *Algorithms and Complexity: Recent Results and New Directions*, pages 21–39. Academic Press, 1976.
55. M.O. Rabin. Probabilistic algorithms for primality testing. *Journal of Number Theory*, (12):128–138, 1980.
56. R. Reischuk. *Einführung in die Komplexitätstheorie.* B.G. Teubner, 1990.
57. H.G. Rice. Classes of recursively enumerable sets and their decision problems. *Transactions of ASM*, 89:25–59, 1953.
58. R.L. Rivest, A. Shamir, and L. Adleman. A method for obtaining digital signatures and public-key cryptosystems. *Comm. Assoc. Comput. Mach.*, 21:120–12, 1978.
59. H. Rogers. *Theory of Recursive Functions and Effective Computability.* McGraw-Hill, New York, 1967.
60. A. Salomaa. *Formal Languages.* Academic Press, 1973.
61. A. Salomaa. *Public-Key Cryptographie.* Springer-Verlag, 1996.
62. U. Schöning. *Perlen der Theoretischen Informatik.* BI-Wissenschaftsverlag, Mannheim, Leipzig, Wien, Zürich, 1995.
63. U. Schöning. *Algorithmik.* Spektrum Akademischer Verlag, Heidelberg, Berlin, 2001.

64. A. Shamir. IP= PSPACE. *Journal of the ACM*, 39:869–877, 1992.
65. M. Sipser. *Introduction to the Theory of Computation.* PWS Publ. Comp., 1997.
66. L.J. Stockmeyer and A.K. Chandra. Intrinsically difficult problems. *Scientific American*, 240(5), 1979.
67. B.A. Trakhtenbrot. *Algorithms and Automatic Computing Machines.* D.C. Heath & Co., Boston, 1963.
68. A.M. Turing. On computable numbers with an application to the Entscheidungsproblem. In *Proceedings of London Mathematical Society*, volume 42 of *2*, pages 230–265, 1936.
69. V. Vazirani. *Approximation Algorithms.* Springer-Verlag, 2001.
70. A. Waksman. A permutation network. *Journal of ACM*, 15:159–163, 1968.
71. I. Wegener. *Theoretische Informatik – eine algorithmische Einführung.* R.G. Teubner, Stuttgart, Leipzig, 1999.

Index

Monographs in Theoretical Computer Science · An EATCS Series

K. Jensen
Coloured Petri Nets
Basic Concepts, Analysis Methods
and Practical Use, Vol. 1
2nd ed.

K. Jensen
Coloured Petri Nets
Basic Concepts, *Analysis Methods*
and Practical Use, Vol. 2

K. Jensen
Coloured Petri Nets
Basic Concepts, Analysis Methods
and *Practical Use,* Vol. 3

A. Nait Abdallah
The Logic of Partial Information

Z. Fülöp, H. Vogler
Syntax-Directed Semantics
Formal Models Based on Tree Transducers

A. de Luca, S. Varricchio
Finiteness and Regularity in Semigroups and Formal Languages

E. Best, R. Devillers, M. Koutny
Petri Net Algebra

S.P. Demri, E. S. Orłowska
Incomplete Information: Structure, Inference, Complexity

J.C.M. Baeten, C.A. Middelburg
Process Algebra with Timing

L.A. Hemaspaandra, L.Torenvliet
Theory of Semi-Feasible Algorithms

Texts in Theoretical Computer Science · An EATCS Series

J. L. Balcázar, J. Díaz, J. Gabarró
Structural Complexity I
2nd ed. (see also overleaf, Vol. 22)

M. Garzon
Models of Massive Parallelism
Analysis of Cellular Automata
and Neural Networks

J. Hromkovič
Communication Complexity and Parallel Computing

A. Leitsch
The Resolution Calculus

G. Păun, G. Rozenberg, A. Salomaa
DNA Computing
New Computing Paradigms

A. Salomaa
Public-Key Cryptography
2nd ed.

K. Sikkel
Parsing Schemata
A Framework for Specification
and Analysis of Parsing Algorithms

H. Vollmer
Introduction to Circuit Complexity
A Uniform Approach

W. Fokkink
Introduction to Process Algebra

K. Weihrauch
Computable Analysis
An Introduction

J. Hromkovič
Algorithmics for Hard Problems
Introduction to Combinatorial Optimization,
Randomization, Approximation, and Heuristics
2nd ed.

J. Hromkovič
Theoretical Computer Science
Introduction to Automata, Computability,
Complexity, Algorithmics, Randomization,
Communication and Cryptography

S. Jukna
Extremal Combinatorics
With Applications in Computer Science

P. Clote, E. Kranakis
Boolean Functions and Computation Models

L. A. Hemaspaandra, M. Ogihara
The Complexity Theory Companion

C.S. Calude
Information and Randomness.
An Algorithmic Perspective
2nd ed.

Former volumes appeared as EATCS Monographs on Theoretical Computer Science

Vol. 5: W. Kuich, A. Salomaa
Semirings, Automata, Languages

Vol. 6: H. Ehrig, B. Mahr
Fundamentals of Algebraic Specification 1
Equations and Initial Semantics

Vol. 7: F. Gécseg
Products of Automata

Vol. 8: F. Kröger
Temporal Logic of Programs

Vol. 9: K. Weihrauch
Computability

Vol. 10: H. Edelsbrunner
Algorithms in Combinatorial Geometry

Vol. 12: J. Berstel, C. Reutenauer
Rational Series and Their Languages

Vol. 13: E. Best, C. Fernández C.
Nonsequential Processes
A Petri Net View

Vol. 14: M. Jantzen
Confluent String Rewriting

Vol. 15: S. Sippu, E. Soisalon-Soininen
Parsing Theory
Volume I: Languages and Parsing

Vol. 16: P. Padawitz
Computing in Horn Clause Theories

Vol. 17: J. Paredaens, P. DeBra, M. Gyssens, D. Van Gucht
The Structure of the Relational Database Model

Vol. 18: J. Dassow, G. Páun
Regulated Rewriting in Formal Language Theory

Vol. 19: M. Tofte
Compiler Generators
What they can do, what they might do, and what they will probably never do

Vol. 20: S. Sippu, E. Soisalon-Soininen
Parsing Theory
Volume II: LR(k) and LL(k) Parsing

Vol. 21: H. Ehrig, B. Mahr
Fundamentals of Algebraic Specification 2
Module Specifications and Constraints

Vol. 22: J. L. Balcázar, J. Díaz, J. Gabarró
Structural Complexity II

Vol. 24: T. Gergely, L. Úry
First-Order Programming Theories

R. Janicki, P. E. Lauer
Specification and Analysis of Concurrent Systems
The COSY Approach

O. Watanabe (Ed.)
Kolmogorov Complexity and Computational Complexity

G. Schmidt, Th. Ströhlein
Relations and Graphs
Discrete Mathematics for Computer Scientists

S. L. Bloom, Z. Ésik
Iteration Theories
The Equational Logic of Iterative Processes